THE REIGN OF RELATIVITY

OXFORD STUDIES IN PHILOSOPHY OF SCIENCE

General Editor
   Paul Humphreys, University of Virginia

Advisory Board
   Jeremy Butterfield
   Peter Galison
   Ian Hacking
   Philip Kitcher
   Richard Miller
   James Woodward

The Book of Evidence
Peter Achinstein

Science, Truth, and Democracy
Philip Kitcher

The Devil in the Details: Asymptotic Reasoning in Explanation, Reduction, and Emergence
Robert W. Batterman

Science and Partial Truth: A Unitary Approach to Models and Scientific Reasoning
Newton C. A. da Costa and Steven French

Inventing Temperature: Measurement and Scientific Progress
Hasok Chang

Making Things Happen
James Woodward

The Reign of Relativity: Philosophy in Physics 1915–1925
Thomas Ryckman

THE REIGN OF RELATIVITY

Philosophy in Physics 1915–1925

Thomas Ryckman

OXFORD
UNIVERSITY PRESS

2005

OXFORD

UNIVERSITY PRESS

Oxford   New York
Auckland   Bangkok   Buenos Aires   Cape Town   Chennai
Dar es Salaam   Delhi   Hong Kong   Istanbul   Karachi   Kolkata
Kuala Lumpur   Madrid   Melbourne   Mexico City   Mumbai   Nairobi
São Paulo   Shanghai   Taipei   Tokyo   Toronto

Copyright © 2005 by Oxford University Press, Inc.

Published by Oxford University Press, Inc.
198 Madison Avenue, New York, New York 10016

www.oup.com

Oxford is a registered trademark of Oxford University Press

Library of Congress Cataloging-in-Publication Data
Ryckman, Thomas.
   The reign of relativity : philosophy in physics 1915–1925 / Thomas Ryckman.
      p. cm.—(Oxford studies in philosophy of science)
   Includes bibliographical references and index.

   ISBN-13 978-0-19-517717-6
   ISBN 0-19-517717-7

      1. Relativity (Physics)—History. I. Title. II. Series.

   QC173.52.R93 2004
   530.11'09—dc22   2004041576

9   8   7   6   5   4   3   2

Printed in the United States of America
on acid-free paper

for Pamela, light of my life

PREFACE

The theories of special and general relativity have been essential components of the physical world picture for now more than eight decades, longer than a generous span of human life. Among physicists, familiarity has not bred contempt. Both theories continue to challenge implicitly held notions in ways that even adepts can yet find surprising. The change in outlook occasioned by relativity theory thus has something of the character of "permanent revolution", continually turning up things new, interesting and possibly disturbing. On the other hand, its revolutionary image would appear to be considerably dulled among philosophers of science, excepting, of course, certain philosophers of physics and others interested in space-time theories. To be sure, Einstein retains the halo of universal genius among the public at large. But today one can easily acquire the impression that it is the quantum theory, the other principal component of the current physical world-view, which has largely captured the contemporary philosophical imagination. No knowledgeable person would seriously question its revolutionary character or inherent philosophical interest. But while philosophers are generally aware of the vigorous epistemological debate that accompanied the quantum theory's rise, was epitomized in the Einstein-Bohr dialogues, and still continues, recognition seems altogether lacking that a corresponding controversy worthy of present philosophical scrutiny occurred in the early years of general relativity. In part this ignorance is traceable to a false, but understandable, impression that such philosophical engagement as took place principally involved supporters and opponents of general relativity, a conflict abating, and justly forgotten, as the opponents of the theory faded away into oblivion. A sallow bill of goods adapted and adopted by logical empiricism, it is still frequently found retailed within the literature of philosophy of science. This book was written to finally inter that insidious narrative, and to recover, if possible, something of the freshness of the philosophical encounter with that most beautiful of physical theories by two of its greatest masters, Hermann Weyl and Arthur S. Eddington.

I am grateful to the National Science Foundation and the National Endowment for the Humanities for grants that relieved me from teaching duties in 1995–1996 in order to begin the project of the book. In relieving me of any further duties on my return, an interim dean at a private university on Chicago's North Shore unwittingly furnished me with the requisite motivation to finish it. I should like to thank her, although readers will have to judge for themselves whether I have succeeded in following her injunction to "write more boilerplate". My largest scholarly debts are to Arthur Fine and Michael Friedman, for innumerable conversations, friendly criticism, and for authoring books in philosophy of science that have not ceased to inspire since I read them as a graduate student in the 1980s. It is largely due to them I became a philosopher of science. It was Howard Stein who awakened my interest in Hermann Weyl, long before this book was conceived. With such an introduction, it is small wonder that Weyl has been on my mind ever since. I owe the warmest thanks to Roberto Torretti, who read the penultimate version with his customary meticulousness, and whose expertise and judicious comments vastly improved it. Carl Hoefer's firm but gentle criticisms of an earlier version played a decisive role in shaping the book's final form and content. Over the years I also received encouragement, advice, or assistance from Guido Bacciagaluppi, Mara Beller, Yamima Ben-Menahem, Michel Bitbol, Katherine Brading, Harvey Brown, Jeremy Butterfield, Elena Castellani, Leo Corry, Steven French, Michel Ghins, Friedrich Hehl, Don Howard, Karl-Norbert Ihmig, John Krois, James Ladyman, John McCumber, David Malament, Paolo Mancosu, Yuval Ne'eman, John Norton, Norman Packard, Itomar Pitowsky, Rob Rynasiewicz, Simon Saunders, Hans Sluga, John Stachel, Rick Tieszen, Thomas Uebel, and Daniel Warren. Heartfelt thanks to all. Sadly, some who helped in meaningful ways are no longer with us. I cannot thank them, but mention them here to record debts that I shall find other ways to pay: Jim Cushing, Zellig Harris, Robert Weingard, and Richard Wollheim.

Chapter 2 draws upon "Two Roads from Kant: Cassirer, Reichenbach and General Relativity" by T. A. Ryckman from *Logical Empiricism: Historical and Contemporary Perspectives*, edited by Paolo Parrini, Wesley C. Salmon, and Merrilee H. Salmon, © 2003 by University of Pittsburgh Press, 159–193. Reprinted by permission of the University of Pittsburgh Press. Chapter 4 includes material from my "Einstein *Agonists*: Weyl and Reichenbach on Geometry and the General Theory of Relativity", in *The Origins of Logical Empiricism*, edited by Ronald Giere and Alan Richardson (Minneapolis, University of Minnesota Press, 1996), 165–209. Chapter 6 incorporates much of my "The Philosophical Roots of the Gauge Principle: Weyl and Transcendental Phenomenological Idealism", in *Symmetries in Physics: Philosophical Reflections*, edited by Katherine Brading and Elena Castellani (Cambridge: Cambridge University Press, 2003), 61–88. I am grateful to the editors and the publishers concerned for their permissions to reuse the material here.

Unpublished correspondence of Einstein was obtained from *Albert Einstein: The Collected Papers*, published by Princeton University Press (reprinted by permission of Princeton University Press). I thank the University of Pittsburgh Library System for permission to quote from unpublished correspondence of Hans Reichenbach, and Frau Dr. Yvonne Vögeli of the *Wissenschaftshistorische Sammlungen* of the Swiss Federal Institute of Technology (Zürich) for providing me with photocopies of the unpublished letters of Eddington to Weyl.

I am grateful to the following sources of the photos on the dust jacket:

Dr. Matthias Neuber for locating the photograph of Moritz Schlick, and to Dr. George van de Velde-Schlick for permission to reproduce it here.

Alain Guillard of the Interlibrary Loan Service at the Bibliothèque universitaire de Paris–XII–Val de Marne for permission to reproduce the photograph of Emile Meyerson from the Ignace Meyerson collection. All rights reserved.

Brigitta Arden at the Archives of Scientific Philosophy, Special Collections, Hillman Library, University of Pittsburgh, for locating the photograph of Hans Reichenbach. Reproduction here is by permission of the University of Pittsburgh. All rights reserved.

Professor John Krois for kindly lending his photograph of Ernst Cassirer and for giving permission to reproduce it here.

Professor Dirk van Dalen of the University of Utrecht, and to Dr. Helmut Rohlfing of the Niedersächsische Staats-und Universitätsbibliothek Göttingen, for locating the photograph of Hermann Weyl. Reproduction by permission of the Niedersächsische Staats-und Universitätsbibliothek Göttingen. All rights reserved.

The Emilio Segrè Visual Archives at the American Institute of Physics for permission to reproduce the photograph of A. S. Eddington.

Norbert Ludwig and Sabine Schumann of the Bildarchiv Preussischer Kulturbesitz, Berlin, for the photograph of Albert Einstein. Permission to reproduce the latter was also granted by the Albert Einstein Archives, Jewish National and University Library, Jerusalem. Many thanks to Barbara Wolff for her assistance.

CONTENTS

THE REIGN OF RELATIVITY

1

INTRODUCTION

> It is only a world embodying the principle of relativity, in the form
> which the doctrine entails, that can be said to exhibit the character
> of mind, with its exclusion of disconnected fragments and relations.
>
> Haldane (1921, 138)

For a brief decade in the early part of the 20th century, an observer with a passing interest in the scene may well have forecast a considerably different course for the subsequent development of 20th century philosophy of science. Einstein's theory of gravitation was announced to the wider world at a joint London meeting of the Royal Society and the Royal Astronomical Society on the 6th of November, 1919. Reporting the British expedition's empirical confirmation of the theory through observations of the solar eclipse six months earlier, the Nobel prize winner J. J. Thomson went on to characterize Einstein's theory, called a theory of "general relativity," as "one of the greatest achievements in the history of human thought".[1] Thompson's proclamation set the stage for the ensuing public clamor. To a world wearied by war, hatred, sickness, and destruction, the British scientific establishment's official endorsement of the theory of a German physicist seemed to beckon a new era of international cooperation and understanding. Yet the excitement occasioned by the theory reached much further into the collective psyche, to an extent that is both difficult to imagine, and as yet unrivaled, for a result of pure science without foreseeable practical or technological application. Much of the commotion was certainly fueled by journalistic sensationalism, mysticism, and even anti-Semitism. But in overthrowing such permanent fixtures of the cognitive landscape as Newtonian gravitational theory and the Euclidean geometry of space, Einstein's theory rather suddenly attained the iconic cultural standing, retained still today, of a revolutionary transformation of outlook.

Naturally the theory became a principal focus of philosophical interest and inquiry. However, its abstract statement in the mathematics of the tensor calculus, not to mention the very fluidity of physical and mathematical meaning attending its fundamental principles, of equivalence, of general relativity, of general covariance, and, finally, of what Einstein, in 1918, had termed "Mach's Principle", greatly complicated the task of coming to a synthetic understanding. Such initial ambiguities are not unusual. Many new scientific theories bring unfamiliar mathematics, and physical theories, if sufficiently robust, are rarely if ever without unproblematic aspects, often taken to say different things at different times. In this situation, it is understandable that there was considerable interpretive latitude for inherently antagonistic philosophical viewpoints, all seeking vindication, confirmation, or illumination from the revolutionary theory. Perhaps only semi-facetiously, Bertrand Russell, at the end of the decade (1926), observed,

> There has been a tendency, not uncommon in the case of a new scientific theory, for every philosopher to interpret the work of Einstein in accordance with his own metaphysical system, and to suggest that the outcome is a great accession of strength to the views which the philosopher in question previous held. This cannot be true in all cases; and it may be hoped it is true in none. It would be disappointing if so fundamental a change as Einstein has introduced involved no philosophical novelty.[2]

Russell himself found general relativity to be a source of "philosophical novelty", for he recast his own analytical metaphysics of "neutral monism" on its basis.[3] But as he implied, the very project of seeking to identify such a revolutionary transformation of thought with any one philosophical viewpoint is suspect from the outset, ignoring the fact that schools of "philosophical interpretation" in turn "evolve" to accommodate or domesticate the prestigious novel conceptions. This is precisely what happened in the case of general relativity.

The contours of philosophy of science since the 1920s testify to a somewhat different perception. Logical empiricism forthrightly admitted the influence of the theory of relativity in shaping the fundamental core of its outlook. Since the rise of logical empiricism, from which stem the main trends in subsequent philosophy of science, if only critically, it has been widely if not universally accepted that relativity theory had shown the untenability of any "philosophy of the synthetic *a priori*". If individual responsibility can be assigned for this assessment, it belongs to Moritz Schlick, to become the *éminence grise* of the Vienna Circle and logical empiricism generally. In the same year, 1922, in which Schlick came to Vienna to occupy the chair of Philosophy of the Inductive Sciences that had been created for Ernst Mach, he addressed an audience of several thousand at the centenary meeting of the German Society of Natural Scientists and Doctors on the topic of "The

Theory of Relativity in Philosophy". No complex or lengthy argument detained him from reaching the conclusion that still has circulation in the curriculum of philosophical instruction:

> Now along comes the general theory of relativity, and finds itself obliged to use non-Euclidean geometry in order to describe this same world. Through Einstein, therefore, what Riemann and Helmholtz claimed as a possibility has now become a reality, the Kantian position is untenable, and empiricist philosophy has gained one of its most brilliant triumphs.[4]

Schlick was not just any philosopher—he had a Ph.D. in mathematical physics under Max Planck in Berlin and was author of both the first philosophical monograph on general relativity and an epistemology book favored by Einstein. In short, Schlick was *the* recognized authority on the philosophical direction of the new theory. As it turned out, the empiricist philosophy whose triumph Schlick celebrated in 1922 scarcely yet existed, for Schlick himself held a *holist* and *conventionalist* account of the metric of space-time in his previous writings on relativity theory. But with some strategic assistance from a recent text of Einstein, and several older ones of Helmholtz, it was fashioned in short order and its influence was far reaching, encompassing the younger philosophers Rudolf Carnap and Hans Reichenbach, who, together with Otto Neurath, were to become the founding fathers of a new and "logical" empiricism.

It will be seen that, however rhetorically useful, the claim that general relativity sounded the death knell of "the Kantian position" follows only if, as Schlick did, one ignored important neo-Kantian developments of Kant's thought as well as many of the most significant developments in relativity theory in the period 1915–1925. Schlick's judgment was narrowly based and by no means universally shared. To sample but one countering opinion, the Nobel prize winner and fellow Planck student Max von Laue stated, in the first actual textbook on general relativity in 1921, that Kantian epistemology was confirmed by the new theory, although "not every sentence of *The Critique of Pure Reason*" could be regarded as sacrosanct.[5] Yet as pious children of this world, to borrow an expression of Hermann Weyl's, we know that if an assertion is repeated sufficiently often, while remaining unchallenged in the forum of debate, it commonly enters into currency as accepted background knowledge. Certainly the claim that general relativity decisively refuted transcendental idealism *tout à coup* is strewn through the literature of logical empiricism, percolating beyond to its prodigal progeny. Nor was it explicitly challenged in philosophical circles by anyone having the *gravitas* of authority possessed by Schlick, and then by Reichenbach, who would take over the mantle of authority on relativity theory within logical empiricism, as Schlick fell under the influence of Wittgenstein and turned away from philosophical investigations of physics. As a result, the allegation has remained unimpeached amidst the triple assault that proved fatal to the rest of logical empiricism: Quine's attack on the analytic–synthetic distinction, Hanson's and Toulmin's on the observational–theoretical distinction, and Kuhn's critique of logical empiricism's inductivism and its method of rational reconstruction. So it was that, when scientific realism began again to stir in late 1950s and early 1960s, as it always will, against the thin gruel of positivism and instrumentalism, there were scarcely any parties to the conflict who grasped the possibility of an alternative to *both* realism *and* instrumentalism or, beginning in the 1970s, to realism and the resuscitated bogey of "relativism".

That alternative already existed, and it assumed several different, but related forms, in the "reign of relativity" from 1915 to 1925 through the efforts of Ernst Cassirer, Hermann Weyl, and Arthur Stanley Eddington. It is a philosophy that exists only in various incomplete realizations having at most a "family resemblance" among themselves. In this book it is called *transcendental idealism*, and although Kant is the paramount figure historically, its development by no means ended with Kant, as Cassirer, Husserl, Weyl, and others have shown. I will therefore use the term "transcendental idealism" far more broadly than is customary in most philosophical discussions. But for present purposes, the core constituent of the doctrine concerns the "transcendental constitution of objectivity" in fundamental physical theory, according to a "transcendental postulate", in broad generality affirming that "[a] nature is not thinkable apart from the coexistent subjects capable of experiencing that nature".[6] The details of the various and differing conceptions of "transcendental constitution" in general relativity are described in detail below in discussions of Cassirer, Weyl, and Eddington.

Of course, serious discourse on the "constitution of objectivity" has long been out of favor in philosophy of science, another legacy of logical empiricism and, indeed, of Schlick. As has become familiar since the work of Alberto Coffa (1991), the young Reichenbach in 1920 held to a conception of the "relativized *a priori*" that attempted to retain the constitutive standing of *a priori* principles while surrendering any claim that such principles are necessarily valid, constitutive of *any* possible experience. But I show in chapter 2 that his "constitutive" discourse was already fatally compromised at the outset by adopting Schlick's language of a "coordination" (*Zuordnung*) between concepts of formal mathematical theory and empirically ascertainable physical objects. For, as it happened, the abstract relation of "coordination" was Schlick's "line in the sand" against the encroachments of Kantian epistemology, and it became, on assimilating the methodology of rods and clocks of Einstein's "practical geometry", the principal weapon of logical empiricism against neo-Kantian interpretations of relativity theory. In chapters 3 and 4, it will be seen that in Schlick's empiricist alternative, initially proposed as an "empiricism with constitutive principles", talk of "constitution" quickly faded from view. In its place came a new empiricist interpretation of physics wherein the ties of theory to observation are explicitly made through "coordinative definitions". The mechanism of "coordinative definitions" was definitively stated just a few years later by Reichenbach and henceforth was associated with his name. But Schlick's influence was instrumental in weaning Reichenbach away from his early neo-Kantian theory-specific and thus "relative *a priori* constitutive principles" to a "consistent empiricism" where "constitutive principles" have become stipulations (in the case of general relativity) about rigid rods and perfect clocks. The ensuing account of the empirical determination of the metric in general relativity would emerge as the logical empiricist paradigm of how the terms of a physical theory, regarded initially signs of an uninterpreted logico-mathematical calculus, received empirical content through connection to observation terms, via conventionally adopted "correspondence rules".

Although present interest in Cassirer is appreciably rising, he was known to several generations of "analytic" philosophers only as a historian of philosophy and so, in view of the low esteem generally accorded to history of philosophy within analytic philosophy (a situation now fortunately slowly changing), largely unread. A number of his books, including the first three volumes of *Das Erkenntnisproblem* that made his reputation in Germany, have never been translated into English. On the other hand, the English translations of his two books on what may broadly be called *Wissenschaftstheorie* (the theory of science), made in the early 1920s, are off-putting for diluting an already diffuse style with Victorian archaisms and scientific illiteracy. Yet the Marburg tradition of neo-Kantianism, within which Cassirer had been educated, long before rejected the original Kantian distinction between the mental faculties of sensibility and understanding, and on this ground Cassirer could reinterpret the doctrine of pure intuition in *conceptual terms* as pertaining only to "the order in general of coexistence and succession".[7] In his 1921 book of "epistemological reflections" on Einstein's theory of relativity, as discussed in chapter 2, he was in a position to grasp what is arguably the most philosophically significant aspect of general relativity, the principle of general covariance, as a "regulative principle" and constituent part of an ideal of physical objectivity from which all traces of "anthropomorphic" subjectivity have been removed. In an enlightened understanding (which is fully in the spirit of Cassirer's discussion), this is the requirement that dynamical laws must be formulated without a "background" space and time, a constitutive requirement of general relativity, but utterly violated in the standard operator formalism of quantum field theory.

The most systematic articulation of the alternative to the new empiricism is to be found in the writings of the mathematician, and interloper in theoretical physics, Hermann Weyl, who looms disproportionately large in the following pages. Weyl was an original. Universally regarded as one of the premier mathematicians of the century, in the decade in question, his contributions to relativity theory ranked second only to Einstein's, and in fact, it is from Weyl that the present mathematical formulation of the theory stems. In the same period, he was a key figure, along with Hilbert and Brouwer, in the debate over the foundations of mathematics. By its end, in 1926, Weyl had produced what he considered to be his "single greatest contribution to mathematics", the theory of representations of semi-simple Lie groups and Lie algebras, *and* written one of the few classics of philosophy of science and mathematics. Just a year later, in 1927, he pioneered the application of group representations to quantum mechanics. For many years, his seminal contribution to physics, originally made in the course of his work on general relativity in 1918, the idea of "gauge invariance" or "local symmetry", was regarded as somewhat peripheral; this changed with Yang and Mills in the United States and, independently, Shaw in the United Kingdom, right around the time of Weyl's death in 1955. With but few exceptions, Weyl has not been systematically read by philosophers (at least not in the English-speaking world) partly, it must be sadly said again, on account of defective or nonexistent translations, but partly also because of his use of a philosophical language almost entirely alien to those interested in philosophical issues in mathematics and physics. That language, at least in his remarkable book on relativity theory, *Raum-Zeit-Materie*, also first appearing in 1918, is the language of transcendental-phenomenological idealism of Edmund Husserl.

Weyl's challenge to the new empiricism is ostensibly a disagreement about the use of rods and clocks as fiduciary measuring instruments in general relativity. To Schlick and Reichenbach, Helmholtz had created, a half-century before, the outlines of a nonnaive geometric empiricism consequent on making stipulations regarding certain bodies as rigid; once adopted, those bodies could be employed to empirically determine the geometry of physical space. Terming this conception "practical geometry", Einstein, in a widely read lecture of January 1921, entitled "Geometry and Experience", stated that it indeed had been instrumental in setting up the general theory of relativity, even though, he admitted, the concepts of "rigid body" and "perfect clock" had to be accepted as posits independent of the theory. But for Einstein, the stipulation about rigid rods papered over a deeper issue. In the spring of 1918, Weyl had proposed a geometric unification of gravitation and electromagnetism, a further step along the road of general relativity. The basis of the unification was a "pure infinitesimal geometry" permitting neither direct comparisons "at a distance" of direction nor, unlike the Riemannian geometry of Einstein's theory, of magnitude. Within such a geometry, Weyl recast Einstein's theory together with electromagnetism on the privileged epistemological basis of fundamental differential geometric notions having immediate validity only in the tangent space attached to each manifold point P, corresponding to a localized space of intuition. In opposition to the scientific realism of his day, and in a characteristically distinctive fashion combining Husserlian "essential analysis" of space and time as "forms of intuition" with mathematical construction, Weyl sought in this way to provide a transcendental-phenomenological account of the constitution of the *sense* of the objective world of relativity theory, the sense of a "being *for* consciousness". However, Weyl's epistemological motivations were expressed in the obscure language of Husserl, and his theory, thus misunderstood, was critically rejected on both physical and general methodological grounds.

The ties of Weyl's theory to observation *are* indirect; and, if we accept Weyl's recognition of the existence of a "natural gauge" of the world, simply presupposed in Einstein's posit of rods and clocks, they are also present in general relativity. The values of the metric at a point can be determined through the use of freely falling neutral "test particles" and by observing the arrival of light at points in the immediate neighborhood of that point. However, neither of these hypotheses, of "freely falling" test particles or of the behavior of light in a gravitational field, is independent of gravitational theory. Both can be derived from the Einstein field equations for particular models of space-time. For this reason, as Weyl repeatedly stressed regarding Einstein's theory, only the theory as a whole, comprising physics, geometry, and mechanics, can be confronted with observation. If that is so, then, as Schlick put it, there is no place for an empiricism worthy of the name to gain a place to stand. A different epistemology of science would have to be found. For without such an empiricist Archimedean point for general relativity, allegedly endorsed by Einstein and therefore to be retained at all cost, there could be no room for subsequent logical empiricist methodology of science to thrive. So too for the fruits of its analysis of science: an empiricist semantics for theoretical terms and sentences, the empiricist criterion of cognitive meaning, and the positivist rhetoric that any nonempirical statement was either analytic or meaningless "metaphysics". When, a full generation later, these invidious doctrines finally faded from the scene under assault from different quarters, the lack of a clear alternative was perhaps noticeable only to those whose horizon stretched back to the philosophically fecund first years of general relativity. In its absence came the inevitable backlash of scientific realism and its several antitheses.

As it happened, of course, such an epistemology of science was developed, in part in bits and pieces of Weyl's mathematical and physical oeuvre and, in broader generality, in his monograph on philosophy mathematics and natural science in 1926. By then, Weyl had returned for good, except for a brief excursion into the new quantum theory, to purely mathematical pursuits. This left the playing field of "scientific philosophy" open to Reichenbach's "constructive axiomatization" of the theory of relativity (1924), where the mechanism of "coordinative definitions" took over from Schlick's still-born "empiricism with constitutive principles", and in this guise the new empiricist analysis of scientific theories acquired its mature form. After the "linguistic turn" of the early 1930s, the discourse became one of two vocabularies, or languages, "theoretical" and "observational", and of defining the former in terms of the latter, eventually through "meaning postulates". Citing Einstein as a guiding spirit, the logical empiricists claimed the authority of philosophical expertise regarding relativity theory, a title they are still perceived in many circles to hold, as it were, from beyond the grave, and despite Einstein's later public disavowals of their core positions. Ironically, Einstein's own philosophical evolution after 1915 carried him further and further away from the empiricism Schlick viewed as present in general relativity and toward neo-Kantian conceptions and the mathematical speculative methodology for which he had once chastised Weyl.

One more figure played a central role in the possible alternative tradition to logical empiricism and its successors that may be loosely associated with "the Kantian position". If one were to name the grand masters of general relativity in the early 1920s, besides the names of Einstein, Weyl, Hilbert, the young Wolfgang Pauli, Jr., and on the mathematical side, Élie Cartan and George D. Birkhoff, only that of Arthur Stanley Eddington remains. Eddington, Plumian Professor of Astronomy at Cambridge since 1914, was already an internationally known astronomer in 1915. He would become, in the assessment of S. Chandrasekhar, "the most distinguished astrophysicist of his time".[8] He was also the first in Britain to have any detailed knowledge of Einstein's new theory during the first World War. With his mathematical skills, he was also a highly creative relativity theorist. In fact, he was so connected to the new theory, as exponent, expositor, and theoretician, that he became known in Britain as "the apostle of relativity", and we have it from no less a source than Paul Dirac that in the early 1920s, his name, not Einstein's, was most closely linked there with the new theory.

Eddington was also heretical enough to accept Weyl's generalization of Einstein's theory and to generalize it further, for epistemological reasons essentially similar to Weyl's. Weyl had reconstructed the objective world of relativity physics within a "purely infinitesimal geometry", corresponding to the phenomenological standpoint of methodological solipsism wherein only such linear relations as could be present to an infinitesimally bounded spatio-temporal intuition were immediately evident. Eddington sought the same goal of constituting the "real world of physics" by reconstructing relativity theory within a differential geometry capable of yielding only objects that are a "synthesis of all aspects" present to all conceivable observers. The external world of physics might be *defined* in this way as a world conceived "from the viewpoint of no one in particular", a standpoint both necessary and sufficient for *objective* representation in physics. The epistemological significance of relativity theory lay in showing that the attempt to portray the physical world from this impersonal perspective resulted in its geometrization. In turn, the physical knowledge captured in such a portrayal is knowledge only of that world's structure. Physics *could be* about *no* other world than that expressly incorporating all viewpoints at once, an "absolute world" as opposed to the "relative" world of each individual perspective, that is, any "conceivable observer". The relation between the relative and the absolute is mathematically captured by the tensor calculus and physical knowledge accordingly must be represented in the form of tensor identities through a method Eddington called "world building".

As we shall see in chapters 7 and 8, Eddington was adamantly convinced that Weyl's "epistemological principle of the relativity of magnitude" (the origin of the modern "gauge principle") was an essential addition to the outlook of relativity of continually incorporating additional "points of view" into physics. But, in the intricacies of Weyl's transcendental-phenomenological framework of constitution, Eddington judged, Weyl had erred. For Weyl had not made clear that his geometry was *ideal* and purely mathematical, a geometrical skeleton for the "graphical representation" of existing physics from "the point of view of no one in particular". Eddington's idea, therefore, was to develop such an ideal geometry independently of physics, basing it on a purely local and *nonmetrical* relation of comparison, a symmetric linear connection. In a geometry based on such an "affine connection", rather than a metric, a more general kind of invariant than tensors can be "built up"; nonetheless, only one of these is mathematically identical to the metric tensor of Einstein's theory. Setting the two equivalent, one can proceed to "graphically represent" the tensorial quantities of existing physical theory, gravitation, and electromagnetism. The ideal geometry of Eddington's affine field theory then shows that Einstein's geometry, not Weyl's, is exact, but this is a demonstration from the most general "the point of view of no one in particular" available to a continuum theory in 1921. Eddington's theory is not a physical hypothesis but an explicit attempt to cast light on the origin and significance of the great field laws of gravitation and electromagnetism. Within the epistemological reconstruction of "world building", the differential geometric invariants appearing in these laws are structures selected from a vast number of other possible invariant structures derivable from given axioms of "primitive relation structure". Mind is the principle of selection; in particular, it is mind's interest in "permanence" that identifies the Einstein curvature tensor, regarded in "world building" as a purely geometrical quantity, with the physical energy-momentum tensor of matter. Hence, Einstein's law of gravitation for "matter" sources is simply a world geometric definition of matter. In the absence of matter, Einstein's law of gravitation for empty space (as amended with the cosmological constant) is a statement that the world is "self-gauging", that rods and clocks, apparatus of course part of the world (and explicitly so, in "world building"), are used in measuring the world. As Eddington pointed out later on, there similarities between his view of physical knowledge and those of Kant. One difference, certainly, is that Eddington's account of the constitution of physical objectivity simply assumes relativity theory, where Kant had assumed Newton. I shall show that the similarities are considerably more noticeable when set in the context of transcendental idealism, more broadly conceived.

In the pages that follow it will be seen that the emergence of logical empiricism in the 1930s as the apotheosis of "scientific philosophy" (a reputation still widely upheld) had little to do with its purported expertise regarding relativity theory but was achieved largely through rhetoric and successful propaganda rather than through philosophical argument. Its most (and still) alluring appeal lay in a self-styled contrast of enlightenment versus reaction, and in its identification of science as the primary instrument of human advance from the dreary annals of superstition, dogma, and fanaticism that permeate human history. Its great myths even today have hardly been questioned: that relativity theory had overthrown any form of "Kantianism"; that "empiricism" stood opposed only to an antiscientific and dogmatic "rationalism"; that logical empiricism, itself modeled on the methodology of relativity theory, was *d'accord* with modern physics (relativity theory and quantum mechanics). The doctrinal triumph of logical empiricist philosophy of science itself, of course, was not lasting. Its employment of a new favorite tool, symbolic logic, as the *organon* of philosophy of science, an *ersatz* for actual knowledge of science, still succeeds to some extent in reviving the desiccated corpse of logical empiricism through the boom-and-bust cottage industry of mainstream philosophy of science. But even symbolic logic could not save "the received view" from the inevitable cognitive discord induced by a glaring awareness of the enormous gap between its rational reconstructive portrait of science and that of a new history of science, reinvigorated by Koyré and, above all, Kuhn, as was recognized by Hempel in his last writings.[9] Rather, these myths live on institutionally, subconsciously continuing in the sclerotic distinction between "analytic" and "continental" philosophy. Surmounting that artificial distinction, the family resemblance among the "transcendental idealisms" of Cassirer, Weyl and Eddington contains the seeds of promise for an actual philosophical understanding of the *non plus ultra* role of abstract mathematics in fundamental physical theory.

In 1931, P.A.M. Dirac prefaced his celebrated paper on magnetic monopoles with several remarks that announce a sea change in the methodology of theoretical physics. Stating that drastic revision of fundamental concepts may be required to address the current problems of theoretical physics, Dirac nonetheless cautioned that such a transformation in outlook is likely to be beyond the power of human intelligence to directly grasp the required new ideas without the assistance of mathematical speculation. In the face of these cognitive limitations, a more indirect approach is suggested, wherein "the most powerful method of advance" would be

> to perfect and generalize the mathematical formalism that forms the existing basis of theoretical physics, and *after* each success in this direction, to try to interpret the new mathematical features in terms of physical entities (by a process like Eddington's Principle of Identification).[10]

Now this principle, as Eddington himself made clear, was directly inspired by Weyl's mathematical identification of the vector and tensor structures of his purely infinitesimal world geometry with those of gravitation and electromagnetism. That being the case, Weyl's 1918 theory can be justly regarded as the locus of the modern revival of the method of *a priori* mathematical conjecture in fundamental physical theory. How such a method can ever be fruitful in constructing well-confirmed fundamental physical theories has long appeared a mystery, for which extreme solutions (such as Platonism) have been seriously proposed. The argument of this book suggests that less desperate measures may have been overlooked. The work of Cassirer, Weyl, and Eddington on general relativity provides a needed "Copernican about-face" on the question, by demonstrating how and why *a priori* constraints of reasonableness can be imposed on nature without proudly (but naively) presuming them to be inherent in nature itself. They did not leave us a fully worked out presentation of an alternative epistemology of science, each going on to other endeavors that effectively removed their work from the sphere of the familiar that so bounds human understanding, even in philosophy. In all likelihood, such a completed account doesn't, or shouldn't, exist except as an ideal guiding inquiry. What they did leave has been allowed here to "speak for itself", a presentation that comes at times at the cost of effusive length, but that appeared necessary in the light of the unfamiliarity, and even inaccessibility, of many of their core writings. Perhaps any further development, any "future music", to quote Weyl again, might be well advised to at least consider what they once had to say.

2

GENERAL COVARIANCE AND THE "RELATIVIZED *A PRIORI*"

Two Roads from Kant

> For the transcendental idealist, the object of knowledge is … neither immanent nor transcendently "given" ["*gegeben*"], but rather "posed as a problem" ["*aufgegeben*"].
>
> Rickert (1921, 316)

> I did not grow up in the Kantian tradition, but came to understand the truly valuable which is to be found in his doctrine, alongside of errors which today are quite obvious, quite late. It is contained in the sentence: "The real is not given (*gegeben*) to us, but put to us [*aufgegeben*]" (by way of a riddle)".
>
> Albert Einstein (1949, 680)[1]

2.1 Introduction

Kantian and neo-Kantian publications comprised a not-insignificant torrent in the "relativity rumpus" following the announced confirmation of the general theory of relativity in November 1919.[2] Of course, many other "schools" of philosophy felt obliged to immediately pronounce upon the theory, often as a stunning vindication of basic principles or outlooks. But it was incontrovertible that general relativity, on corroboration of the dramatic prediction of star images displaced by the sun's gravitational field, minimally required modification or clarification of the necessarily

Euclidean structure of space implied by the Transcendental Aesthetic. Most of this literature, regardless of its provenance, contains little of present interest. But within a few months in late 1920 and early 1921, Ernst Cassirer and Hans Reichenbach published neo-Kantian appraisals of the theory of relativity whose historical and philosophical significance has acquired renewed relevance at the beginning of the 21st century.[3] Looking backward, the superficially similar monographs of Cassirer and Reichenbach[4] originate in distinct vantage points that pose with considerable sharpness two diverging paths for what might be broadly termed the "epistemology of physical science", and so for the subsequent development of 20th century philosophy of science. The roads taken by Reichenbach and Cassirer are not peripheral, resting on differing appraisals of a difficulty *within* Kant, perhaps *the difficulty within Kantian epistemology*. A central problem of the Transcendental Analytic, it concerns the distinct mental faculties of "sensibility" and "understanding" and the nature of the relation between them if, as Kant insisted, these are to be regarded as *independent sources of knowledge*. Carried over to the context of epistemological reflection on general relativity, the issue is transposed into one of the meaning and significance of the theory's mathematical framework, and, most important, of the requirement of general covariance that framework must satisfy. Both Cassirer and Reichenbach drew upon a revisionist conception of the role of *a priori* elements in physical theory as not fixed for all time but "relative", changing with the advance of physical science. Each regarded general covariance as such an *a priori* principle, "constitutive of object of knowledge" for contemporary physics, yet their distinctive treatments of how that role is performed are based upon fundamental differences regarding the respective contributions of "understanding" and "sensibility" to knowledge. That difference marked a watershed for subsequent philosophy of science.

It is a matter of record that, in doctrine as well as institutionally, logical empiricist philosophy of science, and therefore much of mainstream 20th century philosophy of science, emerged from the turn taken by Reichenbach in 1920. In his version of the "relativized" *a priori*, constitutive "principles of coordination" are regarded necessary for the univocal coordination of (as he put it) "reason" to "experience" that comprises physical cognition. These principles themselves collectively represent the "subjective contribution of reason to knowledge", varying with the development of physical theory, whereas the "objective contribution to knowledge" is alone provided by experience. The principle of general covariance is but one of a set of coordination principles for the general theory of relativity, having essentially the meaning of a generalized principle of relativity, "the relativity of the coordinates". In any case, within a few years Reichenbach's coordination principles re-emerged in the new guise of "coordinative definitions". Following Schlick's lead, Reichenbach came to see coordination principles as playing so "thin" a "constitutive" role as to be indistinguishable from stipulations that certain empirically accessible objects and processes are (approximately) described by core relations of the mathematical framework of a physical theory. Yet the most significant aspect of this reevaluation is that it consolidated a fundamental shift of epistemological discussion initiated by Schlick's influential *definition* of cognition as a "univocal coordination" in 1918. Subsequently, philosophers of science would have progressively less and less understanding of the relevance of any account of "physical objectivity" in accordance with conformity to presupposed "conditions of possible experience". Instead, the relevant epistemological issues of interest concerned the applicability of an uninterpreted mathematical formalism to an empirically given concrete subject matter or, in terms more redolent of mature logical empiricism, the semantical rules through which a mathematical framework acquires empirical content, and so "cognitive meaning" in physics.

It is my contention that Cassirer's different articulation of a role for the relativized *a priori* has been rather amply confirmed in the subsequent development of physical theory. Namely, Cassirer expressly pinpointed the specific "meta-empirical" standing of invariance principles in physical theory, in particular, emphasizing that the principle of general covariance significantly transformed the concept of "objectivity" in physics. In this role, principles of invariance have *both* a "constitutive" *and* an ideal "regulative" *a priori* significance. To be sure, in Kant's account of "constitution of the object of knowledge" from the two independent contributions of sensibility and understanding, "regulative" principles, systematic ideals of unity of the "higher faculty" of pure reason, can play no direct "constitutive" role. Hence, one salient division between Cassirer's and Reichenbach's epistemological analyses of relativity theory pertains to just where modification or amplification of original Kantian doctrine is required in order to retain a constitutive but nonconventional meaning of the "relativized *a priori*". Ultimately, the different determinations reduce to opposing answers to the question of *whether there are nonanalytic* a priori *elements in physical theory*. As shown in chapter 3, due in large measure to Schlick's considerable authority regarding the new theory as well as his rhetorical ability to pose the issue in his own terms, this was a very short debate that, in the eyes of "scientific philosophy", Cassirer lost.[5]

In this chapter, and in those subsequent to it, I venture to challenge this received wisdom on grounds internal to various epistemological analyses of the theory of general relativity, all carried out within a broader genus of "transcendental idealism" than is to be explicitly found in Kant. Here, attention is directed toward how, constrained by the resources afforded by their competing revisions of Kantian doctrine, Reichenbach and Cassirer respectively assessed the implications of the principle of general covariance for the "epistemology of physical science". Of course, the matter of precisely what, if any, physical significance this principle may have, has long been perhaps the most controversial issue in the foundations of general relativity. Einstein's considered judgment, that the principle is not physically vacuous but has "considerable heuristic force" in the construction of physical theories, was explicitly recognized by Cassirer, but also has been adopted (or insisted upon) by several leading members of the current generation of theorists of quantum gravity. Following Cassirer's lead, I argue that, in epistemological terms, the "heuristic force" of general covariance is located in the principle's "constitutive" significance in constraining the concept of possible object in field theory to objects that are "background independent". This is to be understood as the expression of an "ideal of reason", namely, that "the objects of which the world is made do not live over a stage and do not live on space-time; they live, so to say, over each other's shoulders",[6] a goal not yet attained in the present state of fundamental theory. But as this conception first emerged in Einstein's resolution of the so-called Hole Argument in 1915, I first consider this argument, together with some recent work that illuminates the regulative idea of background independence. After a survey of these results, I turn back to the Kantian "faculties" of sensibility and understanding, the two roads taken by Cassirer and Reichenbach, and how their differing accounts of objectivity issue in different evaluations of the significance of general covariance.

2.2 The Constitutive Character of General Covariance

In the first complete exposition of general relativity in 1916, Einstein introduced a "postulate of general relativity", deemed an extension of the principle of relativity: *"The laws of physics are to be of such a kind that they apply to systems of reference in any kind of motion"*. A few pages on, a condition of coordinate generality is posed: *"The general laws of nature are to be expressed through equations which are valid for all coordinate systems, that is, are covariant with respect to arbitrary substitutions (generally covariant)"*. Freedom to make arbitrary continuous transformations of the space-time coordinates, Einstein continued, "takes away from space and time the last remnants of physical objectivity" while rendering the laws of nature "suitable for the postulate of general relativity". In support of these contentions, Einstein offered a "reflection" concluding that "all our physical experience" is reducible to "coincidences" of point-events, for whose description alone a reference system is required.[7] No aspect of the foundations of the general relativity has resulted in more controversy and confusion than these puzzling assertions, coupling a generalized principle of relativity for rotating and accelerating frames of reference (apparently expressing the "relativity of all motions"), to a condition of coordinate generality, to reach such striking conclusions. "Eight decades of dispute" ensued regarding the meaning of the principle of general covariance. Yet despite the clarification brought in a recent exhaustive survey,[8] it is a fair assessment to say that the debate rages on into a ninth decade, with the promise of more to follow. Obviously, the many, often intricate, issues involved cannot be resolved here to the satisfaction of all concerned parties. Rather, I wish to outline a case that the fundamental motivation for the principle, arguably Einstein's own, is that it serves as a guiding specification, in ways that are notoriously hard to be completely precise about, of *what is a possible object of fundamental physical theory*. In imposing the condition of general covariance, Einstein sought to legislate that the properties and motions of such objects, as represented in their governing or "constituting" field laws, must be specified without any reference, even implicit, to the setting of a background space-time. In Einstein's ideal conception, fields are *not* properties of space-time points or regions: they *are* those points and regions in whose terms a Machian fully relational dynamics may be implemented.[9] Neither Einstein nor anyone else has been successful in explicitly attaining this ideal;[10] nonetheless, in his later years he returned to the problem in ways that prove helpful in stating our case. Before considering these, I briefly survey the story of Einstein's own confusions over general covariance and how in 1915 he overcame the "hole argument" that had led him astray, the invariable starting point of most contemporary discussions of general covariance.

2.2.1 The Received View

To be sure, the above passage is deeply puzzling, for it appears to intertwine two egregious errors. First is the claim that satisfaction of the condition of coordinate generality is prerequisite to a principle of relativity generalized to include accelerating reference frames. Indeed, Einstein had named his theory of gravitation a *general* theory of relativity because, by application of the principle of equivalence, the behavior of bodies freely falling in a gravitational field (of a highly artificial kind) is indistinguishable from that behavior observed in a frame of reference uniformly accelerated in a gravitation-free region.[11] As there is "an exact physical equivalence" between the freely falling and uniformly accelerated bodies,

> one can just as little speak of the *absolute acceleration* of the system of reference, as in the usual [special] theory of relativity one can speak of the *absolute velocity* of a system.[12]

Leaving aside the (for all practical purposes) *ideal* case of a perfectly homogeneous gravitational field where tidal forces are zero, however, local experiments measuring these forces *can* distinguish between a frame at rest in a gravitational field and an accelerated frame far from any gravitating bodies. To the extent that rigorous validity of such a principle of equivalence fails, a general relativity of motion based on that principle, insofar as it has a clear meaning, must be false.[13] To be sure, unlike in the special theory of relativity, according to the principle of equivalence, in Einstein's gravitational theory there cannot be *global* inertial frames, and so it *must* be formulated with full space-time coordinate generality.[14] But the second error concerns the latter condition. The objection of a young mathematician, Erich Kretschmann, in 1917 is now legend. General covariance, if mere coordinate generality is intended, is merely a formal constraint on the theory's mathematical form, having *per se* nothing to do with a "principle of general relativity" or with the theory of gravitation.[15] In a purely formal sense, an equation is generally covariant just in case it preserves its form (is "covariant") under arbitrary transformation from one coordinate chart to another, $\bar{x} \Rightarrow \bar{x}'$. Moreover, Kretschmann claimed that since, as Einstein affirmed, the totality of physical experience must ultimately refer to coincidences, any physical theory that preserves the lawful connections among coincidences can be written in generally covariant form, subject only to the introduction of additional variables.[16] If it had been Einstein's intent to lend physical significance to general covariance, he had not succeeded in distinguishing that meaning from this purely formal constraint.[17] In a response the next year, Einstein admitted the correctness of Kretschmann's objection, nonetheless maintaining that the "relativity principle", according to which laws of nature find "their sole natural expression in generally covariant equations", has "a significant heuristic force". By contrast, he argued, if one were to write down the equations of Newtonian gravitational mechanics in (four-dimensional) generally covariant form, the result would be seen to be so unnatural as to be readily excluded from theoretical consideration. One can only speculate about Einstein's criteria of theoretical naturalness, but in any case, his reply has been widely judged inadequate, or at least as an essential backpedaling, from his earlier claims on behalf of general covariance.[18]

On the "received view" then, Einstein's vain efforts to give physical significance to general covariance are accounted as due to an overweening philosophical ambition to fully implement Mach's program for the relativization of all inertial effects. In so doing, he inadvertently conflated mathematical technique and physical content.[19] Such assessments may be found even among the close circle of Einstein's collaborators, as the following example from one of Einstein's assistants from the 1930s, Banesh Hoffmann, shows:

> And as for the principle of general covariance, Einstein's belief that it expressed the relativity of all motion was erroneous.[fn.] Worse, as was quickly pointed out, the principle of general covariance is, in a sense, devoid of content since practically *any* physical theory expressible mathematically can be put into tensor form—and this includes not only the special theory of relativity but also the Newtonian theory.[20]

Perhaps the most widely read recent formulation of the received view, that of Michael Friedman (1983), maintains that in upholding general covariance as a principle of general relativity, Einstein illicitly mixed together distinct notions that pertain either to the form or to the content of the theory, but not to both. "We now know that all space-time theories can be given a general covariant formulation: general covariance is only a new mathematical technique, not an expression of new physical content".[21] In fact, from the very first days of general relativity, it has been known that the principle of general covariance, understood as coordinate generality, does place formal constraints on the possible interactions of gravitation and matter. In particular, one consequence of general covariance alone is that not all of the Euler–Lagrange equations, obtained by varying the gravitational action with respect to the field variables $g_{\mu\nu}$ and their first derivatives are independent, as Hilbert first demonstrated in 1915. In this way, a constraint on the set of solutions is introduced that can be precisely formulated: in a generally covariant field theory for n unknown functions, there can exist no more than $n - 4$ independent equations.[22] Furthermore, in conjunction with other requirements, namely, a prohibition against other geometric object fields (than $g_{\mu\nu}$) and that the Euler–Lagrange equations are no higher than second order, the constraints imposed by general covariance on the gravitational action are substantive and highly nontrivial.[23]

In any case, late in his life Einstein made several attempts, to which John Stachel in particular has called attention,[24] to clarify his position in the controversy over the requirement of general covariance. Notably, the context of these last efforts is that of Einstein's unsuccessful unified field theory program. Here is perhaps the clearest example (using our notation):

> On the basis of the general theory of relativity . . . space as opposed to "what fills space". . . has no separate existence. . . . If we imagine the gravitational field, i.e., the functions $g_{\mu\nu}$ to be removed, there does not remain a space of the type [of Minkowski space-time], but absolutely *nothing*, and also not "topological space". For the functions $g_{\mu\nu}$ describe not only the field, but at the same time also the topological and metrical structural properties of the manifold. . . . There is no such thing as an empty space i.e., a space without field. Space-time does not claim existence on its own, but only as a structural quality of the field.[25]

From this passage it appears that the fundamental intent of general covariance is to forbid any principled separation of the metrical and underlying topological structure from space-time itself, or, in Stachel's compression, "no metric, no space-time". Alternately this can be phrased, "position with respect to a 'background' space-time is a meaningless concept".[26] No longer regarded as distinct from the metrical field defined on space-time, space and time have indeed lost "the last remnants of physical objectivity", whereas chronogeometrical relations appear only as "structural qualities of the field". But what, and where, is the argument for this bald affirmation? Stachel first showed that the argument was made some four decades before. The 1916 passage cited above is a reiteration of the "lesson" Einstein learned late in 1915, when extracting himself from a fallacious belief that that his field equations of gravitation *could not be* generally covariant, on pain of indeterminism. In turn, historical consideration has provided a basis for a textually accurate, sympathetic—and instructive—reconstruction of Einstein's claims on behalf of general covariance and hence for a new understanding of its significance. As Stachel and John Norton have emphasized,[27] when Einstein's 1916 remarks are placed in its context, it is possible to see how Einstein could attribute a heuristic physical significance to the requirement of general covariance.

2.2.2 The "Hole Argument"

Recent scholarship, initiated by Stachel, has shown that Einstein's remarks in 1916 on behalf the physical significance of general covariance pertain to an argument whose corrected conclusion only fully appears in Einstein's correspondence with P. Ehrenfest, H. Lorentz, and others.[28] Einstein himself dubbed this the "hole argument" (*Lochbetrachtung*), and its history may be reconstructed as follows. As early as 1912 Einstein posited general covariance (freedom to make "arbitrary" coordinate transformations; alternately, the requirement that the laws of physics have the same form in any coordinate system) as a *sine qua non* for his new gravitational theory. The first comprehensive outline of that theory, written with Marcel Grossmann in 1913, was unable to give a generally covariant weak field approximation to Newtonian gravity.[29] Subsequently, in late 1913, Einstein managed to convince himself that generally covariant field equations of gravitation were in any case not admissible, through an argument alluded to in several publications in 1913 and 1914. The "hole argument" seemed to show that any gravitational theory satisfying the requirement of general covariance would lead to a violation of the *Eindeutigkeit* of physical laws, that is, to a failure of univocal causal determination.[30] Specifically, Einstein considered a hypothetical example where the *same* matter field sources (i.e., the stress-energy-momentum tensor $T_{\mu v}$) outside a "hole" in the space-time manifold (where these sources vanish) gives rise to *two different values of the metric field* $g_{\mu v}(x)$ and $g'_{\mu v}(x)$ at the same point in the *same* coordinate system within the "hole"—that is to say, apparently, to *two different physical situations* at a given point P. Since the values of the metric field in a given region of space-time are not *a priori* or constant (as in both special relativity and Newtonian gravitational theory) but are to be unambiguously determined by field equations for given contingent sources of matter and energy, Einstein concluded that these equations could not be generally covariant. It was only in November 1915, during the feverish rush of presentations to the Prussian Academy culminating in the generally covariant field equations on 25 November, that Einstein came to recognize the erroneous assumption that had made the conclusion of the hole argument appear valid. To see what this assumption was, and why Einstein came to regard it as erroneous, it will be helpful to define some preliminary conceptions from the modern perspective.

The first is to distinguish between coordinate and point transformations on a manifold, here the four-dimensional space-time manifold M^4. A coordinate transformation from one coordinate chart $\bar{x}$ to another $\bar{x}'$, is "passive" in that it leaves the physical system of space-time points alone and merely changes the labels of the various points covered in the intersection of the two charts. In distinction from a coordinate transformation, a point transformation is "active": it leaves the coordinate labels in place and shuffles around the points. In particular, a manifold *diffeomorphism* D is a smooth invertible one-to-one mapping $M^4 \Rightarrow M^4$, carrying points P to other points P'. Correspondingly, there are distinct "passive" and "active" notions of general covariance. In "passive" mode, the requirement that the Einstein field equations are generally covariant states that if S is one solution of the field equations (relating to one another a collection of real-valued metric field functions $g_{\mu\nu}$ and matter-energy field functions $T_{\mu\nu}$ defined on some region of M^4), then any coordinate transformation $\bar{x}^\sigma \Rightarrow \bar{x}'^\sigma$, ($\sigma = 0, 1, 2, 3$) in that region yields another solution S' relating the corresponding transformed field functions $g'_{\mu\nu}(x')$ and $T'_{\mu\nu}(x')$. The primed functions, in the primed coordinate system, are regarded as the same physical state of affairs, the same relational structure among the point-events, viewed from another perspective. This indeed has nothing particularly to do with relativity principles or gravitation, but merely with the definition of a tensor.

"Active" general covariance is more interesting in that diffeomorphisms can generate arbitrarily many more solutions S' from S in the *same* coordinate system by differently spreading the values of the metric field functions $g_{\mu\nu}$ over the space-time manifold of points. To take the scenario of the "hole argument", assume that S contains a "matter hole" H inside of which, by definition, $\forall x \in H, T_{\mu\nu} = 0$. Then the values of the metric field $g_{\mu\nu}$ within H are determined, according to the Einstein field equations, by the matter-energy fields $T_{\mu\nu}(\neq 0)$ outside and on the boundaries of H. Assume the solution set S assigns the distinct points P and P' within H the values of the metric field, $g_{\mu\nu}$ and $g'_{\mu\nu}$, functions of the respective coordinates (in the same chart) of P and P'. Now define a diffeomorphism $D : P \rightarrow P'$, that acts only on points within H, smoothly vanishing on the boundary, while leaving points outside H unchanged. Such a mapping redistributes the metric field functions $g_{\mu\nu}$ within H in that it "drags along" the value of the $g_{\mu\nu}$ at each old point P to each new point P'. The result is that there are now two generally different values of the metric field at P', $g'_{\mu\nu}(x)$, as assigned by S, and the "drag-along" field $D^*g_{\mu\nu}$ of the diffeomorphism D. It must be emphasized that these are *different* field values at the *same* point in the *same* coordinate system. For each of the arbitrarily many diffeomorphisms D, a new solution set S' can be generated from any other solution set in this way. The question is then whether these are physically distinct solutions.

For two years, Einstein believed that they were, hence that the gravitational field equations do not univocally determine the metric field from given matter sources. That two different solutions arise at a single point within the hole, whereas the sources outside the hole have not changed, certainly appears to be a failure of causal determination. Clearly, Einstein's difficulty pertained to the interpretation of the diffeomorphic point transformations rather than to any trivial confusion regarding coordinate transformations, as has been frequently alleged. In late 1915, Einstein realized the faulty presupposition required for the discordant conclusion that these distinct solutions correspond to different *physical* situations. The answer lay in seeing that the coordinates x^σ have no metrical or other physical meaning but serve as essentially arbitrary labels for space-time points, required for the operations of the differential calculus on a manifold. In other words, the points of the space-time manifold (and so also those within the "hole") *do not inherit their individuality, hence physical existence, from the underlying differential-topological structure* of the manifold. To the contrary, their *physically* distinguishing properties and relations derive not from coordinate labels, but from the fields assigned to them by the generally covariant equations of physical theory; in general relativity these include *at least* the metric field functions $g_{\mu\nu}$. In turn, in a generally covariant space-time theory of fields, only the thus-designated events (possible "point-coincidences") and the relations between them are the "true observables", a (topological) structure preserved under one-to-one continuous diffeomorphic point transformations.[31] In the current parlance stemming from Hawking and Ellis (1973), general relativistic space-times are regarded as equivalent if they have isomorphic models $\{\langle M, g_{\mu\nu}, T_{\mu\nu}\rangle, \langle M, D^*g_{\mu\nu}, D^*T_{\mu\nu}\rangle\}$, physically indistinguishable under a manifold diffeomorphism D. The physical equivalence of these models expresses the principle of general covariance, understood actively as diffeomorphism invariance.

Having now recognized his mistake, Einstein in 1916 sought to underscore this new understanding by adopting a programmatic characterization of what is physically observable as, in principle, reducible to the broad category of "point-coincidences" (or intersections of world lines). This ensures that the conclusion of the hole argument can no longer go through, since only a physical process—the metric field (and possibly other physical fields)—can accord physical existence to the events that make up the space-time manifold.[32] For according to this criterion there is truly no "empty space", no space-time points bereft of *at least* the metric field and so no (merely) "topological space". In holding that space has existence "only as a structural quality of the field", Einstein is underscoring his heuristic postulate that "spatio-temporal individuation of the points of the manifold in a general-relativistic model is possible only after the specification of a particular metric field, that is, only after the field equations of the theory (which constitutes its dynamical problem) have been solved".[33] Then the striking statement situating physical reality in "point-coincidences" represents an attempt to distinguish clearly what is required for certain mathematical structures of the theory to have physical significance.[34] It is not the positivist credo that, since the *in-principle observable* is found in the coincidence of points (intersections of world lines), only such coincidences as are actually observed are real. Alas, this was the message received in Machian circles and welcomed as a confirmation of Mach's positivist philosophy.[35] But once the largely hidden context of the hole argument is restored, it is clear that in locating the "physically real" in "point-coincidences", Einstein gave rhetorical force to the *fact* that, in general relativity unlike special relativity, space-time coordinates alone can have no immediate physical—that is, no chronogeometrical—meaning.[36]

2.2.3 Background Independence

Closely related to the distinction drawn above between "passive" and "active" general covariance is a broader distinction between *covariance* groups and *invariance* groups.[37] By the covariance group of a theory is meant the group specifying the admissible coordinate transformations (here mappings from R^4 to R^4) that provide equivalent descriptions of the same physical state of affairs. In general relativity, the covariance group is necessarily the group of all admissible (invertible, suitably continuous) coordinate transformations.[38] This corresponds to coordinate generality or what Reichenbach in 1920 termed "the relativity of the coordinates". But on the view under examination, "general covariance" is in fact misnamed. It is really the expression of a "principle of general invariance" that is, what was referred to above as "active general covariance" or diffeomorphism invariance.

The principle of "general invariance" may appear to be an extension of the principle of relativity, possessing a higher symmetry than the Lorentz (or Poincaré) invariance of special relativity. Still, as Brown and Brading (2002) observe, the proper context for interpreting Einstein's claim that general covariance is an extension of the relativity principle is the special relativistic limit of general relativity, where locally, $g_{\mu\nu} \Rightarrow \eta_{\mu\nu}$ (i.e., the metric of Minkowski space-time) consistent with the existence of space-time curvature of weak fields. However, as they also indicate, when general covariance is understood as a principle of general invariance, even here the difference is clear. General covariance is an exact symmetry of gravitational physics (diffeomorphism invariance of general relativistic models).[39] On the other hand, relativity principles (Galilean, Lorentz), or any symmetry principles associated with the tangent space structure of space-time, are only approximate, being symmetries of isolated subsystems (rotating, translated, boosted laboratories) of the universe.[40] Expressed in terms of coordinates, this pertains to the fact that the Lorentz transformations perform the same operation at all points of space-time (the coefficients of the transformations are constants), whereas general coordinate transformations perform different operations at different space-time points, the coefficients of the transformations being functions of space-time.[41] Alternately, this is the difference between gauge transformations of the "first" (global) and of the "second" (local) kind.[42]

Now the invariance (or symmetry) group of a theory is the transformation group that picks out all the objects of the theory, if any, given once and for all. Such objects are "absolute", acting but not acted upon, unaffected by dynamical laws and so not among the set of state variables distinguishing different physical states of affairs. Among such objects are, Einstein noted in 1924, "the aether of Newtonian mechanics" as well as that of "special relativity", influencing matter and light propagation through inertial effects but not influenced by "the configuration of matter or anything else".[43] Thereby picking out a theory's "absolute objects", the invariance group of the theory identifies and constrains the space-time framework within which the dynamical laws may be formulated. For example, in special relativity the Minskowski metric $\eta_{\mu\nu}$ is an absolute object, picked out by the Lorentz (or Poincaré) group. But consonant with Einstein's Machian-inspired injunction to banish all elements that act but are not acted upon, there are supposed to be no "absolute objects" in general relativity, corresponding to the fact that the invariance group is the group $Diff(M^4)$ of all diffeomorphisms of the space-time manifold. Thus in general relativity, the space-time metric $\eta_{\mu\nu}$ is everywhere to be replaced by the metric (or gravitational) field $g_{\mu\nu}(x)$, not an absolute object but in principle fully determined from matter-energy sources, including the gravitational field itself. Indeed, the prohibition against absolute objects in general relativity may be formulated by the statement that the metric tensor $g_{\mu\nu}$ is the *only* quantity pertaining to space that can appear in the laws of physics.[44]

In this guise, Einstein's requirement of general covariance is therefore the heuristic injunction that any reasonable field dynamics must be formulated without reference to a background of space-time points to which field functions attach as properties; its intent is to eliminate not only the background metric but also the bare manifold itself as an absolute arena for dynamical laws. Reference to space-time is reference to the frame of the dynamical field itself. Essentially, this is a demand for a fully relational field dynamics where the conception of position with respect to a background space-time lacks all physical meaning.[45] In general relativity, the arena for dynamical laws is the configuration space of all degrees of freedom of the gravitational field, that is, the metric *modulo* diffeomorphisms.[46] However diffeomorphism invariance of general relativity alone does not secure Einstein's vision, for that panorama also encompasses the demand that the distribution of mass/energy (or rather energy-momentum) everywhere fully determines the metric. This fundamentally Machian stipulation is not met in general relativity; ironically, in at least one significant way, on account of the requirement of diffeomorphism invariance.[47] But such failure should not detract from the guiding programmatic character of Einstein's requirement of general covariance, in particular, as carried over to his program for unified field theory. To Einstein, the "most essential thing" lay in removing from physical theory, once and for all, the idea of an inertial system, of any notion of background space-time structures that act (in the explanation of the inertial motion) but which in turn are not acted upon.[48] *A fortiori* this holds for a unified field theory, or any "consistent field theory", where "representing reality by everywhere continuous, indeed even analytic functions" means that the very notion of a "particle" does not exist in "the strict sense of the word".[49] In such a theory of the "total field", the dualism of matter and field is fully resolved in favor of the latter. "Particles" are everywhere to be described as "singularity free solutions of the completed field equations", representing localized large concentrations of electromagnetic, gravitational, and perhaps other forms of energy.[50] Moreover, the law of motion of such "particles" must be derivable from the field equations governing the fundamental field variables, a requirement that is only met by a nonlinear theory and, Bergmann has argued, a direct consequence of the theory's generally covariant field equations.[51] In point of fact, if the field equations are nonlinear (as are the Einstein field equations), there is no unambiguous way to separate the total field into the self-field of the particle (notoriously infinite, before "renormalization", in the usual nonalgebraic formulation of quantum field theory), and the finite external "incident" field immediately surrounding the particle's spatially local extended volume, primarily responsible for its instantaneous state of motion.[52] Successful derivation of an equation of motion within such a theory thus entails the nonexistence of causal influences on the particle not mediated through the immediately adjacent field.[53] In any theory of this kind, the very possibility of such "objects of experience" as "particles" with determinate properties, presupposes some criterion for the individuation of distinct physical systems that nowhere relies upon *a priori* structures of a "background" space and time.[54] This *sine qua non* has been carried over by some within the current research program of quantum gravity.[55] But my general claim here is that these multifaceted programmatic considerations underlie Einstein's various pronouncements on general covariance as a guiding heuristic principle. While current commentators have almost exclusively focused on the failure of classical general theory of relativity to satisfy such lofty requirements, emphasis here is rather on highlighting Einstein's ambitions in finding an encompassing theory that did. That similar motivations are still present in fundamental physics merely underscores the interpretation, of both Einstein and Cassirer, of general covariance as an *a priori* constitutive, yet guiding regulative, requirement to be placed on the conception of physical objectivity. Such a role only can be played by meta-level principles, such as principles of invariance of laws.

2.3 The Problem of Dual Origin

The *terminus a quo* of the two roads is stated already in the first sentence of the Transcendental Logic (the transitional section between the Transcendental Aesthetic and the Transcendental Analytic) and indeed in the single word *Vorstellung*, which is translated into English alternately as "representation" (by Kemp Smith) and as "presentation" or indeed even "conception" or "thought" (by Pluhar):

> Our cognition arises from two basic sources of the mind. The first is [our power] to receive (re)presentations [*Vorstellungen*] (and is our receptivity for impressions); the second is our power to cognize an object through these (re)presentations (and is the spontaneity of concepts).[56]

The operative distinction lies in the contrast between passive receptivity and active spontaneity. Kant's view is that these two sources are independent faculties or powers of mind playing distinct "active" and "passive" roles in the synthesis that yields cognition; "sensibility" (*Sinnlichkeit*) is the faculty of receptivity, and "understanding" (*Verstand*), that of spontaneity. The roles of "understanding" and "sensibility" are not only distinct, but also necessary for the experience that is knowledge of objects. That these two faculties are *independent* sources of cognition is stressed by Kant in the section titled "On the Amphiboly of the Concepts of Reflection" (A271/B327), where Kant objects against Leibniz that he *intellectualized* appearances, while against Locke that he *sensualized* all the concepts of the understanding. Perhaps the defining characteristic of Kant's epistemology is that only the combination of the two produces objectively valid judgments.

One might at first ask whether it really essential to Kantian epistemology to stress the *independence* of sensibility and the understanding. Allison, for example, has argued that Kant's emphasis on independence is in part a reflection the "two-front war" that he simultaneously waged with Hume and with Leibniz.[57] On the other hand, it is fundamental to Kantian epistemology that human cognition requires both concepts and *sensible* intuition.[58] Thus, Kant's separation of the faculties is arguably for purposes of an analysis of the distinct contributions each makes to experience. Yet it is difficult to deny that, on the basis of certain Kantian passages, there is an almost irresistible tendency to reify what has thus been separated by analysis and to regard the faculties as *per se* separate.

Whether separate or not, Kant's account of how the two faculties are related, by means of a "transcendental schematism" of the understanding with respect to the faculty of sensibility, is widely conceded to be one of the most difficult chapters of the *Critique of Pure Reason*. In truncated summary, the argument is this. From the previous conclusion of the Transcendental Aesthetic, Kant could claim that sensory experience is possible only inasmuch as it is framed by, or occurs within the forms of, the pure intuitions of space and time. In particular, the pure intuition of space, the necessary form of our outer sensible intuition, has, in virtue of Kant's model of the construction of geometrical concepts in intuition, the mathematical structure of an infinite three-dimensional Euclidean space. The "active" contribution of the understanding to knowledge is the subject of the ensuing Transcendental Analytic. Kant argued first, in the "transcendental deduction" of the *a priori* "categories of the understanding", that these categories are the logical conditions of any (inner or outer) sensible experience at all. The task is then to show that these purely formal "pure concepts of the understanding" can have application to the spatiotemporal conditions of sense experience established in the Aesthetic. That is, they must be related to the manifold of sensible content structured by the pure intuitions of space and time. The pure concepts of the understanding are accordingly given "transcendental schemata" (rules for application) according to the forms of intuition of time and space. Such schemata, as relating the two *independent* sources of cognition, must be double-sided mediating representations that are *both* "intellectual" and "sensible" and so have something of the "nature" of each the two heterogeneous faculties. Kant notoriously posits a mediating "third" to account for schemata that must nonetheless be "homogeneous" with both pure concepts and sensible intuitions. The schemata are "product and as it were monogram of the pure *a priori* imagination [*der reinen Einbildungskraft **a priori**]*" whose workings are "a hidden art in the depths of the human soul" (A136–138; 141–142/B175–177, 181).

Fundamental critique of this section goes back to Kant's earliest critics, J. G. Hamann and S. Maimon (the latter a significant influence on Hermann Cohen's interpretation of Kant).[59] A classic example, as paraphrased in Beiser (1987), is Hamann's objection, voiced in his *Metakritik*, written in 1784 but not published until 1800:

Although Kant says that knowledge arises from the interaction between understanding and sensibility, he has so sharply divided these faculties that all interchange between them becomes inconceivable. The understanding is intelligible, non-temporal, and non-spatial; but sensibility is phenomenal, temporal and spatial. How, then, will they coordinate their operations?[60]

If the two faculties are indeed the only "basic sources" of cognition, while also being so truly heterogeneous as to be independent, what sense can be made of a "mediating third" somehow homogenous to both? Is it itself a "faculty" of mind? Kant himself perplexed about the matter, at one place, famously speculating that the two faculties may have "a common but to us unknown root" (A15/B29), perhaps indeed, the "pure *a priori* imagination". In the second edition, however, the common ancestor thesis is tempered since the imagination itself is generally characterized as "an effect of the understanding on sensibility" (B152).

The interpretive ambiguity between the "A" and "B" editions on this point continued into 20th century philosophy, most notably, perhaps, in the encounter between Cassirer and Martin Heidegger at the "International University Course" in Davos, Switzerland, in March and April 1929. Michael Friedman has recently argued that this exchange of views acquires particular significance in understanding the subsequent split of philosophy into an analytic tradition, taking its lead from logic and the mathematical and physical sciences, and a continental one, stemming from Heidegger's "existential–hermeneutic" variant of phenomenology. Yet one of the few items of agreement between Cassirer and Heidegger is that both favored the "A" edition's account of the relation between sensibility and understanding as having a common origin. To be sure, there the agreement ended. Cassirer shared the Marburg School's view (e.g., Cohen 1902) of the common logical root of sensibility and understanding as a process of synthesis or construction, brought into the light of analysis through the epistemological method of *Erkenntniskritik*. Heidegger, on the other hand, sought in the transcendental imagination, together with its ground in temporality, the basis for an explicitly "metaphysical" rendering of Kant's *Critique of Pure Reason* as an attempt to uncover the fundamental ontology of being, the project of *Being and Time*. To Cassirer, such a "metaphysical" interpretation of Kant was anathema; in making it, Heidegger "no longer spoke as a commentator but as a usurper".[61] As I argue in this chapter, Cassirer is also a principal disputant in another confrontation over the contested relation between sensibility and understanding, where the context is the problem of the "constitution of the object of physical knowledge" as revised in the light of relativity theory.

The different treatments of Cassirer and Reichenbach lead to quite distinct conceptions of the "problem of physical objectivity", that is, to "the constitution of the object" in physical theory. Many points of contact between the two works have tended to camouflage this fundamental difference. Both are agreed that the fundamental epistemological task of transcendental idealism is posed by the "critical question" concerning the possibility of objective knowledge, and that its answer is to be sought not by mere philosophical reflection but "methodologically", through "logical analysis" of accepted theories in the exact sciences. They shared the view that traditional empiricism has no plausible account of this knowledge. Both claimed that, in at least one respect, the general theory of relativity has confirmed transcendental idealism's claim of the transcendental ideality of space and time insofar as coordinates have become arbitrary parameters for representing space-time events. Yet there is also common agreement that fundamental aspects of Kantian epistemology must be revised or reinterpreted in the light of the theory of relativity. In particular, the original conception of the *a priori* as located in inviolable and eternally fixed categories must be abandoned. In consequence, both argued that the *a priori* elements in physical theories are dynamical and "relativized", changing over time, with each such change representing a transformation of the concept of physical objectivity. In the broadest sense, these are "meta-empirical" principles relative to a given physical theory, "constitutive" of theory's objects in the sense of delimiting the space of "possible objects", but nonetheless not immune from experience, changing with the progress of physical science. Although there will be a significant disagreement concerning just how these relativized *a priori* principles exercise their constitutive function, both Cassirer and Reichenbach acknowledge their necessary role as presuppositions of objectively valid knowledge.

Despite all this broad concurrence, however, the fundamental difference was manifest from the outset. In late June 1920, Reichenbach sent a copy of the typescript of his book to his former Berlin teacher Cassirer, now professor in Hamburg. Cassirer's response is illuminating. While informing Reichenbach that he too has just sent a manuscript on Einstein's theory to press, and that "[o]ur viewpoints are almost interchangeable", Cassirer nonetheless continued,

> however, this does not precisely extend, so far as I can now see, to the concept of *aprioricity* and to the interpretation of Kantian doctrine that, in my opinion you still take too psychologically....[62]

The charge of psychologizing Kantian doctrine is key, a familiar accusation made by Marburg neo-Kantians against interpreters of Kant who are accused of holding a "psychological" doctrine of concepts stemming from Kant's account of the schematism. According to Kant's view, for pure concepts (which are "empty") to acquire a nonformal content or significance, they must be first "schematized" by the pure forms of sensibility (space and time). Only then may such concepts obtain their content through relation to the manifold of empirically given intuition, there finding justification as *a priori* presuppositions of empirical knowledge, conditions of possibility of "objects of experience" imposed upon what is "given" in sensibility. But this mode of relation to the sensuously given, the Marburg criticism continues, always threatens to degenerate into a "psychological" account of conceptual content as a general abstractive combination of contingent, individually given, representations, in which superfluous particular details and generic features are indifferently jumbled together. By reifying as independent the distinct contributions to knowledge that Kant properly separated only for purposes of analysis, it fails to recognize that "in intuition itself the function of the concept is already effectively demonstrated".[63] The pure forms of sensibility *as well as* the pure concepts of the understanding are but different moments or modes of the fundamental synthetic function of unity, a regulative *demand* imposed by pure thought that cognition continually strives to satisfy but can never complete.

We may therefore infer that Cassirer's objection pertained, above all, to Reichenbach's conception of physical cognition as a mere coordination (*Zuordung*) of abstract mathematical representations to the concrete sensible objects they purportedly describe. Similar objections had been, and would continue to be, lodged by Cassirer against accounts of empirical cognition by critical realists like Külpe and Schlick based upon a mere "coordination" of concepts to objects given in empirical intuition.[64] To be sure, the problem does not in itself lie in mere use of the term *Zuordung*, of mapping or correspondence, to describe the fundamental act of cognition, a practice that, by 1920, had become quite common within epistemologies ranging from positivism to realism.[65] In point of fact, a "general law of coordination" (*allgemeines Gesetz der Zuordnung*) lay at the heart of the "functional theory of concepts" Cassirer opposed to the traditional view of concepts as abstracted from sensuously given particulars. However, Cassirer had inherited the Marburg neo-Kantinan view of Hermann Cohen and Paul Natorp that sensibility and understanding are not to be regarded as two completely independent capacities. Instead, the very forms of sensibility are actively attained spatial and temporal representations of the cognizing subject, having "ultimately the same standing as the discursive structures of the understanding". For the Marburg school, representation of what is experienced as *given* is not fundamentally different from representation of how it is *thought*, although not as completely conceptually determined.[66] Accordingly, the Schlick–Reichenbach model of empirical cognition as a mere coordination between two independent faculties was judged as placing a one-sided, and so misleading, emphasis upon one moment of a complex cognitive integration of concepts and already actively structured empirical intuitions. But then, by the same token, Reichenbach's conception of the constitutive role of relativized *a priori* "principles of coordination" first enabling the distinct faculties to be interrelated, appears correspondingly misplaced.

2.4 Reichenbach's *Relativitätstheorie und Erkenntnis A priori*

2.4.1 Cognition as Coordination

At the beginning of a chapter relating how the concept of *a priori* must be revised in the light of relativity theory, Reichenbach helpfully offered a brief *précis* of the Kantian account of the "constitution" of the object of knowledge from the distinct contributions of perception and the categories:

> According to Kant, the object of knowledge, the phenomenal thing, is not immediately given. Perception does not give the object but only the material [*Stoff*] of which it is constructed through an act of judgment. In judgment a subordination [*Einordnung*] into a determinate schema is carried out, according to the choice of scheme a thing or a determinate type of relation develops. Intuition [*Anschauung*] is the form in which perception presents the material of knowledge; accordingly, intuition contains a synthetic moment. However, only the conceptual scheme, the categories, creates the object [*Objekt*]; the object [*Gegenstand*] of science is therefore not a "thing-in-itself" but rather an intuition-based reference structure [*Bezugsgebilde*], constituted through categories.[67]

The basic idea of Kantian epistemology, that constitution of the object of knowledge of science requires both "conceptual scheme" and intuition, is fully accepted as a permanent contribution to epistemology, to be retained even in the light of relativity theory. Accordingly, Reichenbach presupposed that physical knowledge is comprised of the distinct and independent contributions of "reason" (*Vernunft*) [interestingly, not "understanding" (*Verstand*)], and "reality" (*Wirklichkeit*) [not "sensibility" (Sinnlichkeit)]. However, the intricacies of the Kantian account of their interrelation, involving the "schematization" of the intellectual faculty of understanding by application to the pure forms of sensible intuition, is completely bypassed, as is indeed any detailed treatment of the categories or mention of pure intuition. Instead, Reichenbach offered an immensely streamlined version of the Kantian doctrine of cognition, intended to give a general overview of what is essential, while recognizing that "Kant's own conceptual constructions belong to an era distinguished more by grammatical, rather than mathematical, precision".[68] In tandem with Schlick's influential work on "general epistemology" (1918), Reichenbach regarded cognition in general as *defined* as a univocal coordination (*eindeutige Zuordnung*) of conceptual and nonconceptual elements.[69] The sought-for mathematical precision in characterizing the Kantian essentials is found in the set-theoretic notion of a "coordination" (*Zuordnung*) or mapping between two sets, "the most general concept describing the relation between concepts and reality [*Wirklichkeit*]".[70] On account of similarities that conceal significant differences, it will be instructive to briefly contrast Reichenbach's account of cognition as coordination with the earlier, and better known, one of Schlick.

In his grandly conceived *Allgemeine Erkenntnislehre*,[71] Schlick influentially proposed a definition of cognition (*Erkennen*) as a univocal coordination (*eindeutige Zuordnung*) of concepts to "reality". His starting point is a radically formalist doctrine of concepts, explicitly inspired by Hilbert's axiomatization of Euclidean geometry. Both mathematical and physical and, indeed, all properly *scientific* concepts are to be precisely defined implicitly, through their relations to one another in the deductive system developed from the axioms of the respective theory.[72] Against both Machian positivists and neo-Kantians, Schlick argued that as so defined, concepts are merely designative signs, while "designation" is not constitution but presupposes, he emphatically stated, a reality already fully formed. In short, the "bridges are down" between "concepts and intuition, thought and reality".[73] "Designation" itself is nothing more than the act of coordination of a conceptual sign to an object of an external reality, leaving every object as it is, existing independently of concepts and completely individuated. "In its very essence [*Wesen*], coordination [*Zuordnung*] is independent of standpoint and organ", while the act of coordination itself, the relating of one object to another, is a fundamental, irreducible act of consciousness, underlying all thinking.[74] Accordingly, a judgment is *true* if its signs (concepts) unambiguously (univocally) designate objects within that part of reality under consideration; however, the criterion of univocality (*Eindeutigkeit*) can be satisfied by different systems of judgments, all equally "true". Singling out one of these can be achieved only through the adoption of methodological principles (e.g., greatest overall simplicity) that have the standing of conventions.[75] In expressed intent, Schlick's account of knowledge as univocal coordination is emphatically a "critical realism" wherein one speaks of knowledge of "things in themselves".[76] There is neither "pure intuition" nor nonanalytic "pure forms of thought". Relations are simply "forms of the given". Since "reality is already formed", there is no need for "thought to form reality". Arguing explicitly against the neo-Kantianism of the Marburg school, Schlick thus insisted upon the independence of what is given in experience from all forms of thought, thereby endorsing "the Kantian assumption that thought already finds present in intuition a material [*Stoff*] independent of it". The Marburg school's opposition to Kant's independence thesis, summarized in their "striking formula" that in cognition "objects and facts are not *given*, but *posed* as a problem" [*nicht "gegeben", sondern "aufgegeben"*], merely conflates the real with its conceptual wrapping.[77] But in a fundamental departure from Kant, epistemology is to be based entirely on the fundamental concept of coordination—conceived as a relation of designation between concept and object. This renders nugatory, not to mention misleading, the Kantian transcendental machinery of synthesis linking concepts and perception. As a *merely designative* account of knowledge, Schlick proudly declared that his "semiotic" analysis of cognition "thoroughly disposes" of any Kantian concept of knowledge (*der kritizistische Erkenntnisbegriff*).[78]

Regarding cognition as just the reduction of one thing to another, Schlick deemed the task of philosophy, or rather epistemology, as providing clarification of the basic concepts that appear in the advancing processes of explanation in science, an enormously influential view in subsequent philosophy of science.[79] But for Reichenbach in 1920 the task of epistemology is still to answer the "critical question" regarding the possibility of knowledge. This difference in orientation has a profound impact on their respective conceptions of cognition as coordination. To be sure, in stated agreement with Schlick, Reichenbach considered objects of pure mathematics as completely conceptually determined, through implicit definition from axioms (hence in Reichenbach, too, there is no role for "pure intuition" in accounting for mathematical knowledge). Among these systems of interconnected mathematical propositions are to be found the fundamental equations of physics, such as Newton's second law, Maxwell's theory, and Einstein's gravitational field equations. Considered simply as systems of mathematical relations, these are to be regarded as purely formal mathematical expressions. But whereas Schlick regarded the method of implicit definition as an ideal model of definitional precision for all systems of scientific concepts, Reichenbach insisted that the *meaning* of physical concepts is only determined through the character of their connection to experience. After all, unlike in mathematics, where the relation of truth is immanent within the axiom system, the equations of physics are required to have validity for reality.[80] But then the significance of the laws and fundamental equations of physics, "axioms of connection" lying on the "conceptual side of the coordination", must be determined by a specific manner of coordination to perceptual reality.

Now the great advantage of reformulating cognition as a coordination is that it exploits the precise sense of a mapping between two sets. However, Reichenbach recognized that, strictly speaking, the analogy is misleading in the case of empirical knowledge. For there is a "notable fact" that "we carry out a coordination of two sets, of which one not only conserves its order through the coordination, but *whose elements are first defined through the coordination.*[81] That is to say, in "the determination of knowledge through experience",

> the defined side first determines the individual things of the undefined side, and conversely, the undefined side prescribes the order of the defined side. In this reciprocity [*Wechselseitigkeit*] of coordination is expressed the existence of the real [*des Wirklichen*]. It is entirely indifferent whether one speaks of a thing-in-itself or whether one opposes doing so. That mutuality of coordination means that the real exists; this is for us its conceptually graspable sense, and in this way we are able to formulate it.[82]

"Coordination" is accordingly not a mere designation since "the real" is expressed by the reciprocity of the relation between the defined and the undefined sides of the coordination. As in Schlick, a particular conceptual structure may be coordinated to a given perceptual reality in many possible ways, while cognition requires that the coordination be *univocal* (*eindeutig*). Yet "Schlick's psychologizing method" (*Schlicks psychologisierende Methode*) is criticized in that it led him, incorrectly, in Reichenbach's assessment, to consider univocal coordination as an essentialist notion of cognition rooted in a necessary human capacity, while denying "the correct part of Kantian doctrine, namely, the constitutive significance of coordination principles".[83] For Reichenbach, then, such principles, a "subjective contribution of reason" comprising a mediating third between the defined and the undefined side, are required to define a univocal coordination. The criterion of univocality (*Eindeutigkeit*) itself, however, lies not in reason, but in perception, and in any case, is a conceptual fiction that can only be approximated.[84] In contrast to Schlick, perception in itself does not afford a definition of the real.[85] Nonetheless, through a system of "axioms" or "principles" (both terms are used) of coordination, in physical cognition parts of abstract mathematical theory are univocally coordinated to the manifold given in perception, individual elements of reality defined through a coordination to individual equations. In this way, coordination principles "are *constitutive* [**konstitutiv**, original emphasis] of the real object [*wirklicher Gegenstand*]".[86] As constitutive of the object of knowledge, these "principles of coordination" retain the primary meaning of Kant's synthetic judgments *a priori*. On the other hand, Reichenbach will demonstrate that in fact there can be inconsistent sets of such principles, contrary to Kant's (considerably reconstructed) claim that a single set of coordination principles is inherently valid, the permanent contribution of reason to knowledge. Thus, coordination principles do not retain the sense of the *a priori* as possessing universal validity, for they are fallible, theory specific, and relative to a given stage of physical knowledge. In the theory of general relativity, general covariance belongs to the set of coordination principles; its significance is "the relativity of the coordinates".

Thus, unlike Schlick, for whom the "bridges are down between thought and reality", Reichenbach insisted that the mathematical concepts of the fundamental equations of physics are not mere "designations" of the real but require a "mediating third" to first define, and so "constitute", "objects of experience" in physics. In this way there is a univocal determination of the perceptually real (*Wirklichkeit*) in terms of concepts. Every fundamental physical theory presupposes a system of such principles in making a connection to experience. These principles are therefore *a priori* and are purely conceptual in origin, the produce of "reason" and a "subjective contribution to knowledge". It will be apparent, however, that the set theoretic language of "coordination" ill serves to elucidate the *a priori* role of constitutive principles; as Schlick recognized, this language is inherently designative and so realist.[87] It is not at all surprising that Schlick would strongly resist Reichenbach's attempt to doctor up the precise (and "purely semiotic") definition of cognition as a coordination with "constitutive principles" of coordination, opening the floodgates to neo-Kantian conflation of the concept of reality with reality itself.[88]

2.4.2 *Revising the Kantian* A Priori

Recasting of the dynamics of critical (i.e., Kantian) epistemology in the set-theoretic language of "coordination" is thus the setting for Reichenbach's central claim regarding how this epistemology must be revised in the light of the theory of general relativity, namely, that theory contradicts Kant's implicit assumption that there is a unique consistent system of such principles of coordination, and in two different ways. First, the theory of relativity demonstrates that an inconsistent system of such principles exists; that is, in the growth of experience, a given system no longer yields a univocal coordination of concepts to reality. For general relativity, the system comprised by the principles of special relativity, normal induction, general covariance ("relativity of the coordinates"), continuity of laws and physical magnitudes, homogeneity of space, and the Euclidean character of space, has in its totality been shown to be incompatible with experience.[89] Second, from the theory of relativity follows the existence of equivalent descriptions of physical reality, each of which is a univocal coordination of concepts to reality, by the existence of a group of transformations from one reference system to another.[90] Each of these lessons from relativity occasions a necessary revision in Kant's doctrine of the unique nature of the *a priori* rooted in the structure of human cognitive capacities. In response, Reichenbach proposed to eliminate the meaning of the *a priori* as "valid for all time" while retaining that of "constitutive of the object". It is of particular interest that he also regarded relativity theory as itself showing *how* the required modifications may be effected while still remaining within the (now revised) framework of *a priori* constitutive principles. Within relativity theory, these changes are wrought by special application of two general epistemological methods, that of "successive approximation" and of "analysis of science". The epistemological aim of these methods is to restore to a physical theory a consistent system of coordinating principles. This will enable the univocal coordination of concepts to perceptual reality that constitutes the object of physical knowledge and so *defines* cognition. Such a system is, of course, only empirically discoverable, not derived from the nature of reason. It is accordingly fallible and not absolute. But it is *a priori*, owing to its office in defining knowledge of objects in terms of mathematical concepts, that is, in attaining the concept of object in physics.[91]

Reichenbach's illustration of the two methods within the general theory of relativity is concerned to show how, in the face of conflicting experience, a system of coordination principles is recovered that gives a univocal coordination of concepts to perceptions, defining the theory's objects. But each of the aforementioned methods pulls in the direction of realism. The result is that Reichenbach is inexorably led to a realism quite close to Schlick's whereas the notion of objects of physical science as constituted through concepts or conceptual structures is then utterly rejected. The two methods together undermine Reichenbach's idea of the "relative *a priori*" as a set of "constitutive principles" first enabling this coordination. To show this, we turn to the two methods in question.

2.4.3 The "Method of Successive Approximation"

While individual coordination principles are determined by the "subjective" nature of reason, and thus are "arbitrary", establishing a *system* of them is not independent of experience, for the system as a whole is required to univocally connect the mathematical concepts of a physical theory with concrete empirical phenomena. In the face of recalcitrant experience, there is no logical compulsion to single out a particular member of the system as invalidated; rather, a conflict with experience testifies only to the presence of mutually inconsistent members. It would seem that Reichenbach will face here Duhemian problems of empirical underdetermination that can be resolved only by adopting further subjective ("arbitrary") considerations in revising such an inconsistent system of coordinating principles. That he does not is due to "the method of successive approximations ... representing the essential point in the refutation of Kant's doctrine of the *a priori*" since "it shows not only a way of refuting the old principles, but also a way of justifying new ones".[92] The "method" itself involves the meta-level application of a single coordination principle, that of "normal induction", to *systems of coordinating principles as a whole*. "Normal induction" is just the injunction that, among all extrapolations and interpolations from experience, the "most probable" hypothesis is to be chosen. In fact, the pre-eminent standing of the principle of "normal induction" has already been presupposed in any univocal coordination to experience, since "univocality of a cognitive coordination" is simply defined to mean that different empirical measurements may be taken to represent the same value of a given physical state variable.[93] The "method of successive approximations" has a similar normative directive. With its use, it is both "logically admissible and technically possible to inductively discover new coordinating principles that represent a successive approximation of the principles used until now".[94] That is to say, an older constitutive principle can be regarded as an approximation to a new one for certain simple cases. In this way, the method of successive approximations has the standing of a inductive maxim guiding the arrow of disconfirmation to a single coordinative principle. Within the limits of the inductive uncertainties of measurement and observation, that principle can be seen to represent a limiting case of a new, more general, principle. The strongly normative character of the method is stated as a methodological meta-theorem:

> For all imaginable principles of coordination, the following statement is valid: For every principle, however it may be formulated, a more general one can be indicated that contains the first as a special case.[95]

One essential ingredient in Reichenbach's refutation of Kant's "dogmatic" sense of *a priori* (as "valid for all time") is thus a maxim governing scientific change in which advance occurs through monotonic successive generalizations of particular coordination principles.

Kuhnian historiography of science will object that this is not at all a plausible description of the actual mode of conceptual change in science. But the meta-theorem also presents something of an internal consistency problem in view of Reichenbach's claim to have purged Kantian epistemology of its absolute elements. For how can an assertion that "there are no most general (coordinating) principles" and "no most general concepts", indeed, that even the concept of "coordination" itself may prove to be too narrow a definition of cognition,[96] mesh with a meta-theorem asserting that modification of coordinating principles (hence, change in the object of scientific knowledge) always proceeds in the determinate direction of successive approximation? "Relativizing" the *a priori* at the constitutive level of the coordination of equations to reality is thus compensated by an absolutist methodology proscribing a determinate direction to scientific change. It does not appear that the "method of successive approximation" is itself a fallible generalization from the history of science, but neither is it a conventional criterion for theory choice, as with Schlick's principle of "greatest overall simplicity". Nor is it simply the expression of a neo-Kantian "regulative ideal", regarding the task of constitution of the objects of science as an "infinite task", never to be completed. Rather on its basis, Reichenbach asserted that

> even our concepts of the objects of science in general, of the real [*Realen*] and how it can be determined, can only proceed to a gradual process of becoming more and more precise [*einer allmählich fortschreitenden Präzisierung*].[97]

The "method of approximation" accordingly codifies and implements the scientific realist intuition of the growth of scientific knowledge in terms of a logic of approximate truth. In the very course of establishing his central claim that coordinating principles *can* have only an inductive warrant, Reichenbach erected the outline of a classically convergent realist textbook account of scientific progress.

As mentioned above, Reichenbach alleged that relativity theory provides an exemplary instance of the "method of successive approximations" for epistemology. It will be instructive to briefly consider the example, since it does not support either the meta-theorem or its suggestion of a logic of approximate truth. According to Reichenbach, the general theory requires a successive approximation to the special theory despite the fact that it is, at the same time, "a far-reaching extrapolation" of the special theory because it necessitates modification of the system of principles of coordination. One such principle, general covariance ("general relativity of all coordinate systems"), a requirement "obvious from the standpoint of critical philosophy", has led to a rejection of another *a priori* principle, the globally Euclidean nature of space. Reichenbach's concern, then, is to pinpoint a general epistemological significance for the extrapolation of Einstein, based upon the principle of equivalence.[98] His statement of how the general theory arises from its special relativistic limit closely follows similar presentations of Einstein, made as late as 1951[99] arguing that the Minkowski metric can be seen as an instance of the more general metrical structure characteristic of gravitational fields. He does this by showing that a nonlinear coordinate transformation can represent the transition from an inertial frame of reference to a rigidly and uniformly accelerated frame of reference. The argument begins with the expression for the line element in empty Minkowski space,

$$ds^2 = dx^2 + dy^2 + dz^2 - dt^2 \tag{1}$$

which can also be written as

$$ds^2 = g_{\mu\nu}dx^\mu dx^\nu \tag{2}$$

(here, and throughout the book, the Einstein summation convention is employed) allowing for a nonlinear transformation of the coordinates. Obviously, (2) is the expression more suitable for accelerating and rotating frames. But as $g_{\mu\nu}$ are functions of the coordinates, they describe, by the principle of equivalence, a very special, indeed, limiting kind of gravitational field where, in general coordinates, the Riemann–Christoffel curvature tensor is defined to be null, $R^\sigma_{\mu\nu\tau} \equiv 0$. Of course this holds for Euclidean coordinates as well. Relaxing this field law ($R^\sigma_{\mu\nu\tau} \neq 0$) describes the presence of general space-time curvature and so of an *arbitrary* gravitational field in the source-free case ($T_{\mu\nu} = 0$), whose field law requires only the vanishing of the Ricci tensor, $R^\sigma_{\mu\nu\sigma} \equiv R_{\mu\nu} = 0$. It is then natural to suppose that the ds as given in (2), expresses the line element also in the case of a general gravitational field.

Unfortunately, Reichenbach mixed into his own discussion a classic statement of what has been termed "the infinitesimal principle of equivalence" ("at every point of the field, the theory should pass into the special theory of relativity for infinitesimal domains", see Norton [1985]). It is this principle that is regarded incompatible with the constitutive principle that "in all circumstances, the theory permit choice of Euclidean coordinates". As John Norton has pointed out, the problem with "the infinitesimal principle of equivalence", Einstein already remarked to Schlick, who had also employed it, is that in the infinitely small, every continuous line is a straight line, rendering the principle vacuous.[100] In a subsequent admission, the argument of the 1920 book is admittedly "not quite correct".[101] Furthermore, as there presented, the following objection was forcefully made by Hugo Dingler: If true that the coordination (that defines knowledge) is itself only possible by means of the coordinating principles, how can these principles themselves be contradicted in the facts of observation?[102] In a response to Dingler, written in April 1921, the argument is recast.[103] It is essentially this: The three following presuppositions are collectively inconsistent, according to general relativity.

(A) the (global) validity of Euclidean geometry in "natural coordinates"
(B) equality of gravitational and inertial mass
(C) validity of the (laws of) special relativity in *small* (but finite) domains where gravitational effects are negligible

Of these three, now only A is readily recognizable as an *a priori* constitutive principle. But from all three presuppositions, relativity theory, regarding B and C as inductively warranted, draws the conclusion ¬A. This is not a circular inference. A affirms the global validity of Euclidean geometry; the validity of B and C requires only that, in physical measurements, despite the Euclidean presuppositions of the theory of measurement and instrumentation, the departures from Euclidean geometry be too small to be of consequence. The principle of equivalence

(C) thus manifests relativity theory's method of approximation for transiting between the general theory and the special theory.

In fact, it is not really possible to view the principle of equivalence as a legitimate example of an approximative method, in the desired sense that the general case is to contain the special case as a limit. For in the special case, the Riemann–Christoffel curvature tensor necessarily vanishes, while it is necessarily nonvanishing in the general case. The two cases are mathematically inconsistent. In any case, Reichenbach acknowledges that the inference is not logically compulsive. It is possible (as e.g., Dingler) to retain A and challenge, for example, C, in order to avoid the conclusion ¬ A. But given the empirical validity of special relativity, such a move is hardly distinguishable from the postulate of *a priori* philosophy that A is necessarily valid as a condition of possible experience, a claim here precisely at issue. Thus, the "method of successive approximations", via the principle of equivalence, is the essential tool in reaching the conclusion that physical space does not have a global Euclidean structure. The positive case that the pseudo-Riemannian metric of general relativity is an "objective property of reality" is however established through the application of the other method, that of "analysis of science".

2.4.4 The "Method of Analysis of Science" (Wissenshaftsanalytische Methode)

Reichenbach maintained that Kant was correct in arguing that the concept of the physical object "is determined through reason [*die Vernunft*] just as much as it is through the reality [*das Reale*] that it would conceptually formulate".[104] However right about the necessity of these two components of knowledge, Kant was nonetheless wrong in thinking that the coordination of reason and experience produces only a single univocal coordination. Indeed, the theory of relativity shows that any such coordination contains arbitrary or subjective elements and herein lies another lesson of the theory of relativity to epistemology. Indeed, there is an arbitrariness inherent in any univocal coordination that is rooted in "the relativity of the coordinates". Every such coordination of equations to perceptual reality produces an "equivalent description" of that reality within some admissible coordinate system. However, the theory of relativity has shown how to eliminate this arbitrariness of description through coordinate transformations.

> The theory of relativity teaches that the four space-time coordinates can be chosen arbitrarily, but that the ten metric functions $g_{\mu\nu}$ must not be arbitrarily assumed; rather, they have entirely definite values for every choice of coordinates. Through this procedure, the subjective elements of knowledge are eliminated and its objective meaning [*Sinn*] is formulated independently of the special principles of coordination.[105]

This lesson of relativity theory is generalized by Reichenbach into a new "method of analysis of science" ("*wissenschaftslicheanalytische Methode*")[106] whose purpose is the elimination of subjective modes of description from the objective meaning of physical statements. It is "a sort of invariant theoretical method" to distinguish "that part of our scientific knowledge which stems from reason" and so is "subjective", from "the objective content of science, a content which, in the present form of science, is no longer clearly visible".[107] General covariance, or "the arbitrariness of admissible systems", merely gives expression to the "structure of reason" (*die Struktur der Vernunft*), and in fact, what Kant affirmed regarding "the ideality of space and time" has only now been exactly formulated through "the relativity of the coordinates". On the other hand, metric relations in space-time are invariant, objective properties in nature that prescribe determinate limits to the subjective form of physical descriptions. It is this invariance with respect to coordinate transformations that characterizes the "objective content of reality" [*objecktiven Gehalt der Wirklichkeit*]".[108]

Now Kant's "analysis of reason" in the Transcendental Analytic was concerned to demonstrate that knowledge results only from a synthesis of the different sources of cognition. That account, together with its core chapter on the "transcendental schematism", has been jettisoned as leading to inconsistent systems of *a priori* principles, and in doing so, the Kantian model of cognition has been revamped or, rather, generalized. The relation of concepts to experience is no longer established through the murky machinery of the schematism but is generalized into the minimal notion of a coordination. Only then does the "method of analysis of science" come into play, analyzing each coordination into its "subjective" and "objective" constituents by reference to invariance under coordinate transformations. In this way, the method has replaced Kant's analysis of reason:

> The procedure of eliminating from the subjective form of description the objective meaning of a physical statement through transformation formulas, has, by indirectly characterizing this subjective form, taken the place of the Kantian analysis of reason.... This is the sole way that affords us an insight into the cognitive function of our own reason.[109]

The physical object of knowledge, conceptually structured through the coordination of mathematical equations to concrete empirical phenomena, is first constituted as an object of experience only within a particular coordinate system describing the perceptual reality of measurement and observation. As thus constituted, it must be then refined through "the method of analysis of science" in order to determine what within it pertains to "the objective content of reality". As that method is implemented in general relativity, the metric is deemed subjective in as much as the ten independent functions $g_{\mu\nu}$ are functions of arbitrarily chosen coordinates. These metric coefficients cannot in general have Euclidean values (as they would "if the metric were a purely subjective matter"), for a (pseudo-)Euclidean metric requires that, in some admissible coordinate system, the $g_{\mu\nu}$ assume the special values o or ± 1 (Minkowski space-time). But that the metric describes an "objective property of the physical world" lies in its invariance under the admissible transformations of the coordinates.

> If the metric were a purely subjective matter, then the Euclidean metric would have to be suitable for physics; as a consequence, all ten functions $g_{\mu\nu}$ could be selected arbitrarily. However, the theory of relativity teaches that the metric is subjective only insofar as it is dependent upon the arbitrariness of the choice of coordinates, and that independently of them it describes an objective property of the physical world. Whatever is subjective with respect to the metric is expressed

in the relativity of the metric coefficients for the domain of points, and this relativity is the consequence of the empirically ascertained equivalence of inertial and gravitational mass.[110]

The statement that "the relativity of the metric coefficients" is a consequence of the equivalence of inertial and gravitational mass (B above) requires amendment since that equivalence, the "weak principle of equivalence", is also at the base of classical Newtonian gravity.[111] It must therefore be understood as a reference to some stronger version of the principle of equivalence, such as that expressed in C above. Exception may also be taken to the statement that the suitability of Euclidean geometry for physics has the consequence that "all ten functions $g_{\mu\nu}$ could be selected arbitrarily". But the general point is clear enough. By "relativity of the metric coefficients for the domain of points" Reichenbach intends, as discussed in §2.2.2 above, that if the metric field in a region of space-time is represented in one such chart $\bar{x}$, then a representation of that field is admissible in any chart $\bar{x}'$ obtained from $\bar{x}$ through an arbitrary, continuous transformation of coordinates.

But we have still to understand the central contention that freedom (or arbitrariness) in the choice of admissible coordinate systems expresses a *subjective contribution of reason to physical knowledge*, entering into the description of the physical world in that observations and measurements presuppose reference to particular coordinate systems.

> It is obviously not inherent in the nature of reality that we describe it by means of coordinates; this is the subjective form that enables our reason to carry through the description.[112]

The description of reality in terms of coordinates is a subjective contribution of reason, having to do with the nature of our minds, not "reality". From the perspective of a modern formulation of space-time theory, this is entirely trivial, since coordinate freedom is automatically ensured in the standard setting of a differential manifold M^4.[113] In 1920, however, the usual setting for space-time theories was not a modern differential manifold but a number manifold R^4, or one of its open subsets. Number manifolds have too much structure, structure that must be "transformed away" by enlarging the covariance group of the theory.[114] With this in mind, the separation procedure of Reichenbach's method of analysis of science appears more understandable.

Even so, parsing general covariance, or "the relativity of coordinates", as "the subjective contribution of reason", is not really appropriate. To see this, recall that general covariance allows that once any solution to the Einstein field equations is found, any number of other physically equivalent solutions may be derived, "passively", by changing coordinates, or "actively", by diffeomorphism, the lesson of the hole argument. Nonetheless due to general covariance, a constraint on the set of solutions of the Euler-Lagrange variational equations is introduced. In particular, in a generally covariant space-time theory for n unknown functions, there can exist no more than $n - 4$ independent field equations, an ostensible problem for determinism (as Hilbert first noticed) and an essential desideratum in formulating the Cauchy initial value problem of general relativity.[115] As discussed in §2.2.1, general covariance (in giving rise to the Bianchi identities) also plays a vital role in linking the two sides of the Einstein field equations, a formal restriction but hardly an eliminable "subjective contribution of reason". Reichenbach's subtractive procedure appears guided by the thought that it is always possible to identify, and then eliminate, the "subjective contributions of reason" from physical descriptions, rendering physical theories capable of describing the "objective content of reality" without any "subjective" contamination. But this would appear to be only a short step from Schlick's account of cognition as mere designation, and so to endorsing a particular thesis regarding the cognitive representation of the world, namely, that the equations of fundamental physics portray a fully structured mind-independent physical reality, an article of faith that only a realist might adopt. In 1920, Reichenbach is not yet ready to take this step, readily asserting that experience contains rational elements, while denying that these constitutive elements, the principles of coordination, are independent of experience.[116]

Within a few years, Schlick's nagging criticism will sink in: Any invocation of constitutive principles is a remnant of neo-Kantianism having no place in an essentially realist conception of cognition as a coordination. With Reichenbach's growing realization that his coordination principles actually do *not* play a significant constitutive role in anything like the Kantian sense, the result is that all further reference to the epistemological problem of "constitution of the object" in logical empiricism is henceforth dropped.[117] As discussed in detail in chapter 3, by 1924, in place of constitutive principles and axioms of coordination Reichenbach will adopt "coordinative definitions", stipulated assignments of physical objects to certain mathematical concepts, first enabling empirical interpretation of physical theory. Reichenbach's neo-conventionalist treatment of physical geometry, allowing different choices of material standards implementing the mathematical concept of "congruence", would be the first, and most successful, fruit of a general epistemological method requiring the analysis of scientific theories into disjoint factual and a definitional parts. Ironically, in view of this ever more explicit realism, the cost of surrendering his conception of the "relative *a priori*" is that Reichenbach must then refrain from the conclusion, obtained from the "method of successive approximation", that global Euclidean geometry is *not* true of physical space. Instead (as also shown in chapter 3), his epistemological analysis of physical geometry will always manifest a striving to view the theory of general relativity through the more empiricist prism of special relativity.

2.5 Cassirer's *Zur Einsteinschen Relativitätstheorie*

2.5.1 *From Concepts of Substance to Function Concepts*

In an earlier work of 1910,[118] Cassirer encapsulated "the problem of knowledge" from the vantage point of the physical and mathematical ("exact") sciences in the first decade of the century. Looking backward, Cassirer identified *the* central epistemological trend as a transformation of the nature of concepts in the exact sciences, from an abstractive or picture theory, characteristic of empiricism and naive realism, to ever-growing reliance on purely functional, relational, and series concepts. For Cassirer, too, the concept of "coordination" (*Zuordnung*) is all-important, for it is the heart of this conceptual transformation. The development of these sciences has repeatedly illustrated that construction of concepts does not involve a procedure of abstraction of a common property from a collection of individuals, but rather that the individuals are connected through some "general *law of coordination*" (*Gesetz der Zuordnung*).[119] In the very ascription of order to a manifold, such a coordination is presupposed, if not in its completed form, then in its basic function. In accordance with the Marburg rejection of an independent "passive" or "receptive" faculty of intuition, cognition is the resultant of a complete "interpenetration" of sensibility and the conceptual. The "matter" of cognition is not independent of "form" but constantly *is* only in relation to form, while "form" is *valid* only in relation to "matter".[120] For Cassirer, this mutual dependency is encapsulated in the transcendental-logical object-constituting relation of *Zuordnung*. Resisting the pull of mentalism, Cassirer refrained from characterizing anything but the logical form of this "intellectual *coordination*" (*gedankliche* **Zuordnung**) through which diverse elements are connected into a systematic unity.[121] The object of knowledge does not arise from the mere application of formal concepts to sensible experience but is "an expression for the form and mode of conceiving itself".[122] Hence, on the functional theory of the concept, only a relative distinction can be made between the "form" and the "content" of cognition. These are not completely independent realms of existence, but only reciprocal "moments", as concept and as intuition, of a basic process of cognitive synthesis that determines the concept of object. "Content" *is* only as determined through the serial relations of space and time, and the forms of magnitude and number. In physics, the epistemological high point of this (pre-relativistic) development had been attained by Hertz and especially Duhem, who stressed that concepts are pure symbols for relations and functional connections, not in any sense copies, or images, of the real.

In this genealogy of the doctrine of the concept, no particular principle of form or order characteristic even of the present state of science can be taken as immutable or having *apodictic* validity. What remains unchanged through the successive changes in scientific knowledge is merely the "objectifying function" itself, the "supreme law of objectification". A fundamental axiom is Kant's claim that "objective validity and necessary universality (for everyone) are interchangeable concepts".[123] Cassirer's guiding analogy is Felix Klein's *Erlanger Programm* program where a geometry is characterized by the group of transformations under which given relations between points of the space are invariant.[124] Similarly, the method of "transcendental philosophy" is to be a "*general invariant theory of experience*" (*eine "**allgemeine Invariantentheorie der Erfahrung**"*—original emphasis), isolating and investigating the most general elements of form that persist through all change in the material content of experience. Among these are the "categories" ("*Kategorien*") of space and time, of magnitude and functional dependence between magnitudes, presupposed in any empirical judgment or system of judgments. The aim of critical philosophy is to provide a complete inventory of the ultimate *logical invariants* (*die letzten **logicschen Invarianten***) common to all possible forms of scientific experience, persisting from theory to theory as necessary and constitutive factors of any theory. That this is a goal neither completely attained nor attainable at any stage of knowledge is readily admitted. Rather, the significance of this aim is that it is a *"demand"* that a fixed direction is prescribed to "the continuous unfolding and development of systems of experience".[125]

In this distinct account of the relation of "sensibility" and "understanding", intuition and concept, the meaning of the *a priori* is accordingly different. A cognition is *a priori* not because it is prior to experience but because, and only insofar as, it is contained as a necessary premise of valid judgments concerning the "facts" of science. These "logical presuppositions" of physical theory may be seen either as *a priori* elements or as conventions. For Cassirer, the denomination of these ideal conceptual creations as "conventions" is apt only in that it merely acknowledges their spontaneous character; otherwise, it is inappropriate, ignoring that this spontaneity is not unlimited and unrestrained but bound up with the order and connection of the system of perception. The developmental trend of these ideal elements is clearly marked within the physical world-picture (*Weltbild der Physik*) of 1910. All accidents of judgment that are unavoidable from the standpoint of the individual observer are to be excluded in striving toward "that element of necessity universally comprising the kernel of the concept of object".[126] Just as "the most characteristic thesis of critical idealism" lies in the statement that the object is not given, but only attained on the basis of "intellectual necessities" (*Denknotwendigkeit*),[127] so physical objectivity is not given but arrived at, through successive stages of the concept of physical object. The *a priori* standing of the "logical presuppositions" of physical theory accordingly rests upon recognition that the process of perception is not entirely distinct from judgment, and that judgments, distinguishing and systematically ordering the separate contents of a manifold, are the very form of objectifying determination.[128]

Toward the end of *Substanzbegriff und Funktionsbegriff* Cassirer, in a footnote, greeted the recent appearance in print of Max Planck's December 1908 lecture titled *"Die Einheit des physikalischen Weltbildes"*.[129] Planck's lecture was the opening broadside in what became a vitriolic polemic with Mach and his followers over the nature of physical theory and the aim of physical science. In it, Planck pointed to the "unity" of the "physical world picture" as what remained of constant value despite the comings and goings of particular physical theories. In Planck's vivid expression, this is the ideal of "unity of all separate parts of the picture, unity of space and time, unity of all researchers, all nations, all cultures".[130] Some ten years later, in *Zur Einsteinschen Relativitätstheorie*, Planck's thesis of unity through deanthropomorphizing the "physical world picture" reappears as Cassirer's salient theme. The further development of physical theory, with the theory of relativity and especially in the general theory, falls comfortably within the framework of the transition to purely "functional thinking" described in *Substanzbegriff und Funktionsbegriff*.[131] The two further steps along this path taken by the theory of relativity in the interim are qualitatively different. First, in the transformation of the doctrine of measurement, relativity theory has shown that certain concepts ("length", "mass") are not properties of objects but of relations of objects to frames of reference, an additional "de-anthropomorphic" step in the concept of the physical object. But it is the successive step taken by the general relativistic requirement of general covariance, of "the general invariance of laws of nature", that Cassirer sees as bringing a decisive advance in the concept of physical objectivity. For this demand illustrates, as he later put it, that "the ultimate stratum of objectivity" lies in "the invariance of such relations and not in the existence of any particular entities".[132]

2.5.2 General Covariance: A Principle of Objectifying Unity

The task undertaken in Cassirer's relativity monograph is quite specifically marked out: to determine the significance for epistemology of Einstein's claim that his theory has removed from space and time "the last remnants of physical objectivity". "What are we to understand by physical objectivity", Cassirer asks, "here denied to the concepts of space and time?"[133] It is not sufficient to merely observe that space in itself and time in itself do not satisfy Planck's often-invoked formulation of the criterion of physical objectivity—"What can be measured, exists". While this may be adequate for physics, measurement itself rests on presuppositions that require epistemological elucidation. Nor is it enough to understand Einstein's remark in the sense that space and time are forms of phenomena and not things, in the sense of naive realism. "That physical objectivity is denied to space and time by this theory must signify... something other and something deeper than the knowledge that the two are not things in the sense of 'naive realism'".[134] For none of the genuine concepts of physical objects—energy, mass, momentum, and so forth—are such naive "thing-concepts". What is left still unaccounted is the "logically special position" (logische Sonderstellung) occupied by the concepts of space and time. Space and time are a further abstractive step away from most physical concepts, "representing, as it were, concepts and forms of measurement of higher than the first order". Hence, any attempt to provide an answer to the question concerning the loss of "physical objectivity" by space and time is constrained to recognize the more fundamental character of these concepts. As befits the method of Erkenntniskritik, admitting no superior epistemic authority outside of science itself, the answer must be sought in terms of the changing manifestations of the concept of physical object within physical science. So the specific task Cassirer has set himself is an examination of how "physical objectivity" is to be construed from within the physical perspective of the new theory such that it is denied to space and time.

In chapter 2 of his book, Cassirer (1921) already arrived at a preliminary result: the requirement of general covariance—namely, that laws be stated in a form valid for all frames of reference—represents a further, but qualitatively different, advance in the line of conceptual development (Begriffsbildung) extending from classical mechanics through the special theory of relativity. In the latter instance, the validity of the general laws of nature was still restricted by reference to a class of determinate reference bodies; with general relativity, this restriction is altogether removed. Although some determinate reference system (Bezugsystem) is implied in testing these laws, "the meaning and value [Sinn und Wert] (of the laws) is independent of the particularity of these systems and remains self-identical, whatever changes experience may bring to them". This is to assert that "independence from the arbitrary standpoint of the observer" (Unabhängigkeit vom zufälligen Standort des Beobachters) is just what is meant in speaking of an object of "nature" and of "laws of nature" as determinate in themselves.[135] In unknowing agreement with the thesis of Eddington's *Space, Time and Gravitation* of the same year (see chapter 7), measurement in one system, or in any of the unrestricted plurality of "justified" systems in the end yields only particularities (*Einzelheiten*), but not the genuine "synthetic unity" of the object. With reference to Planck's Leiden lecture, these new requirements of physical objectivity mean that "the anthropomorphism of the natural sensuous world picture, the overcoming of which is the task of physical knowledge, has been compelled to take a further step back".

The interpretation of general covariance as a further development of the methodological principle of "objectifying unity" is the central theme in the remainder of Cassirer's essay. Where experience had unexpectedly failed to find the preferred reference frame posited by Galilean–Newtonian mechanics for the motion of the solar system or the motion of the earth in Michelson's experiment, the theory of general relativity made a virtue out of necessity by requiring that there *cannot* and *must* not be such a preferred system. The general theory of relativity thus adopts the principle (*Prinzip*) "that for the physical description of the processes of nature [*Naturvörgange*] no particular reference body should be distinguished above all the others".[136] The requirement of general covariance ("that all Gaussian coordinate systems are of equal value for the formulation of the general laws of nature") is designated a "rule of the understanding" ("*Regel des Verstandes*") adopted within physics not only as a formal requirement on mathematical representation, but as a "principle that the understanding uses hypothetically, as a norm of investigation, in the interpretation of experience". The sole meaning and justification of such a principle rests upon the fact that, through its application, it will be possible to attain the "synthetic unity of phenomena in their temporal relations" ("*synthetische Einheit der Erscheinungen nach Zeitverhältnissen*"), that is, lawful explanation of all observed phenomena. The guiding norm itself is unconditioned, and so only ideal: it is just the "idea of unity of nature, of univocal determination itself".[137] Nonetheless, with the requirement of general covariance, the general theory of relativity has given a new meaning to the Kantian idea of unity of nature as a "unity of determinate functional relations", assimilating under arbitrary transformations of the coordinates, all measurement results obtainable in particular reference systems. The concept of object of physics has become the concept of what remains invariant under such arbitrary transformation, and dynamics is more and more resolved into geometry (*reine Metrik*), a tendency, Cassirer observed, most clearly evident in Weyl's treatment of general relativity".[138] In this regard, Cassirer's brief comments on the principle of equivalence are also telling. The equivalence between a uniformly accelerating frame in a gravity-free region and one falling freely in a static gravitational field is "a precept [*Vorschrift*] for the development of our physical concepts [*physikalische Begriffsbildung*]: a requirement made not of experience but only for our manner of intellectually representing it".[139]

Cassirer recognized, with an eye toward Kretschmann's "correction" of Einstein, that general covariance ("that the general laws of nature are not changed in form by arbitrary changes of the space-time variables"—termed here the "principle of general relativity") may appear to be an analytic assertion, specifying the meaning of a *general* law; nonetheless, that in general there be such invariant laws is a *synthetic* demand.[140] As such, the principle of relativity is "a general *maxim* set up for the investigation of nature" (*eine allgemeine Maxime der Naturbetrachtung*): a formal restriction but also (here, Cassirer quotes from Einstein [1917a]) "a heuristic guide in the search for the general laws of nature".[141] Similarly, citing the Kantian formulation of the object of knowledge as a "concept, with reference to which presentations have synthetic unity", Cassirer judged the requirement that a physical theory be generally covariant (in "the form of... systems of equations, which are covariant with respect to arbitrary substitutions") to be a purely logical and mathematical relativization (*Relativierung*). Yet it is through this relativization that the object of physics is determined as a "phenomenal object", although no longer connected with "subjective arbitrariness and subjective contingency". Such an object is not "objective" because all subjective contributions of reason have been eliminated. Rather, general covariance is one of the "ideal forms and conditions of knowledge upon which physics rests as a science, that secures, and at the same time grounds, the empirical reality of all that physics regards as 'fact' and to which it accords the name objective validity".[142] Cassirer later on underscored Einstein's emphasis that this new ideal of physical objectivity is but a norm, a "methodological maxim" or "regulative principle" for the intellectual treatment of nature.[143]

Not until Cassirer's chapter 5 ("The Concepts of Space and Time of Critical Idealism and the Theory of Relativity") is what is usually posed as the primary obstacle to Kantian or Kantian-derived interpretations of the theory of relativity, the doctrine of pure intuition, straightforwardly confronted. Drawing upon Marburg revisionism regarding pure intuition, Cassirer argued that the general theory of relativity, whose fundamental feature is characterized as having removed from space and time "the last remnant of physical objectivity", has improved on Kant in bringing about a clarification of the role of pure intuition in empirical cognition. While following the broadly critical idealist injunction that space and time are "forms of phenomena" and not "things", Cassirer also enjoined that they are *conceptual* "sources of knowledge", pure ideal *concepts* of the relational orders of "coexistence" and of "succession", as they were indeed for Leibniz but not for Kant. A "coincidence" of two world lines, presupposing nothing concerning the metrical relations of space and time, involves only topological relations, the "serial forms of the relations of coexistence and succession". As such, the notion of "point-coincidence" gives the most general meaning for that "synthesis of the manifold" for which Kant formulated the term "pure intuition". In this regard, general relativity, in robbing "pure intuition" of its chronometrical background structure, has indeed *clarified* the Kantian meaning of the term, whose "most general sense... was certainly not always maintained by Kant equally sharply on account of his involuntary substitution of special meanings and applications".[144] In this regard, the general theory of relativity exhibits "the most determinate application and carrying through within empirical science of the standpoint of critical idealism".[145] Kant's *intention*, with regard to the use of the term "pure intuition", was simply to express the "methodological presupposition" of characteristic "thought-forms" (*Denkformen*) of connection and of ordering entering into all scientific knowledge; these are the concepts of number, of function, and of space and time. Such forms are not to be conceived as "rigid" but rather as "living and moving"; none is given to thought "at one stroke" but is only revealed through the process of "coming to be" in the concrete manifestations of scientific thought. But in the continual attempt of physics to bring these changing forms into a mutually determinative relation with the manifold of sensibility, the latter "progressively loses its 'fortuitous' ["*zufälligen*"] anthropomorphic character and receives the impress of thought, the impress of systematic unity of form".[146] The loss of "physical objectivity" by space and time, triumphantly announced by Einstein, refers precisely to the appearance, in the general theory of relativity, of the concepts of space and time solely as functional forms of succession and of coexistence.

Cassirer thus represented the principle of general covariance as a qualitatively new stage in the continual development of the conception of physical objectivity stretching back to the birth of modern science. In that process can be documented a progressive "movement of thought" (*Denkbewegung*), an unmistakable trend of the replacement of "substance" or "thing" concepts, uncritical "anthropomorphic" modes of representation, by functional and relational concepts. A yet further step, and a decisively higher stage of "de-anthropomorphization", has been taken with general relativity, for in its wake, the concept of "physical objectivity" incorporates the *methodological norm* of general covariance: that the laws of nature find their only natural expression in generally covariant equations. Although "objects of experience" require the choice of a suitable coordinate system (through the concrete calculation of a result to be compared with experimental data), there can be no general *preferred* set of coordinates (reference frames, or foliations of the space-time manifold). Singling out any reference frame for such distinction violates the spirit, and the letter, of general covariance, according to which *any* adopted reference object is itself a dynamical, not an absolute, object. As so "relativized", the fundamental concept of "object of nature" is not a picturable but a "pure structure" entity identifiable only in relation to other structures of the field. Deprived of the anthropomorphic stage of a background space-time that is always presupposed picturable or visualizable "thing-concepts", such a dynamical object is completely resolved into the pure measure relations (*reine Maßbeziehungen*) of a fully relational dynamics.[147]

In this latter sense, general covariance is, as noted above, the most thoroughgoing refinement yet of the *normative* methodological principle of "unity of determination". Having the standing of a Kantian "concept of reason", it is concerned "solely with the use of the understanding", but proposes "to extend the synthetic unity thought in the categories, all the way to the absolutely unconditioned", so prescribing to the understanding a "direction towards a certain unity of which the understanding has no concept". While a "transcendental idea ... considering all experiential cognition as determined through an absolute totality of conditions", it is not "arbitrarily invented, but posed as a problem (*aufgegeben*) by the nature of reason itself" (A326–327/B383–384). As "the concept of a maximum to which nothing congruent can ever be given *in concreto*", it is, certainly, a "subjective contribution of reason" in Reichenbach's parlance, yet far from being *eliminable* from physical description, it is deemed *essential* for restructuring that description toward the goal of a conception of objects of a completely impersonal nature. Hilbert, apparently independently, came to the same view of the significance of general covariance as the regulative idea of a "radical elimination" from physical description of the "anthropomorphic slag" contributed by the senses and intuition.[148] So understood, general covariance is, for both Cassirer and Hilbert, *the* epistemologically salient aspect of the general theory of relativity.

2.6 Conclusion

Stemming from a fundamental difference in the interpretation of Kantian epistemology, the 1920 monographs of Cassirer and Reichenbach on the theory of relativity point in diametrically different directions for subsequent philosophy of science. Each proposed a conception of the "relativized *a priori*" as meta-level constitutive principles governing empirical laws. However, the "relative" standing of Reichenbach's principles of coordination is counterbalanced by two epistemological methods consonant only with the commitments of scientific realism, a fact evidenced in his later writings, even as Schlick retreated from realism to a Wittgenstein-inspired positivist distaste of all "metaphysics". To philosophers not independently persuaded of the virtues of realism or positivism, Cassirer, the "historical" philosopher, proposed a significantly richer appreciation of the epistemological innovation of the theory of general relativity. While lacking the language and mathematical tools of symmetry readily available today, Cassirer nevertheless succeeded in grasping that the revolutionary epistemological idea of general relativity lies in general covariance, the regulative idea of all fundamental physical objects interacting through dynamical laws completely without reference to a background space-time. Such a conception of general covariance as an "idea of reason" constraining fundamental physical theory is no longer constitutively *a priori* in Kant's sense.[149] That *regulative* ideals can play a heuristic but still *constitutive* role in physical cognition is then not Kantian orthodoxy. But it is universally agreed that general relativity needs occasion some revision or clarification in those deep, and often murky, waters.

3

1921

"Critical or Empiricist Interpretation of the New Physics?"

"O Kant, wer rettet dich vor den Kantianern?"
Hans Reichenbach to Arnold Berliner, 22 April 1921[1]

3.1 Introduction

The appearance early in 1921 of Cassirer's "epistemological considerations on Einstein's theory" prompted the editors of the *Kant-Studien*, the official organ of the venerable *Kant Gesellschaft*, Germany's largest and most notable professional association of philosophers, to ask the philosopher Moritz Schlick once again to consider the viability of a Kantian philosophical understanding of the theory of relativity. Already in 1915, just following the appearance of the general theory in November, Schlick had published an assessment of the philosophical significance of the (special) theory of relativity, arguing that Kant's doctrine of time as an *a priori* form of intuition had been too closely modeled on Newtonian time to be compatible with the new Einstein kinematics. Hence, any claim that space and time are necessary *a priori* forms of intuition could pertain, at most, to purely qualitative and subjective properties of space and time, not to the quantitative measurable relations of physics. While "not abolishing the core of the Kantian doctrine", the (special) theory of relativity showed "the necessity of modifying essential parts of it".[2] Now, with the general theory of relativity, Kantian and neo-Kantian epistemological analyses of relativity theory seemed, *prima facie*, in further difficulty. The Kantian claim regarding the necessarily Euclidean character of space in the doctrine of pure intuition of the Transcendental Aesthetic appeared straightforwardly refuted. Nonetheless, Cassirer, a leading "neo-Kantian" (although Cassirer himself rejected the label as suggesting a dogmatic attachment to orthodoxy[3]), had concluded that the theory of general relativity, as evidenced by Einstein's claim that in the new theory space and time lost the "last vestige of physical objectivity", exhibited "the most determinate application and implementation of the standpoint of critical idealism within empirical science".[4] What were philosophers, excluded from first-hand knowledge by the theory's highly abstract mathematics, to think?

Cassirer's book merited the editors' special attention for a number of reasons. Recently called, in 1919, as professor of philosophy to the University of Hamburg, newly created by the Weimar government, Cassirer was one of the leading philosophers in Germany.[5] He had first made his name in the philosophical world as a historian of philosophy, tracing the "problem of knowledge" from the early Renaissance up through Kant in two large volumes of the same title, in 1906 and 1907. A third volume in the series, covering the post-Kantian developments of the 19th century, followed in 1920.[6] A ten-volume edition of Kant's works, edited by Cassirer, appeared in 1912; his intellectual biography of Kant completed the edition as eleventh volume in 1918.[7] But it was above all the publication of his first book of systematic philosophy in 1910, *Substanzbegriff und Funktionsbegriff*, that made Cassirer the recognized leader of the second generation of the neo-Kantian Marburg School. Founded by Hermann Cohen and Paul Natorp, this branch of neo-Kantianism was most closely associated with epistemological questions of the mathematical and physical sciences.[8] As noted in chapter 2, Cassirer's "epistemological considerations" on relativity theory continued and extended the broad theme of that earlier work in which Cassirer tracked the transformation of the concept of object in mathematics and physical science since the 17th century. In the "Forward" to his relativity book, Cassirer reported that Einstein himself had read his book through in manuscript, and "encouraged (it) through several critical comments".[9] Einstein's criticism, expressed in a letter to Cassirer of 5 June 1920, was rather benign, especially when contrasted with his disparaging comment to Schick, in autumn 1919, regarding "how eagerly the philosophers are already striving to cram the general theory of relativity into the Kantian system".[10] To Cassirer, Einstein wrote that he could understand "your idealistic mode of thought" (*Ihre idealistische Denkweise*) regarding space and time, and even believed it to be free of contradiction, and he likewise agreed that "conceptual functions [*begrifflichen Funktionen*] must enter into experience in order for science to be possible". However, he cautioned, the choice of such functions could by no means be thought to be "compelled by the nature of our intellect".[11] Cassirer, as is evident from chapter 2, could only agree.

Schick's credentials for this assignment were impressive. Schick was the first philosopher in Germany to be recognized as a competent authority on the theory of relativity. He had received his Ph.D. in physics in 1904 under Max Planck in Berlin with a thesis on the reflection of light in inhomogeneous media that demonstrated Schick's facility in carrying out detailed scattering calculations.[12] Years later, at a celebration on the occasion of his eightieth birthday in 1938,

Planck would single out, as most notable among his many students, Schlick, together with the Nobel prize winner Max von Laue.[13] Still *extraordinarius* professor in Kiel in 1921, Schlick would go to Vienna in 1922 to become the fourth occupant of the chair in the Philosophy of the Inductive Sciences at the University of Vienna created for Ernst Mach in 1895, and subsequently occupied by Ludwig Boltzmann and the physical chemist L. Stöhr. Besides the 1915 paper on the philosophical significance of (special) relativity principle, Schlick had also written an exposition of the general theory for the scientific laity, *Space and Time in Contemporary Physics* [*Raum und Zeit in der gegenwärtigen Physik*], first appearing in the pages of the scientific weekly *Die Naturwissenschaften* in two installments in March 1917.[14] Reprinted as a separate monograph, the third (1920) edition was, in 1921, still on the booksellers' shelves. (As will be seen, the fourth and final edition appeared in 1922, with a significant change.) Finally, Schlick had known professional and personal connections to Einstein; there was considerable correspondence between them in the period between 1915 and the early 1920s.[15] Their first correspondence, upon Einstein's reading of Schlick's initial philosophical appraisal of special relativity in 1915, revealed Einstein to be an enthusiastic supporter of Schlick's philosophical writings on relativity theory. In late 1919, he was Schlick's house guest in Rostock, following which, he wrote to Max Born, "Schlick is a clever person [*ein feiner Kopf*]; we must try to obtain a professorship for him", a task that would not be easy, Einstein judged, because Schlick "does not belong to the philosophical church of the Kantians".[16] It is likely that Einstein played some role in obtaining Schlick's post in Vienna. In any case, Schlick's large 1918 book on general epistemology, *Allgemeine Erkenntnislehre*, the first number of a series of monographs and textbooks in the natural sciences published under the editorial direction of the editors of *Die Naturwissenschaften*, was well known to Einstein. Preparing to travel to Holland in October 1919, Einstein wrote to Schlick that the only reading for the journey was to be his "epistemology".[17] This reading had a visible influence in Einstein's widely read essay "Geometry and Experience" ("*Geometrie und Erfahrung*"), originating as a rare public lecture on 27 January 1921. There, Einstein not only commended the book's epistemological emphasis of the method of implicit definition, he also wrote of a "geometrical-physical theory" as "necessarily unintuitive, a bare system of concepts", the geometrical and some of the physical laws of which are conventions, whose relation to "experiential objects of reality (experiences)" is one of "coordination", all views found in Schlick's book.[18] Perhaps having this reference in mind, the mathematician Hermann Weyl lamented to Edmund Husserl in March 1921 (see chapter 5, §5.2.1), that Schlick's epistemology book had a considerable resonance with "the leading theoretical physicists". The editors of the *Kant-Studien* could not possibly have regarded Schlick as neutral toward the Kantian or "critical" ("*kritizistische*") philosophy, for his epistemology book pointedly defended, in explicit opposition to all varieties of Kantianism (including Husserlian phenomenology) and Machian positivism, a form of scientific realism. Nonetheless, Schlick, as both philosopher and *Fachmann* regarding the theory of relativity, could be regarded as uniquely placed to assess what would be the most philosophically sophisticated attempt to link the general theory of relativity with the broad trend of Kantian thought.

In the event, Schlick's review essay appeared in mid-1921 in the *Kant-Studien*. Its title, echoed in that of this chapter, directly revealed his understanding of his assignment. It was to render an answer to an either/or interrogative, *tertium non datur*. Posed in this way, there could be little surprise regarding his verdict, although garden-variety empiricists might not have recognized Schlick's "empiricism" as falling quite within any of the known species of that doctrine. Let us now turn to a recent, and authoritative, assessment of the significance of Schlick's answer:

> It is the first clear statement of the inconsistency between Kantian philosophy and relativity. This remarkable article may well be regarded as the point of departure of a new direction for scientific philosophy.[19]

This appraisal, by the late Alberto Coffa, can be accepted as the "received view" of Schlick's essay. There are two claims here, that the "Kantian philosophy" *is* inconsistent with the theory of relativity, and that Schlick opened up a new direction for scientific philosophy. Certainly both claims are of interest. Although in this chapter I am principally concerned with the latter claim and, in particular, the nature of the empiricist interpretation of the new physics that Schlick offered in place of Cassirer's *kritizistiche* interpretation, a few words of reminder about the former claim are in order. First of all, thanks in large measure to Schlick's authority and rhetorical ability to pose the issue on his own terms, the debate between "empiricist" and "critical" philosophy over relativity theory effectively ended with Schlick's essay. Within a few years, the postulate that physical theories neither require nor contain any "constitutive" or synthetic *a priori* elements would become a cornerstone of the new "scientific philosophy" of logical empiricism, as would also the polemical tarring of the "philosophy of the synthetic *a priori*" with the broad brush of "metaphysics". Second, if consideration is limited to those few who in 1921 had demonstrated an innovative expertise with the new theory that Schlick himself certainly did not possess, this outcome, so influential for philosophy of science in the 20th century, was by no means inevitable. Subsequent chapters will consider at length the philosophical standpoints of Weyl and Eddington, both kindred spirits to the "philosophy of the synthetic *a priori*". Here I simply note the assessment of Schlick's cohort, Max von Laue, made also in 1921 in the first edition of his well-regarded text on general relativity:

> It is, frankly, an identifying characteristic [*Kennzeichen*] for a correct epistemology, that it remains invariant against all transformations that the physical world picture experiences in the course of time. We would not conceal our conviction that Kant's critical idealism (although not every sentence of the "Critique of Pure Reason") satisfies this requirement even against the general theory of relativity.[20]

Third, and finally, on a careful reading, Schlick's argument bears not upon Cassirer's understanding of the "synthetic *a priori*" as regulative principles or "rules of the understanding" governing the development of concepts of physical objectivity, but upon a more traditional Kantian conception of apodictically certain and unrevisable principles. Perhaps to avoid tiresome discussions of Kant interpretation, perhaps because he considered it the identifying characteristic of all Kantian philosophy, perhaps for rhetorical purposes, Schlick located "the essence of the critical viewpoint" in the claim that the constitutive principles of physical knowledge

> are to be *synthetic judgments a priori* in which to the concept of the *a priori* inseparably belongs the characteristic of *apodeicticity* (universal, necessary and inevitable validity).[21]

The "critical" (i.e., neo-Kantian) philosopher must maintain this understanding of the synthetic *a priori*, or else, in Schlick's lexicon, he is no longer a "critical" philosopher. Still more, continuing a reading of Kant given in his *Allgemeine Erkenntnislehre* (1918), Schlick refused to allow that the Kantian doctrine of "pure intuition" could ever be purged of its psychological trappings and so could not be revised or refined to be a "method of objectivification" in the manner Cassirer had adopted. The gauntlet thus laid down, Schlick had little trouble in dispatching such claims of Cassirer's book as could be represented in this fashion, misleadingly, since Schlick completely ignored the genetic character and historical evolution of the "regulative principles" and "rules of the understanding" that comprised the core of Cassirer's account of the development of the concept of physical objectivity culminating in general covariance. Accordingly, with this declaration the issue is no longer joined, for Cassirer had been denied any possibility of distinguishing his conception of constitutive *a priori* principles from an orthodoxy that Schlick could easily show was rendered obsolete by the new physics. In one recent assessment, Schlick's "challenge" to Cassirer to produce examples of such unrevisable synthetic *a priori* principles "represents a fundamental misconstrual of Cassirer's conception of the *a priori*".[22] In any case, Schlick's traditional reading of the synthetic *a priori* was not a necessary one, as he himself already knew. The principal thesis of a 1920 monograph from the neo-Kantian perspective of Schlick's logical empiricist colleague-to-be, Hans Reichenbach, denied that apodictic certainty is inseparably attached to synthetic *a priori* principles. Schlick had reviewed Reichenbach's book and subsequently expended considerable effort, in correspondence with Reichenbach in late November, 1920, arguing that Reichenbach's theory-relative conception of synthetic *a priori* principles did not suffice to distinguish them from conventions in the sense of Poincaré.[23] Showing some understandable sensitivity on this interpretive point, in his essay Schlick still insisted on this view of the *a priori*, while stating that his was "an inquiry directed to systematic rather than historical questions". Thus, he gave himself an easy target indeed.

However, this chapter focuses on Coffa's second claim, although factually it is not in dispute. Rather, my concern here is with the nature of the empiricist interpretation of the new physics pointing a "new direction for scientific philosophy" that Schlick offered in place of Cassirer's "critical idealism". While polemically counterposed to Cassirer as an interpretation already extant in the literature, in fact Schlick's empiricist interpretation, namely, his empiricist account of the metric of space-time in the general theory of relativity, was a work in progress, hardly then existing. At the time of Schlick's diatribe against Cassirer it had merely been hinted at in several of the notes and elucidations Schlick appended to two of Helmholtz's papers on physical geometry, republished in 1921 in a new edition of

Helmholtz's epistemological writings. In turn, these crucial notes and elucidations sought to recast Helmholtz's views on physical geometry in the bright light of Einstein's recent paper on "Geometry and Experience" so as to make Helmholtz appear to be an empiricist precursor of both Einstein and Schlick. By themselves, they provide evidence of a fundamental shift in Schlick's account of the interpretation of the geometry of physical space, differing from that presented in his previous writings on relativity theory. For this reason, 1921 draws our attention as a "pivotal year" for "scientific philosophy" because it brought Schlick's crystallization of the new empiricist interpretation of physics. Based on a highly selective reading of these texts of Einstein, and of Helmholtz, Schlick's new empiricism is an almost "on-the-spot" improvisation, seeking to find the resources for an empiricist interpretation of the metric of space-time in observable facts about measurement bodies and light rays whose fiduciary behavior has been fixed by conventional stipulation.

In what follows, three central aspects involved in the emergence of Schlick's "empiricist interpretation" of the new physics are identified and treated severally. First and foremost is Einstein's well-known lecture "Geometry and Experience" that, as Schlick is writing against Cassirer, had just recently been published as a *separatum* from the Proceedings of the Berlin Academy of Sciences. Following Schlick's lead, this article would become virtually a founding hymn of logical empiricism. But Schlick chose to ignore the *pro tem* character of Einstein's defense of rigid rods and ideal clocks as metrical indicators in the general theory of relativity, a hypothesis Einstein knew to be inconsistent with the spirit, if not the law, of his field equations of gravitation. On the other hand, Einstein's treatment of the line element of the space-time interval as physically defined by measurements of rigid rods and clocks was his principal weapon against Hermann Weyl's theory of "gravitation and electromagnetism". As discussed in chapter 4, and in further detail in chapter 6, from Weyl's epistemological vantage point such a stipulation regarding rigid bodies represents the last vestiges of Euclidean "distant geometry" in Riemann's infinitesimal geometry and so is unjustifiable even if it is in accordance with the observed behavior of rigid rods and ideal clocks as metrical indicators in weak gravitational fields. Completely disregarding this argumentative context of Einstein's essay, Schlick interpreted Einstein's provisional endorsement of rods and clocks in the present state of physics as instead a methodological affirmation of the conventionalist underpinnings of a new empiricist realism in physics, resting upon a *stipulation* regarding rigid bodies.

Schlick employed this selective assessment of Einstein's essay in order to refurbish Helmholtz's epistemological project of attempting to base the geometry of physical space upon "facts" about rigid bodies. Blithely overlooking Helmholtz's own attempt to salvage a modified version of the Kantian theory of space as a form of "outer intuition" in his account of the "facts" underlying geometry, Schlick interpreted Helmholtz as an occasionally naive geometric empiricist whose "greatest epistemological achievement, his theory of space", once corrected in the light of Einstein's supposed definitional treatment of rigid bodies, is not only plausible for the new physics of general relativity but also "quite certainly *true*".[24] Yet while celebrating Helmholtz as the Elijah of the new empiricism, Schlick also found it necessary to modify his previous unqualified endorsement of the holist conventionalism he had associated with Poincaré, a change that occasioned wider reaching ramifications within his general epistemology. Thus, in order to present the "consistent empiricism" he opposed to the neo-Kantians, Schlick eliminated the gray area recognized in the first edition of that work between "hypotheses" and "definitions". In the book's second edition (1925), the classification of types of judgment was revised to feature a sharp distinction between definitions and empirical judgments. This new discrimination was the prototype for the particular version of the analytic/synthetic distinction that became a defining characteristic of logical empiricism and the principal target, in an ironical turn of the wheel of fortune, of Quinean holism.

3.2 A New Empiricism?

What is the central tenet of Schlick's new empiricism? The official view of the Vienna Circle identified it as a renunciation of the conception of synthetic judgments *a priori* in all "scientific philosophy".[25] However, excoriation of the synthetic *a priori* can be but a necessary, not a sufficient, characterization, for it does not distinguish the new empiricism from what Schlick called the "extreme empiricism" of Mach, in criticism of which Schlick largely agreed with Cassirer. But although unwilling—as the title of his essay reveals—to allow a third way in the choice between empiricism and the strictly Kantian synthetic *a priori*, Schlick insisted upon such with respect to the choice Cassirer had posed in 1921 between Machian empiricism (*Sensualismus*) and critical idealism. There is yet another alternative: it is an empiricism *with constitutive principles*.

> Between the two remains standing the empiricist view, according to which these constitutive principles are either *hypotheses* or *conventions*; in the first case they are not *a priori* (since they lack apodeicticity), and in the second they are not synthetic.

Because of this *tertium quid*,

> a thinker who in general perceives the unavoidability of constitutive principles for scientific experience should not yet on that account be designated a critical philosopher (*als Kritizist*). An empiricist can, for example, very well recognize the presence of such principles; he will only deny that they are synthetic and *a priori* in the sense described above.[26]

As we know, Schlick's novel idea of an "empiricism with constitutive principles" created the broad mold for the logical empiricist or logical positivist analysis of scientific knowledge. To invoke only Coffa's evaluation, "there was no doctrine more central than this to the development of logical positivism in the late 1920s and early 1930s".[27] In a brief time, the language of "constitution" would fade from view, particularly after Carnap's ambitious sketch in 1928 of a "constitution theory" for empirical science based on the type theoretic logic of *Principia Mathematica*.[28] What remained of Schlick's holist conventionalism was the idea that a convention or stipulation, Schlick's surrogate for constitutive principles, must be made in order for a physical theory to acquire empirical content. The requirements of empiricism mandated that such conventions, in the form of physical or "co-ordinative definitions" concern observable objects and processes. In short order, the core thesis of logical empiricism emerged, that all cognitive statements could be factored into purely tautological (or analytic) and empirical (or synthetic) components. This revitalization of Humean empiricism was above all due to Schlick, and from it nearly all the subsequent currents of 20th century philosophy of science might be traced, if only in dialectical opposition.

But consider, for a moment, Schlick's claim that the constitutive principles of scientific theories are located within the alternatives "either *hypotheses* or *conventions*". In order to gauge the significance of this disjunction, some further details of Schlick's general epistemology of science are worth noting, as that position is represented in the first edition of his *Allgemeine Erkenntnislehre*. For Schlick, the process of cognition is essentially unifying and explanatory; its "great task" is to find out how to use more and more general concepts to designate individual or particular objects. This is all the more the case in the sciences posing the ultimate aim of cognition as bringing the totality of phenomena under a minimum of explanatory principles.[29] As I showed in chapter 2, §2.4.1, Schlick's account of the process of cognition is entirely erected upon the concept of "coordination" (*Zuordung*), the basis of the "merely designative (semiotic) character of thinking and cognition". In Schlick's exposé, concepts are signs coordinated to objects, taken in the wide sense as including perceived or inferred qualities, properties, and relations. With a gesture to Hans Vaihinger's "Philosophy of the As If", Schlick maintained that, strictly speaking, concepts are not real, they are "mere fictions", valuable only for their instrumental role in designation, which is the essence of cognition. Since they are "not real mental forms of any kind", concepts can be precisely defined through definitions, particular judgments that set up relations among concepts.

What is wanted is a mode of definition compatible with the "purely semiotic" character of scientific cognition and with the character of concepts as mere signs. Here Schlick took Hilbert's axiomatization of Euclidean geometry, in particular, to have indicated "a path that is of the highest significance for epistemology" in emphasizing *implicit definition* of concepts in mathematics as a method that frees concepts from any nonexplicitly expressed (and so, nonconceptual) trappings of meaning. In mathematics, primitive concepts appearing in the axioms of a theory are implicitly defined by their occurrence in the deductive consequences of the axioms, the sole requirement being the mutual *consistency* of the axioms that relate concepts to concepts. All verbal concepts may bring with them other, more or less vague, semantic connections and psychological associations. But in mathematics, the method of implicit definition has epistemological significance precisely because whatever *intuitive* meaning (*anschauliche Bedeutung*) thereby attaches to such concepts (like "point" or "line") is "completely unimportant" for the deduction of mathematical theorems. The fundamental innovation of Schlick's general theory of knowledge is then to suggest that the method of implicit definition "is by no means restricted to mathematics but is in principle just as valid for all scientific concepts as for mathematical ones". In this way, rigorous exactness of thinking is purchased at the cost of "a radical separation" of concepts from intuition, and of thinking from reality (*Wirklichkeit*).[30] On the one side lies a system of scientific concepts, precisely defined within a (hopefully) consistent axiomatic system; on the other side lies a reality composed of the "forms of the given", objects and relations of scientific experiment and observation. At this juncture, the "bridges" between the two spheres, of axiomatic scientific theories and "reality", "are down". But they are restored by the coordination of judgments, affirming relations among concepts, to facts, always concerning at least two objects and a relation between them.[31] The target of Schlick's insistence upon this radical separation is clearly marked out; it is any version of synthetic *a priori* constitutive principles: "Thought never creates the relations of reality [*Wirklichket*]; it has no form which could imprint it, and reality allows no imprinting, for it is already formed".[32] Those of the neo-Kantian school, Schlick declared, "commit the error of taking the conceptual wrapping for reality itself".[33]

In Schlick's account, judgments, if not tautologies, explicit definitions, or false, are thus signs for facts of the world.[34] *Cognitive* judgments, propositions representing new knowledge claims, are *new* combinations of *old* concepts occurring in other propositions; some of these concepts are previously known, for example, from an explicit definition that is based on a convention ("A yard is three feet"). In general, four different classes of possible judgments are distinguished:[35] (1) definitions, a coordination completed through an arbitrary stipulation; (2) empirical judgments, designations of facts of experience; (3) hypotheses, judgments formed from known concepts for a *provisional* designation of facts, in the hope of attaining a univocal coordination; and (4) synthetic judgments *a priori* in the sense of Kant, noted above. Schlick will always deny the actual existence of the latter.

Of relevance here is the opposition between definitions or conventions and hypotheses. A reader of the first (1918) edition of the *Allgemeine Erkenntnislehre* might be inclined to think that there is little at issue in this distinction, since Schlick's view is that this only a *relative* difference, relative essentially to what is already known, that is, to the state of the system of scientific cognition at a given time. The more self-contained and developed is the deductively connected scientific system of concepts and judgments, the more "genuine judgments" differ from definitions only in a "practical or psychological sense, not in a purely logical or epistemological one".[36] An axiomatized mathematical theory provides the illustrative example: it is to some extent arbitrary whether certain sentences are derived as theorems, or treated as axioms from whose consequences other judgments may be derived that ordinarily serve as definitions of the concepts. In the less deductively developed and self-contained empirical sciences, the difference between definitions and "genuine judgments" appears to be clearer and better founded. For in the empirical sciences a definitional judgment first allots a given meaning to a concept, but then concepts designating real objects continually acquire "an ever richer content" through the process of inquiry and so the judgments containing them appear as instances of knowledge, "genuine judgments", rather than definitions. But in principle the situation is no different; the difference in kinds of judgment is merely a relative one, and the same "linguistic formulation" may serve in either role:

> Once a science has developed into a rounded-out, more or less closed, structure, what is to count in its systematic exposition as definition and what as knowledge [*Erkenntnis*] is no longer determined by the accidental sequence of human

experiences. Rather one will accept as definitions those judgments that resolve a concept into such characteristics that one can construct from the *same* characteristics, many—perhaps even all—concepts of the given science in the simplest possible way.[37]

This fluidity between cognitive judgments and definitions is ultimately solidified through appeal to the principle of simplicity, in which, following Poincaré, Schlick located the operative criterion for conventional choice, even among judgments, or systems of judgments (theories) that can each be considered to be "true". For if a judgment, formed within the interconnected scheme of concepts that is the axiomatized theory, designates a fact univocally, it is called "true". Truth, the "only virtue" of judgments, is just the "univocal designation" of facts by judgments which means that different conceptual systems containing judgments univocally designating all the facts in question, may equally be deemed "true".[38] Underdetermination of theory by empirical evidence is expressly recognized; unlike contemporary scientific realists, Schlick regarded it as posing no particular obstacle for his entirely semiotic conception of truth.

It is in precise accord with this fluid distinction between hypothesis and definition in the first edition of *Allgemeine Erkenntnislehre*, that Schlick assessed the philosophical significance of the (special) principle of relativity in 1915. In this essay, published in December 1915, shortly after Einstein presented his generally covariant gravitational theory to the Prussian Academy on 25 November, Schlick allowed that both the Einstein "view" (without the ether) and the Lorentz "view" (with a substantial ether), can be reckoned as "true" because each gives a univocal designation of the kinematical facts of space and time measurement. A decision in favor of Einstein's theory of (special) relativity can be made only if one further accepts "the principle that the simplest theory, the one least encumbered with hypotheses, is to be regarded as a 'true copy' of reality". Indeed, for Schlick in 1915, "the real" or "reality", as that concept is presupposed in science, is to be *defined* through the principle of simplicity:

> We can simply assert that among the possible assumptions the simplest should be designated as the one "corresponding to reality". "Reality" is then just a word for that unknown reason which "brings it about" that certain theories yield the simplest type of natural regularity.[39]

Any assertion that *nature is simple* cannot be based on experience but must be a mere stipulation. Complex theories can always be thought up that equally provide a univocal designation of all the relevant facts. The prototype for this kind of approach to theory-choice in the face of empirical underdeterminism, Schlick made quite clear, is Poincaré's conventionalist preference for Euclidean geometry.

Thus, although Schlick posed the issue to Cassirer in 1921 as "empiricist or Kantian", it is very difficult to construe Schlick in 1915 as supporting anything like an empiricist interpretation of the geometry of space-time. Indeed, there he subordinated empiricism to a conventionalism that nonetheless has realist aspirations, appealing to Poincaré's geometric conventionalism to secure his semiotic conception of truth as "univocal designation", while tacitly criticizing Helmholtz, along with Gauss, for holding that an "empiricist" conception of physical geometry was possible:

It is therefore no contradiction, but lies, rather, in the nature of the matter, that under certain conditions several theories may be true at once, in that they provide to be sure different, but still in each case completely univocal designation of the facts. One of them, indeed, will do this more skillfully and simply than all others, and one may therefore work with it alone, and even agree to call it the sole "correct" one, but a logically compelling reason for this may not at first be apparent.... As [an] example we may refer to the possibility of using different geometries in the physical description of the world, without doing any harm to the univocality [*Eindeutigkeit*]. Henri Poincaré has shown with convincing clarity (although Gauss and Helmholtz still were of the opposing opinion), that no experience can compel us to lay down a particular geometrical system, such as the Euclidean, as a basis for depicting the physical lawfulness of the world. Rather, one can choose entirely different systems for this purpose, though in that case we also have at the same time to adopt other laws of nature.... We always measure, as it were, only the product of two factors, namely the spatial and the, in the narrower sense, physical, properties of bodies, and we can arbitrarily assume one of the two factors, so long as we merely take care that the product agrees with experience, which can be achieved by a suitable choice of the other factor.... The theory must now make it its task to so choose *both* factors, that the laws of nature are given the simplest possible expression. As soon as it succeeds in this, it appears to us with great persuasive power as the "correct" one. In the case of space, it is known that all experience teaches that it is by far the most convenient thing to base it on Euclidean geometry; physics can then be founded on the simplest assumptions of all (e.g., that a body retains its shape unaltered during a uniform translation]. We therefore absolutely [*schlechthin*] designate our space as Euclidean, although strictly speaking there is nothing that compels us to put nature's laws into Euclidean dress. That happens, as Poincaré expressed it, on the basis of a *convention*, and his view has therefore been given the name of conventionalism.[40]

With the general theory of relativity, Schlick understandably backed away from Poincaré's view that the simplest theory combining the two factors of geometrical and physical properties of bodies, will cast "nature's laws into Euclidean dress". But the most noteworthy aspect of this passage is Schlick's unequivocal endorsement of geometric conventionalism, not empiricism.

Nor did a recognizably "empiricist" interpretation of physical geometry emerge in Schlick's 1917 monograph on the general theory of relativity, "Space and Time in Contemporary Physics", although here, inexplicably, Helmholtz is now aligned with Poincaré, as a conventionalist. Once again, Schlick's *relative* distinction between hypotheses and definitions appears very clearly in his discussion of spatial measurement in the context of the general theory of relativity. There the reason guiding choice of the key stipulation—that certain bodies are to be regarded as rigid—is located in a so-called Principle of Continuity, namely, that we "maintain continuity with the physics that has hitherto proved its worth".

Comparing measuring rods and observing coincidences result in a measurement, as we have seen, only if they are founded on some idea, or some physical presupposition [*Voraussetzung*] or, rather, stipulation [*Festsetzung*]; the choice of which, strictly speaking, is essentially of an arbitrary nature, even if experience points so unmistakably to it as being the simplest that we do not waver in our selection.

The "physical presupposition" or rather "stipulation" is that we are to regard the length of a rod as remaining constant, so long as its place, position, and velocity change only slightly. In other words,

> we stipulate that, for infinitely small domains, and for systems of reference, in which the bodies under consideration possess no acceleration, the special theory of relativity holds.... The equations of the general theory of relativity must be, in the special case mentioned, transformed into those of the special theory.[41]

Now it is only in the absence of a firm distinction between physical "hypothesis" or "presupposition" and "definition" or "convention" that the validity of the special theory (in which Euclidean measure determinations are employed) in "infinitesimal regions" of the variably curved space-times of the general theory, a claim often underwritten by the "infinitesimal principle of equivalence" criticized in chapter 2, §2.4.3, could be described as a stipulation. Such a characterization of the principle of equivalence can serve only to ease the assimilation of Schlick's 1917 treatment of geometry in the general theory of relativity to the conventionalism account he associated with Poincaré.

However, in the second (1925) edition of *Allgemeine Erkenntnislehre*, the fourfold classification of judgments of 1918 has become a threefold one.[42] Hypotheses are no longer distinguished from other empirical judgments, but both are sharply differentiated from the broad class of definitions, analytic not synthetic judgments.[43] "Concrete definitions", the association of a name with a given particular object that can be "a quite arbitrary stipulation", obviously belong here. But the class of definitions also contains conventions *per se*, definitions that enable a concept to apply to reality by attaining an univocal designation of the real (*eine eindeutige Bezeichnung von Wirklichem*).[44] Spatiotemporal relations are "the true domain" of conventions, in particular, conventions asserting an equality of spatial or temporal intervals. In any case, they are quite different from "concrete definitions" and nontrivial in that it is "one of the most important tasks of natural philosophy" to investigate their nature and meaning.[45] Yet in 1921, in his essay on Cassirer, Schlick still clearly recognized a relative distinction between definitions and hypotheses, and as discussed above, this relative distinction, together with its solidification through a principle of simplicity, is the core of his conventionalist account of truth as univocal correspondence. Why, then, did Schlick in 1925 efface this relative distinction between hypotheses and definitions?

Here is a clue, and it comes, most familiarly, from Quine: to the extent that a physical theory is regarded as confronting "the tribunal of experience" not statement by statement but only "here and there" through connections that implicate, sometimes in ambiguous fashion, large blocks of theory, to that extent do the individual concepts and judgments of the theory lack an individual empirical (or "cognitive") meaning. And this is the case if there is only a relative, "practical or psychological, not a purely logical and epistemological" difference between judgments that are definitions and those that are "genuine", empirical judgments designating, or purporting to designate, "facts. By the same token, the more indirect the empirical warrant of fundamental theoretical concepts may be allowed to be, the less suitable becomes any "empiricist" semantic analysis of the theory in question, given empiricism's mandate that observational evidence alone provides the meaning of physical concepts, or can serve as sole justification for affirming or denying statements about the physical world. While Quine and others might still insist that a more liberal doctrine of empiricism can survive even these holist ties of theory to observation, Schlick in 1921 needed a more clearly identifiable variety of empiricism to counter the threat posed by Cassirer, who attempted to tether general relativity to critical idealism. Don Howard has recently put his finger on the difficulty here:

> [T]he neo-Kantian can argue that since the coordination does not occur empirical proposition by empirical proposition, since only whole theories are coordinated with reality, and since, therefore many different theories can equally well be coordinated with reality ... which is to say that experience alone does not determine unambiguously our choice among possible theories, it is the function of synthetic *a priori* judgments to resolve the ambiguity.[46]

Then in order to save his purely designative conception of truth and to pose an empiricist alternative to the neo-Kantians, Schlick needed to rein in the holist conception of physical theory as a system of statements whose central concepts are implicitly defined by the axioms of the theory. By fortuitous circumstance, the prototype of an empiricism suited to the theory of relativity had just appeared on the scene, promoted, no less, by Einstein himself. Or so Schlick apparently thought.

3.3 "Geometry and Experience"

At the Berlin Academy's Leibniz-day public celebration on 27 January 1921, Einstein gave an address entitled "Geometry and Experience". This short lecture, also issued separately in expanded form, would be regarded by logical empiricism as a paradigm-defining text, fixing key parameters of logical empiricist philosophy of science. Prominently reprinted (in part) in the Feigl and Brodbeck reader in philosophy of science that virtually defined Anglo-American philosophy of science when it appeared in 1953, Einstein's essay perhaps is best known for its clear statement of the distinction between the modern axiomatic conception of geometry (Einstein observed approvingly, in a manner "Schlick in his book on epistemology has thus very aptly characterized as 'implicit definitions'") and "practical geometry", a contrast given a sharply terse formulation in the dictum: "In so far as the propositions of mathematics refer to reality [*Wirklichkeit*], they are not certain; and in so far as they are certain, they do not refer to reality".[47]

Three strands of Einstein's argument readily stand out. First, he introduced and defended a conception there called "practical geometry" ("*praktische Geometrie*"), a geometry arising from the "empty conceptual schemata" of axiomatic geometry through a coordination of the latter to "practically rigid bodies", that is, measuring rods and clocks that behave "as do solid bodies in Euclidean space of three dimensions". The presupposition of such bodies, in turn, renders geometry an empirical science.

> It is clear that the system of concepts of axiomatic geometry alone cannot make any assertion as to the behavior of those objects of reality which we designate as practically rigid bodies. To be able to make such assertions, geometry must be

stripped of its merely logical-formal character by the coordination of experienceable objects of reality [*erlebbare Gegenstände der Wirklichkeit*] with the empty conceptual schemata of axiomatic geometry. To accomplish this, we need only add the proposition: fixed bodies are related, with respect to their possible situations, as are bodies in Euclidean geometry of three dimensions. Then the propositions of Euclidean geometry contain assertions concerning the behavior of practically-rigid bodies.

Geometry thus completed is evidently a natural science; we may regard it in fact as the most ancient branch of physics.... We would call the thus completed geometry "practical geometry" and distinguish it in the following from "pure axiomatic geometry".... To this portrayed conception of geometry I attach special importance, because without it, it would have been impossible to set up the theory of relativity. Without it the following reflection would have been impossible: in a system of reference rotating relatively to an inertial system, the laws of situation of rigid bodies do not correspond to the rules of Euclidean geometry on account of the Lorentz-contraction; therefore in the admission of non-inertial systems as equally justified systems, Euclidean geometry must be abandoned.[48]

With the assumption of "practically rigid bodies" (and "perfect clocks"), the metrical relations of space (space-time) are *not* a matter of convention but can be empirically determined from measurements made with rods and clocks; in this way, a clear decision can be made regarding the Euclidean or non-Euclidean character of physical space. Second, the position contrasted to that of "practical geometry" is a holist form of geometric conventionalism that Einstein apparently regarded as stemming from Poincaré's rejection of the concept of *actually* rigid bodies. But with this denial, Einstein remarked, practical geometry's "original, immediate relation between geometry and physical reality [*Wirklichkeit*] is destroyed".[49] Third, Einstein nonetheless admitted, against his "practical geometry", that the position of Poincaré is, *sub specie aeterni*, in principle correct. These different aspects of Einstein's argument shall now be scrutinized to reveal the dialectical interplay between them.

The ostensible main point of Einstein's lecture is the argument that the metric of the space-time continuum is empirically determinable (and non-Euclidean) against a conventionalist view of geometry of physical space identified with Poincaré. One might well wonder why Einstein believed it necessary to uphold the viewpoint of the empirical determinability of the geometry of space-time at this time, that is, just a little over a year since the results of the British expedition confirming Einstein's theory of gravitation were announced (on 6 November 1919) from the podium of the Royal Society of London. One reason is certainly a concern to portray the historical development of the theory, which is alluded to at the end of the above quotation. As John Stachel has shown in detail, Einstein tacitly refers here to the case of a uniformly rotating disk, the simplest example of a stationary gravitational field.[50] In particular, the assumption that rigid measuring rods correspond to distances on the rotating disk enabled Einstein to conclude (in 1912) that, due to the Lorentz contraction of the rods placed upon the circumference of the disk, the geometry of the disk could not be Euclidean, a crucial step down the road to the curved space-times of general relativity.

But a more general reason is this: although fully aware that the concept of a rigid body is not really permissible in his gravitational theory of variably curved space-time, Einstein considered the empirical basis of his theory as lying in the connection of the line element ds to the rod and clock measurements of "practical geometry". In particular, Einstein used the supposition that the "segment" or rather "tract" (*Strecke*) between two neighboring points of space was empirically definable independently of the theory by the extension of an "infinitesimal" rigid rod connecting them, "normed" as a unit interval.[51] Similarly for timelike curves, a unit of duration was normed by considering the periods of two ideal clocks, running always at the same rate, no matter when and where they are brought together and locally compared. Then distances, angles, and durations in the theory could be presumed to be read off directly from the measurements of rods and clocks, paving the way to the British Expeditions' observations confirming the theory's successful prediction of the "bending" of light rays passing through the strong regions of the solar gravitational field. The idealization of rigid rods and regular clocks—that is, measuring appliances reckoned as unaffected by the presence of a surrounding gravitational field or other fields—served as a "bridge" to link the phenomena of gravitational mechanics to Einstein's non-Euclidean Riemannian geometry of curved space-time. For example, assuming "practical geometry", the metrical "distance" between finitely separated points P and Q corresponds to the measure obtained by the number of times an "infinitesimal" rod could be laid down along a "straight line" (as given by a light ray) joining the two points. On the other hand, the "distance" can be theoretically computed by integrating the invariant interval ds^2 between all neighboring points P', P'', ..., Q along a path connecting P and Q. The intervals ds^2 are found from the components of the metric tensors that are functions of the coordinates of the respective points. In turn, these components can be calculated from the Einstein field equations and so depend on the amount of "matter" (momentum, energy, stress, etc.) surrounding the region containing P and Q. According to the principle of general covariance, the coordinate patch covering the region of the two points need not be Cartesian or Galilean (i.e., a rigid grid); no metrical significance is immediately attributable to coordinate differences. But in the supposition that the measured and the theoretical value for the "distance" between the two points are the same, Einstein located the empirical basis of his gravitational theory and so rendered its non-Euclidean geometry a part of physics.

However, as discussed in chapter 4, since the spring of 1918 Einstein had continually inveighed, in private and in public, against Hermann Weyl's theory of gravitation and electromagnetism, on the grounds that it doesn't permit rods and clocks to exhibit the behavior that makes them the suitable instruments of measurement that they, in fact, are. In that theory, rods and clocks (or rather the radii of atoms and their spectral frequencies of vibration) are not independent of their position in space and time but rather depended, Einstein argued, on their "prehistory", that is, the electromagnetic fields through which they had passed. Thus, two atoms of, say hydrogen, should display different spectra if one, but not the other, had passed through a strong electromagnetic field. For the time being, I leave aside whether this is an adequate rebuttal of Weyl's theory; what is important here is that Weyl's theory lacked the direct connection to experience as did general relativity via the above-mentioned "norming". Moreover, to Einstein, Weyl's theory wrongly predicted that the spectral frequencies of the chemical elements should not be constant and independent of position as in fact they are observed to be.

Although Einstein's objection and its subsequent elaboration by Pauli (1921) persuaded nearly all interested parties, Einstein felt it incumbent upon him to raise the cudgel once again against Weyl in this Berlin Academy lecture of January 1921. Without mentioning Weyl's theory by name, Einstein reiterated his principal objection, that the demonstrable existence of sharp spectral lines of the atoms of the chemical elements, no matter what the prehistory of the atom had been (i.e., no matter what electromagnetic fields it had passed through), provided "compelling empirical proof" (*überzeugenden Erfahrungsbeweis*) for the "basic postulate of practical geometry", that is, of the existence of infinitesimally rigid rods and clocks. Indeed, it is this assumption, Einstein continued, on which ultimately rests the physical meaningfulness at all of speaking of a metric in Riemann's sense within the four-dimensional space-time continuum.[52] Thus, Einstein argued that the very applicability of Riemannian geometry to the physical world presupposed the existence of infinitesimal rigid rods and ideal clocks, and so the possibility of the empirical confirmation of the general theory of relativity rested upon the supposition that these idealized bodies give physical meaning to the concepts "unit measuring rod" and "unit clock (period)". Measurements with these instruments physically attest to the metric field of gravitation. This posited direct connection with experience was, to Pauli, "the most beautiful achievement of the theory of relativity", even though "logically, or epistemologically, this postulate does not admit of proof".[53]

To "practical geometry's" conception of the direct linkage of the Riemannian geometry of general relativity to experience, Einstein opposed what he termed the view of Poincaré that it is a matter of convention which geometry we take to obtain in physical space. Recall that Poincaré had argued that in the absence of truly rigid bodies, purely geometrical statements affirm nothing about experience until they are combined with statements of physics.[54] But this meant that in answer to the question as to the nature of the geometry of physical space, any geometry can be chosen since thereby one commits oneself only to a set of ideal propositions, as long as the supposed laws of the behavior of physical objects can be adjusted to be in agreement with what is actually observed. According to Einstein, this view is concisely represented in the formulation

$$\text{Total theory} = G + P$$

In other words, a geometry G can be chosen arbitrarily and also part of the system of physical laws P, as long as the remainder of P enables the total theory to be brought into agreement with experience. As just shown, Schlick both formulated and endorsed this holist and conventionalist conception of physical geometry in his writings on relativity theory prior to 1921. Moreover, for Schlick and for Einstein's Poincaré *sub specie aeterni*, it is not really germane that the actual Poincaré additionally thought that Euclidean geometry would always be adopted as the simplest geometry, with the corresponding adjustments to physical laws (e.g., maintaining that light rays no longer traversed Euclidean straight lines). But even while arguing against geometric conventionalism, Einstein conceded Poincaré's point that there are, in fact, no "actually rigid bodies"; indeed, Poincaré had maintained that this is a confused conception, a kind of categorical mistake. The concepts of the fixed body and clock, Einstein admitted, "do not play the role of irreducible elements in the conceptual edifice of physics but rather of composite structures which should play no independent role in the construction of theoretical physics".[55] What Einstein meant is that general relativity, which in principle is capable of encompassing all matter fields into the geometry of space-time (in 1921, this was only electromagnetism), should, again in principle, explain material structures (e.g., rods and clocks) as composite structures whose structure and behavior are derivable from the theory's field equations. Hence, connecting such a theory to experience by means of concepts treated as independent of the theory (in particular in the assumption of rigidity), when in fact they are not, is a less than consistent procedure.

This was not really a new concession by Einstein. A year before completion of the general theory of relativity in November 1915, Einstein had expressed views on the connection of geometry to experience in a manner similar to that which he designated in 1921 as "practical geometry". However, in that context, his attention was limited to pointing out that the advent of field theories, with their prohibition of action-at-a-distance, presented a new and critical perspective on Euclidean geometry.

Before Maxwell, the laws of nature were, in *spatial* relation, in principle *integral laws*; this is to say that distances between points finitely separated from one another appeared in the elementary laws. This description of nature is grounded upon Euclidean geometry. The latter signifies at first nothing other than the system of consequences of the geometrical axioms; they have, in this respect, no physical content. However, geometry becomes a physical science by adding the requirement that two points of a "rigid" body must be separated by a determinate distance, independent of the position of the body. Propositions supplemented through this stipulation [*Festsetzung*] are (in the physical sense) either applicable [*zutreffend*] or inapplicable. In this extended sense, geometry forms the foundation of physics. From this viewpoint, the propositions of geometry are to be considered as integral physical laws, since they deal with distances of *finitely separated* points. . . . Through and since Maxwell, physics has undergone a through-going radical change in gradually carrying through the demand that distances of finitely separated points may no longer appear in the elementary laws, that is, "action at a distance theories" [*Fernwirkungs-Theorien*] are replaced by "local-action theories" [*Nahewirkungs-Theorien*]. In this process it was forgotten that also Euclidean geometry—as employed in physics—consists of physical propositions that from a physical viewpoint are to be set precisely on the side of the integral laws of the Newtonian point mechanics. In my opinion, this signifies an inconsistency from which we should free ourselves.[56]

Einstein went on to introduce his readers to the novel idea that his gravitational theory, in treating coordinates as arbitrary parameters in the space-time continuum, reduced the integral laws of Euclidean geometry to differential laws, removing the mentioned inconsistency, and appearing as a natural extension of the requirement of a "local-action theory". But if set against the 1921 essay "Geometry and Experience", what is particularly striking about this 1914 passage is the lack of a contrast to the method of (what is later termed) "practical geometry" or any anticipation of another tie of geometry to experience *other than through a stipulation* about the invariant distance between points on a "rigid" body. In other words Einstein did not recognize in 1914, as he did in 1921, Weyl's point (see further in chapters 4 and 6) that reliance upon the rigid bodies of geometrical measurement is itself an *inconsistency* with the character of the general theory of relativity as a "local action theory" (*Nahewirkungstheorie*) but represented a remnant of the *Ferngeometrische* past of Euclidean geometry.[57] In 1921, Einstein additionally admitted that, in principle, the "practically rigid" behavior of such complicated physical structures as rods and clocks (manifesting what Weyl would call the "natural gauge of the world"; see chapter 6, §6.4.2.1) should be explicable as a remote consequence of the field equations. In a theory such as general relativity, such empirically direct ties of geometry to experience via "practical geometry" are unexplained explainers. This, and not a conventionalist freedom to always choose Euclidean geometry, is the sense in which Poincaré is right *sub specie aeterni*. Nonetheless, Einstein stated his belief that at the present stage of knowledge, that is, in the absence of a field theoretic account of matter, these concepts must be provisionally accepted.

Einstein's admission, in 1921, of the *sub specie aeterni* correctness of Poincaré's point of view is not a concession to conventionalism, that is, to freedom to choose any geometry we like, but to the inevitable epistemological holism of a theory in principle capable of explaining its own measuring applicances, and so its ties to observation. Poincaré's position is valued for its principled unwillingness to consider certain physical objects as "geometrical", that is, as *ideal*, and so as independent from the field laws that, in principle, are accountable for the behavior of all material structures. But this is precisely the point of view of what Einstein later termed "a consistent field theory". It is accordingly important to keep sight of the *qualified* character of Einstein's methodological analysis in favor of "practical geometry". By 1921, Weyl had shown that there are other, less direct means of connecting the geometry of space-time to observation through the paths of freely falling "test particles" and of light rays, avoiding the inconsistent assumption of even infinitesimal rigid bodies in a theory of the gravitational and, perhaps, electromagnetic fields. In such a procedure, the tie of the total theory, geometry plus physics, to experience involves quite intricate and theory-internal complexions: tracks of force-free neutral test particles of negligible mass are taken to manifest geodesics of the affine structure of space-time; the equation of motion of such particles is itself (at least in the case of a pure gravitational field) derivable from the theory's field equations. The totality of the affine geodesics endows the space-time manifold with a projective structure. Paths of light rays, in turn, provide the space-time manifold with a causal-conformal structure; it can then be shown that the paths of light rays (conformal geodesics) are a limit case of the projective structure and that together they give enough information to determine a metric at a given point. (The projective and the conformal structure determine the metric, up to a factor of scale; see chapters 4 and 6.) This construction of the metric of space-time shows that it is neither necessary nor desirable to posit a theory-independent (i.e., of gravitational theory) definition of the distance between two neighboring points.[58] Such a construction, to be sure, involves its own assumptions, but these are assumptions compatible with the field equations of the theory. In so many words, without the assumption of rigid bodies, the empirical foundation of general relativity acquires the epistemological and semantic complexity that fall under the rubric of holism.

It is therefore worth noting that within just a few years Einstein took repeated pains to distance himself from an unqualified commitment to "practical geometry" on grounds that are explicitly holist in nature. Here just two instances are cited. First, in the prominent context of his "Nobel lecture" (1923e), delivered to the Nordic Assembly of Naturalists at Gothenburg (serving in lieu of a Nobel prize lecture), he brought up the "deficiency of method" in the stipulation that measurement bodies are rigid. Before doing so, Einstein stated the kind of empiricist meaning criterion often deemed characteristic of the method of the theory of relativity:

> [C]oncepts and distinctions are only admissible to the extent that observable facts can be assigned to them without ambiguity (stipulation that concepts and distinctions should have meaning). This postulate, pertaining to epistemology, proves to be of fundamental importance.

A bit further on, however, Einstein returned to consider this empiricist "meaning stipulation" in regard to the notions of rigid body and uniform clock of chronogeometrical measurement:

> The concept of the rigid body (and that of the clock) has a key bearing on the foregoing consideration of the fundamentals of mechanics, a bearing which there is some justification for challenging. The rigid body is only approximately achieved in nature, not even with desired approximation; this concept does not therefore strictly satisfy the "stipulation of meaning". It is also logically unjustifiable to base all physical consideration on the rigid or solid body and then finally reconstruct that body atomically by means of elementary physical laws which in turn have been determined by means of the rigid measuring body. I am mentioning these deficiencies of method because in the same sense they are also a feature of the relativity theory in the schematic exposition which I am advocating here. Certainly it would be logically more correct to begin with the whole of the laws and to apply the "stipulation of meaning" to this whole first, that is, to put the unambiguous relation to the world of experience last instead of already fulfilling it in an imperfect form for an artificially isolated part, namely, the space-time metric. We are not, however, sufficiently advanced in our knowledge of nature's elementary laws to adopt this more perfect method without going out of our depth. At the close of our considerations we shall see that in the most recent studies there is an attempt, based on ideas by Levi-Civita, Weyl, and Eddington, to implement that logically purer method.[59]

Eddington was the first to build up a "world geometry" of the curved space-time continuum by beginning with an affine connection, making the metric of secondary fundamental importance (see chapter 8). Obviously, in such a theory (in 1923 envisaged as a field theory of gravitation and electromagnetism), originally nonmetrical, there can be no immediate connection—requiring both the supposition of a metric and a "realization" of distance through rigid rods—of metrical notions with experience.

Another, even more explicit, methodological endorsement of holism in the ties of geometry to experience is given the essay "Non-Euclidean Geometry and Physics" (1925a). Although Einstein's main thesis in this article is that the geometry of space-time in general relativity is empirically determined to be non-Euclidean, the methodological issues between "practical geometry" and holism are posed very clearly.

> According to [the] more refined conception of the nature of the fixed body and of light, there are no natural objects which correspond *exactly* in their properties to the basic concepts of Euclidean geometry. The fixed body is not rigid [*starr*], and the light ray does not rigorously embody the straight line; of course in general, it is not a one-dimensional structure. According to modern science, geometry by itself [*allein*] anyway corresponds to no experiences, but rather only geometry together with mechanics, optics, and so on.

A few paragraphs later Einstein returns to the issue of the tie of geometry to experience, remarking that "one must take up either one of two consistent standpoints". The first is the standpoint of the "practical physicist":

> Either one accepts that the "body" of geometry in principle is actualized through the fixed body of nature, if only certain regulations are imposed regarding temperature, mechanical demands, and so on. Then, to the "tract" ["*Strecke*"] of geometry corresponds a natural object, and with this all propositions of geometry attain the character of expressions about real bodies. This standpoint was represented especially clearly by Helmholtz, and one can add that without it, the setting up of the theory of relativity would have been practically impossible.

The other standpoint, as noted above, is however that of "modern science":

> Or, one in principle denies the existence of objects which correspond to the basic concepts of geometry. Then geometry by itself contains no expressions concerning objects of reality [*Wirklichkeit*] but only geometry together with physics. This standpoint, which may be more perfect [*vollkommenere*] for the systematic representation of a completed physics, was represented especially clearly by Poincaré. From this standpoint the total content [*Inhalt*] of geometry is conventional; which geometry is preferred depends upon, through its use, how "simple" a physics can be set up that is in agreement with experience.[60]

Just as in "Geometry and Experience", Einstein then remarks that he chooses the first standpoint, that of the "practical physicist" and so, of Helmholtz, "as better agreeing with the current position of our knowledge". If this is done, the issue between the Euclidean or non-Euclidean character of the geometry of space-time is plainly posed and, Einstein argued, clearly answerable. The other standpoint, of "modern science", is that of Poincaré and holism, with the overall simplicity of physics together with geometry as the determining criterion in the choice of a particular geometry. But even from this standpoint, although unlike in Poincaré, choice of Euclidean geometry within "modern science" is really nonadmissible. This point is tacitly conceded at the end of the lecture in referring to the further generalizations beyond the Riemannian geometry of general relativity by Weyl and Eddington, based upon Levi-Cività's concept of the "infinitesimal parallel displacement" of a vector (see chapters 4, 6, and 8). These theories have shown, Einstein concluded, that "the ideas which have developed out of non-Euclidean geometry have proven eminently fruitful in modern theoretical physics".[61] Such a fruitfulness of geometric ideas in physics would seem to belie Einstein's claim that from the standpoint of "modern science", "the total content of geometry is conventional". Moreover, in these theories, the question of the geometry of space-time within the total system comprised of geometry plus physics centers on a choice among different *non*-Euclidean geometries.

In sum, the holism of the standpoint that Einstein saw Poincaré as representing "especially clearly", is independent of Poincaré's own conventionalist choice of Euclidean geometry. Einstein's conjunction of the distinct issues of holism and conventionalism is understandable, rooted as it is in Poincaré's own discussions of geometric conventionalism. But the dynamical character of the space-time metric of general relativity provides ample grounds for disentangling epistemological holism from conventionalism. That geometry could be dynamical was a possibility not taken seriously by Poincaré in the light of his remark that Riemann's geometries of variable curvature, which are "incompatible with the motion of a rigid figure", "could never therefore be other than purely analytic".[62] But in any case, these two texts, of 1923 and 1925, suffice to show that Einstein's account of the connection of space-time geometry to experience via the expedient of "practical geometry" was considerably more nuanced and provisional than Schlick, and subsequently Reichenbach, would take it to be. Far from simply choosing a stipulation through which a metrical concept acquires physical meaning, the issue of the empirical character of geometry can be assessed from two different and complementary standpoints, either of which has distinct advantages. "Helmholtz" and "practical geometry" had proved invaluable in the heuristic genesis and initial confirmation of general relativity and, moreover, provided a lens through which to focus attention on the speculative and unphysical character of Weyl's theory. Yet "Poincaré" and "completed physics" (and, although unstated, "Weyl" and "Eddington") are necessarily the perspective of a "unified field theory" such as Einstein, after 1921, sought with unrelenting determination. Within unified field theory, the standpoint of epistemological holism in linking space-time geometry with experience is no longer merely an option.

3.4 Helmholtz and "Schlick's Helmholtz"

In 1921, Schlick and Paul Hertz, a physicist in Göttingen and relative of the famous Heinrich Hertz, editorially collaborated to publish a centenary collection of Helmholtz's *Epistemological Writings* annotated with extensive "elucidatory" footnotes.[63] Significantly, this collection of four papers included Helmholtz's two classic articles on the foundations of geometry, "On the Facts Underlying Geometry" (*"Über die Tatsachen, die der Geometrie zugrunde liegen"*, 1868) and "On the Origin and Significance of the Geometrical Axioms" (*"Über den Ursprung und die Bedeutung der geometrischen Aziome"*, 1870), as well as "the free and untrammeled statement of his philosophical position",[64] "The Facts in Perception" (*"Die Tatsachen in der*

Wahrnehmung", 1878). Schlick wrote the elucidatory notes to the 1870 and 1878 papers, and it is here that the immediate impact of Einstein's lecture shall be discerned.

Helmholtz published "On the Facts Underlying Geometry" in 1868 in the *Nachrichten* of the Göttingen Academy of Sciences; the previous year, 1867, Riemann's celebrated *Habilitationsrede* of 1854 "On the Hypotheses Underlying Geometry" had been published after Riemann's death by Dedekind in the *Abhandlungen* of that Academy. Indeed, the publication of Riemann's essay came as a revelation to Helmholtz, who, in 1866, had taken the pre-Riemannian position that free mobility of rigid bodies implied the validity of the Euclidean axioms.[65] But as still appears from the echo in his chosen title, Helmholtz's 1868 essay is ostensibly an attempt to replace the "hypotheses" Riemann saw as underlying geometry with "facts"; it will emerge, however, that Helmholtz's essay treats only a considerably more restricted conception of geometry. At the very beginning, Helmholtz posed the fundamental question he would attempt to answer: to determine to what extent the propositions of physical geometry have "an objectively valid meaning" and to what extent are they dependent on definitions or the form of descriptions. Despite his subsequent reputation as a "geometric empiricist", Helmholtz immediately continued that in his opinion "this question is not to be answered all that simply". The principal difficulty facing an empiricist account of geometry is then squarely pinpointed:

> [I]n geometry we continuously deal with ideal structures [*idealen Gebilden*] whose bodily representation in actuality [*Wirklichkeit*] is always only an approximation to requirements of the concept, and we only decide whether a body is rigid [*fest*], whether its surfaces flat, its edges straight, by means of the same propositions whose factual correctness the examination is supposed to demonstrate.[66]

The difficulty is that the concept or idea of a rigid body must be already be operatively legitimate in order to be in a position to know whether any given body is approximately rigid. For this reason, the concept of a perfectly rigid body is not itself acquired from experience. Nonetheless, it must be presupposed in the practice of geometric measurement, particular instances of which serve to confirm or disconfirm whether geometrical-physical space has a Euclidean structure.

In this 1868 paper, Helmholtz sought to derive Riemann's central hypothesis, that the length *dl* of an infinitesimal Pythagorean line segment is expressed by a quadratic function of the coordinate differentials (written here in the modern way),

$$dl^2 = g_{ik}dx^i dx^k$$

from certain hypotheses that expressed "facts" primarily about the observable behavior of the standard bodies of geometrical measurement, in particular, the rigid measuring rod. As he made clear at the outset, Helmholtz's concern, unlike Riemann's more general investigation, was limited to the consideration of "actual space" (*wirklicher Raum*), a space satisfying the requirement that in it, finite systems of fixed points (rigid bodies) could move around without distortion. With the constraint imposed by the free mobility of finite fixed bodies, he could show that Riemann's hypothesis was derivable from four less restricted assumptions.

Helmholtz's characterization of actual space as allowing free mobility of rigid bodies stemmed from his supposition that all geometrical measurement, and so the very possibility of physical geometry, is based on the observation of the relation of congruence between spatial magnitudes. Observations of congruence between bodies, according to Helmholtz, presupposed the possibility of motions of fixed finite bodies up to and adjacent to one another, as well as that congruence of spatial magnitudes is independent of all motions. So the facts about congruent spatial magnitudes are facts about rigid bodies, and their motions, a considerably different view of measurement from that of Riemann, for whom geometric measurement presupposed only the total or partial superposition of one ideally thin and perfectly flexible "measure string" or thread on another.[67] These facts are expressed in the last three of the four hypotheses underlying Helmholtz's investigation; the first is taken directly from Riemann. They are 1) that a space of n-dimensions is an n-fold extended manifold (meaning a point in it is specified by n independent and continuously varying coordinates), and the motion of a point in such a manifold is accompanied by a continuous change in at least one of the coordinates; 2) that there exist mobile finite rigid bodies, or fixed point systems; 3) that there is no constraint on these motions, but completely free mobility of these bodies; and finally, 4) that space is monodromous (i.e., that two congruent bodies are still congruent after one of them has undergone a complete rotation about any axis of rotation); Sophus Lie, using the language of continuous (Lie) groups, later showed this hypothesis (for spaces of three or more dimensions) to be redundant.[68] As clarified by Lie, justification for Riemann's hypothesis of an "infinitesimal Pythagorean metric" on an n-dimensional manifold involved showing that the isometries of this metric (its congruences, or equality of measures) are precisely captured by an infinite parameter (continuous) group of motions acting on the manifold such that, for each motion carrying a rigid body from any given point P to any other point Q, there is exactly one transformation in the group.

One immediate difference between Helmholtz and Reimann is that for Riemann, free mobility is not assumed for finite bodies but only for infinitely small bodies, corresponding to a Riemannian manifold's assumption of "flatness in its smallest parts". There are two further restricting conditions: (a) unlike Riemann, Helmholtz was concerned only with the case of a three-dimensional manifold, corresponding to actual space, and (b) since the aim is to drive Riemann's result, expressed in terms of differentials of the coordinates, Helmholtz considered "only points having infinitely small differences in the coordinates". So a "congruence independent of limits will be presupposed only for infinitely small spatial elements",[69] ruling out applicability of his hypotheses to bodies of arbitrary size. As thus pertaining only to infinitesimal displacements rather than finite motions, Helmholtz apparently believed that his hypotheses were weaker than Riemann's axioms. In fact, they are completely different; Lie showed that thereby they have been radically changed since it is then unclear what Helmholtz's axioms, as interpreted infinitesimally, assert about observable finite motions and, conversely, what the finitely interpreted axioms state about his infinitesimal displacements, since neither the finite nor the infinitesimal cases are inferable from one another.[70] Moreover, precisely because Helmholtz's derivation pertains only to infinitesimal displacements, it does not furnish what Helmholtz claims for it (already in the title of his paper) as its significance, for pending some physical understanding of the extent of the spatially "infinitesimal", such as relativity theory later provided, the validity of his axioms in the infinitesimal cannot be connected, except at most indirectly and hypothetically, with the "observable facts" about the free mobility of rigid finite bodies, that is, as an integral and not a differential law. Helmholtz's proof therefore does not, and cannot, accomplish what it claims to do, a derivation of Riemann's hypothesis from observable facts. This remarkable lapse is alluded to briefly in notes of P. Hertz to Helmholtz's 1868 paper.[71] But with the exception of the second, there is no reason to linger over the other hypotheses. So consider the salient issues Schlick raised with Helmholtz's hypothesis about the existence of rigid bodies.

The second hypothesis notably requires a definition of rigid body, but to all appearances the definition given is circular, for it invoked a conception of points already fixed in space. The definition, cited from the 1870 geometrical paper, reads:

> [T]the definition of a rigid body can now only be given by the following characteristic: Between the coordinates of any two points belonging to a rigid body, an equation must exist that expresses an unchanged spatial relation between the two points (which finally turns out to be their separation) for any motion of the body, and one which is the same for congruent point pairs. Such point parts however are congruent, which can successively coincide with the same fixed point pairs in space.[72]

The apparent circularity was noted by both P. Hertz and by M. Schlick in their respective "elucidations" to Helmholtz's two papers on the foundations of geometry reprinted in the centenary collection. Schlick's comment to the above passage (n. 31 to the text) is worth quoting at length:

> This definition reduces congruence (the equality of two tracts [*Strecken*]) to the coincidence of point pairs in rigid bodies "with the same fixed point pairs in space" and thus presupposes that "points in space" can be distinguished and held fixed. This presupposition was explicitly made by Helmholtz..., but for this he had to presuppose in turn the existence of "certain spatial structures which are regarded as unchangeable and rigid". Unalterability and rigidity...cannot for its own part again be specified with the help of that definition of congruence, for one would otherwise clearly go round in a circle. For this reason the definition seems not to be logically satisfactory.
>
> One escapes the circle only by stipulating by convention that certain bodies are to be regarded as rigid, and one chooses these bodies such that the choice leads to a simplest possible system of describing nature [here Schlick refers to Poincaré, *The Value of Science*, p. 45 of the German edition]. It is easy to find bodies which (if temperature effects and other influences are excluded) fulfil this ideal sufficiently closely in practice. Then congruence can be defined unobjectionably (as by Einstein in "*Geometrie und Erfahrung*", p. 9] as follows: "We would call a tract [*Strecke*] the embodiment set out by two marks on a practically rigid body. We imagine two practically rigid bodies with a tract marked on each. These two tracts shall be called 'equal to each other' if the marks on the one can constantly be brought into coincidence with the marks on the other".[73]

A noticeable symptom of this alleged difficulty with Helmholtz's definition is that, although he *appeared* to recognize that the notion of a rigid body is an idealization (as in the remarks at the very beginning of his 1868 paper), in point of fact his text several times gives evidence of a stubborn belief in the existence of "actually rigid bodies". Now, as highlighted above, Helmholtz clearly thought that the axioms of geometry also make assertions about the behavior of physical measuring bodies. But he also indicated that he considered it meaningful to assert of two bodies once held to be congruent that they might, at some later time, and perhaps other situation, "have changed in the same manner" although they still are observably congruent. The reason is that Helmholtz, while recognizing the intertwining of geometry and mechanics in the concept of congruence, consistently maintained that the notion of a rigid body is *constitutive* of the concept of congruence on which geometrical measurement rests. This was expressly indicated earlier in his paper, where he noted,

> So all our geometrical measurements rest upon the presupposition that the measuring instruments which we take to be rigid, actually [*wirklich*] are bodies of unchanging form.[74]

Schlick inserted a footnote at the word "actually" that reads:

> In the little word "actually" there lurks the most essential philosophical problem of the whole lecture. What kind of sense is there in saying of a body that it is *actually* rigid? According to Helmholtz's definition of a rigid body . . . , this would presuppose that one could speak of the distance between points "of space'" without regard to bodies; but it is beyond doubt that without such bodies one cannot ascertain and measure the distance in any way. Thus one gets into the difficulties already described in note 31. If the content of the concept "actually" is to be such that it can be empirically tested and ascertained, then there remains only the expedient already mentioned in that note: to declare those bodies to be "rigid" which, when used as measuring rods, lead to the *simplest* physics. Those are precisely the bodies which satisfy the condition adduced by Einstein (compare note 31). Thus what has to count as "actually" rigid is then not determined by a logical necessity of thought or intuition, but by a convention, a definition.[75]

In fact, this "correction" completely transforms Helmholtz's views on space, as I will show. But according to Schlick, Helmholtz, either unawares or disingenuously, in seeking to base geometry upon "facts" about "actually rigid bodies", has given a circular definition of such bodies. The solution, in the light of Schlick's understanding of Einstein's "Geometry and Experience" is to break the circularity by a stipulation that certain bodies are "rigid" when, using those bodies as measuring rods, the "simplest physics" results. In this way, truths concerning geometrical-physical space are part of such of system of "simplest physics" and are determined by measurements made with the fiduciary rigid bodies. The geometry that enables such a system of physics is then empirically ascertainable.

However, if Helmholtz's geometrical papers are set in the context, as he himself did, of his prior "investigations on spatial intuitions (*räumlichen Anschauungen*) in the visual field", then another, and more internally consistent, reading of this seemingly circular definition of rigidity emerges.[76] In a recent examination of the relationship between Schlick's semiotic epistemology, epitomized above in §3.2, and Helmholtz's "theory of signs" (*Zeichentheorie*), Michael Friedman has described how Helmholtz's researches in the psychology and physiology of perception lay behind the most elaborate statement of his epistemology, "On the Facts in Perception" (1878), and its principal conclusion, which is critical of the causal realist theories of perception.[77] According to his "theory of signs", perceptions, our sensory representations of objects external to our body, are symbols or signs for these objects that need not resemble them in any way (indeed, the idea of a comparison is not even thinkable). For Helmholtz, the external cause standing behind the play of our sensations is not an object in another realm existing behind the veil of our perceptions but is just the lawlike relation governing the patterns of sensations themselves. As Friedman documents, Helmholtz's investigations of geometry were a key step in his rejection of the causal realist view of perception he had previously entertained. In particular, his researches on visual perception in the late 1850s and early 1860s had led to the conclusion that our ability to localize things in space stem from an innate capacity to imaginatively construct, from given sense impressions of an object or objects, lawlike sequences of the sense impressions that would or could be obtained through the voluntary movement of our bodies toward them, away from them, around them, and so on. Subsequently, in his researches on geometry, Helmholtz saw that these patterns of lawlike sequences of sensations, connected with the actual and possible motions of our bodies, provide the means for a representation of space itself.[78] In this sense, space is not some substantive arena in which objects behind the veil of perception are sporting about, but rather, through the anticipated possible motions of our bodies,

> a *given form* of intuition, possessed *prior to all experience*, in so far as its perception is connected with the possibility of motor impulses of the will for which the mental and corporeal capacity had to be given us, by our organization, before we could have spatial intuition.[79]

That we have such a form of "outer intuition" for spatial relationships is *a priori* in the sense that its origin lies the physiological and psychological makeup that affords the possibility of an *anticipation* of the patterns of sensations stemming from imaginatively projected voluntary motor impulses of our bodies; the perception of the space of intuition is thus bound up with the possibility of the volitional motor impulses. In turn, a mathematical representation of this form of "outer intuition" can be constructed from the possible lawlike sequences of sensations that stem or would stem from an imaginative free mobility of fixed bodies throughout the space. Such a conception of space is the actual space of physical objects, for the concept of congruence, underlying the possibility of geometrical measurement, itself rests on possibility of free mobility of rigid bodies of measurement, and judgments of congruence are based on the lawlike uniformity in sensations in the perception of superimposed bodies.

Contrary to Kant, this space of intuition is not necessarily described by the axioms of Euclidean geometry; all that follows from the condition of free mobility of fixed bodies is that this space has the mathematical structure of a three-dimensional space of constant curvature. If it is further supposed that this space is infinite, that two bodies can be continuously moved indefinitely far apart, then this space can be either a hyperbolic space of constant negative curvature or Euclidean flat space. In either case, the particular axioms that describe geometrical–physical space are only to be discovered from the spatial measurements made with rigid measurement bodies. Different systems of axioms of geometry (Euclidean, elliptic, hyperbolic) are compatible with the *a priori* form of spatiality that is mathematically describable as a space of constant curvature, but it is an empirical discovery which set of axioms accurately characterize geometrical figures in this space. This relation between space as an *a priori* form of outer intuition and the axioms of geometry is summarized in Helmholtz's well-known aphorism, "Space can be transcendental without the axioms being so".[80]

It is now possible to see how Helmholtz could have given the definition of "rigid body" that in fact he did. Space as a form of "outer intuition", spatiality as such has the structure of a space of constant curvature, as it must if the free mobility of fixed bodies is possible. In such a space, systems of points can be represented as rigid in supposing an *ideal* extension between any pair of such points, ideal as not occupied by any material body. For purposes of mathematical representation, coordinates can be assigned to the points in such a way that to any difference in coordinates between pairs of points, there corresponds an ideal dematerialized fixed extension. Geometrical measurement is then possible on the presupposition that the measurement bodies of geometry are actually bodies with such a constant fixed extension. It is this *meaning-constituting* presupposition, according to Helmholtz, permits speaking of "actually rigid bodies".[81] Regarded as a condition of the possibility of geometrical measurement at all, such measurements being restricted to a space of three dimensions in which free mobility of fixed bodies can be intuitively represented, it is not a conventional stipulation that some bodies are to be considered rigid nor is it a naive confusion regarding actual, physical bodies that, as Helmholtz knew very well, are never more than approximately rigid.

The Helmholtzian view of physical geometry is thereby distinct from a "geometric empiricism" holding that the metrical relations of space can be straightforwardly determined from facts produced by the use of measurement bodies naively regarded as actually rigid. It is also a different view than Schlick's holist conventionalism, based on a stipulation of the rigidity of certain bodies that, when employed as measuring rods, lead to the "simplest physics". Certainly, for Helmholtz, too, the axioms of geometry are propositions not regarding spatial relationships alone but also "the mechanical behavior of our most rigid bodies during motions". But the sense in which a measurement body is considered "rigid" is that provided by notion of an ideal imaginative extension belonging to spatiality itself as a form of "outer intuition". In the last analysis, Helmholtz's "geometric empiricism" boils down simply to his view that the propositions of *pure* geometry by themselves make no determinate assertions about space but only in connection with the instruments of geometrical measurement.

> Then such a system of propositions is given an actual content, which can be confirmed or refuted by experience, but which for just that reason can also be obtained by experience.[82]

This has been rightly regarded as "a very powerful argument against the Kantian philosophy of geometry and is perhaps the main reason why the latter could not survive the discovery of non-Euclidean geometries: *a priori* knowledge of physical space, devoid of physical contents, is unable to determine its metrical structure with the precision required for physical applications".[83] In a word: Helmholtz argued against the Kantian philosophy of geometry while retaining an inherently Kantian theory of space.

Some care must be taken here, for Helmholtz also described (and then rejected) the possibility of a "transcendental geometry", based on a "transcendental" or "inner intuition" of ideal *geometrical* spatial structures. The concept of such "absolutely unalterable and immobile" geometrical figures is a "transcendental concept", formed independently of any experience of actual bodies and to which the behavior of actual bodies need not correspond.[84] Relations of congruence and likeness of such figures to one another could be given in a "transcendental intuition" without the figures ever being brought into coincidence through motion, for motion "belongs only to physical bodies".[85] Now "a strict *Kantian*" (*ein strenger* **Kantianer**, original emphasis), Helmholtz allowed, could maintain that the axioms of geometry were therefore "propositions given *a priori* through transcendental intuition". But, Helmholtz then pointed out, the axioms of such a "transcendental geometry" would no longer be synthetic, instead following analytically from the transcendental concept of the immobile fixed geometrical structures.[86] To be sure, the procedure of comparison of magnitudes in "transcendental intuition" (and so, without mobility) is an idea of doubtful physical meaningfulness; our scientific and practical interest is irrevocably attached to measurable relations of spatial alikeness, not to what appears as spatially alike in "this inner intuition" (*diese innere Anschauung*).[87] The differentiation from "a strict Kantian" is therefore instructive: Helmholtz, a nonstrict Kantian, recognized an *a priori* form of spatial or "outer intuition" as a condition of the possibility of geometrical measurement, but not the "transcendental axioms" governing the geometrical relations of ideal and immobile spatial figures given within an "inner" or "transcendental intuition". Space ("outer intuition") is transcendental without the axioms being so.

We are now in a position to see how Schlick tried to appropriate Helmholtz as an empiricist precursor for his own "empiricist interpretation of the new physics". The talk of an *a priori* spatial intuition, Schlick cautioned, is to be understood only in "a non-epistemological and psychological sense, pertaining only to the psychic makeup of the cognizing consciousness". On the other hand, in the proper "transcendental-logical exegesis" of Kant,

> the essence of the *a priori* consists in its comprising the ultimate axioms which alone form the foundation for all rigorous cognition and guarantee the latter's validity.[88]

As he did with Cassirer, Schlick once again set the ground rules of what is, and what is not, suitable to be designated as "Kantian" (or "neo-Kantian"). As just shown, Helmholtz had rejected the "strict Kantian's" account of a "transcendental geometry". So, Schlick concluded, Helmholtz's "epistemology thoroughly deviates from Kant's". To be sure, Helmholtz considered physical geometry, characterized by a set of axioms, as an empirical theory—there is no necessary geometry. But he limited the available options for empirical selection to spaces of constant curvature because of the "fact" of free mobility of fixed bodies, a restriction imposed by the *a priori* form of spatiality itself. In failing to give recognition to Helmholtz's attempt to retain a Kantian account of space as a necessary form of "outer intuition", Schlick could only judge Helmholtz's discussion of "actually rigid bodies" to rest on a definitional circularity, broken only by following the example of Einstein and making a stipulation regarding "rigid bodies".

3.5 Conclusion

In the ensuing year, Schlick's empiricist sanitizing of Helmholtz continued apace; even the charge of "circularity" in Helmholtz's definition of rigid bodies was laid aside. Thus, in a 1921 talk, published in 1922, commemorating Helmholtz as an epistemologist, Schlick lauded Helmholtz's theory of space as *true*, locating the kernel of his empiricism in the recognition that

> the content of geometrical principles...is thus at bottom, a *physical* claim: something is thereby stated about the observable behavior of bodies, light rays and so on. And if bodies had behaved *differently*, in certain ways, from what we actually observe, we should have adopted from the outset and assumed as correct another theory of space than that of Euclid, without being prevented from doing so by any *a priori* form of intuition. Helmholtz's reasons for this are chiefly founded on the indissoluable union of the spatial and the physical in experience....His arguments are irrefutable....His theory, widely contested at the time, has been brilliantly confirmed, of course, by the progress of science. What Helmholtz declared possible is now known to be the case: through Einstein's general theory of relativity, contemporary physics has in fact reached the conviction that natural phenomena, established by most accurate observation, compel us to attribute non-Euclidean properties to real space.—It is a great satisfaction to the philosopher to observe that even in epistemology there is such a thing as confirmation by the advancement of science.[89]

The statement that *"natural phenomena...compel us to attribute non-Euclidean properties to real space"* signals Schlick's new empiricism, and the end of his geometric conventionalism. To be sure, Schlick could point to Helmholtz's insistence on "the indissoluble union of the spatial and the physical in experience", a demand also made by Poincaré. But nothing remains of the conventionalist strategy of choosing measuring bodies in such a way as to produce "the simplest physics"; rather, "geometrical principles" make "physical claims", indeed, "about the observable behavior of bodies, light rays, and so on". No longer can the geometrical principles be chosen independently and the physical laws adjusted accordingly. Instead, these principles are directly implicated in the "observable behavior" of the objects and processes of geometrical measurement, for the fundamental notions of metric and congruence are tied, by stipulation, to such instruments of measurement, which can therefore be considered to be independent of the basic postulates of the physical-geometrical theory.

Schlick's attempt to situate Helmholtz as a precursor of Einstein, assimilating both to the new empiricism he needed to oppose to Cassirer's critical idealism, can only be viewed as inspired. As shown above, Einstein himself was of two minds about the procedure of tying geometry to experience through the expedient of "practically rigid bodies". In Schlick's eyes, however, the methodology of "practical geometry" furnished a much needed weapon against the Kantian interpretations of relativity theory. His "empiricism with constitutive principles" quickly faded from view, for its bite against the neo-Kantian holism was toothless. In its place, indeed, came a new empiricist interpretation of physics wherein the ties of theory to observation are explicitly made through "coordinative definitions". In chapter 4, I show that the mechanism of "coordinative definitions" was definitively stated just a few years later by Reichenbach and henceforth was associated with his name. But as will be seen there, Schlick's influence was instrumental in weaning Reichenbach away from his early neo-Kantian theory-specific, and so, "relative *a priori* constitutive principles" to a "consistent empiricism" where "constitutive principles" have become stipulations (in the case of general relativity) about rigid rods and ideal clocks. The ensuing account of the empirical determination of the metric in general relativity would become the logical empiricist paradigm of how the terms of a physical theory, regarded initially as signs within an uninterpreted logico-mathematical calculus, received empirical content through connection to observation terms, via conventionally adopted "correspondence rules".

4

EINSTEIN *AGONISTS*

Weyl and Reichenbach

It is important to remember that the physical interpretation of the
mathematical notions occurring in a physical theory must be
compatible with the equations of the theory.

Trautman (1980b, 4)

4.1 Introduction

Most contemporary readers of Hans Reichenbach's works on the philosophy of
space and time have not considered them in the scientific context of their origin,
that is to say, against the background of activity in the small but vigorous community of general relativists in the decade or so after the inception of the theory of
general relativity in November 1915. Of these researches, perhaps the best known
to the history of science are the papers of W. De Sitter, A. Friedmann, G. Lemaître,
and others on the cosmological implications, among them models of an expanding
universe, of Einstein's field equations of gravitation.[1] But this was also a period
characterized by bold attempts to extend general relativity's "geometrization" of
gravity to encompass also the electromagnetic field and thereby to geometrically
represent all (known) physical interactions within space-time geometry. The first,
and historically most significant, of these efforts was the "theory of gravitation and
electromagnetism" initially put forward in 1918 by the mathematician Hermann

Weyl. Weyl himself had a rather complicated relation to his theory. As has now been well documented by physicists and historians of physics, the theory was the birthplace of a requirement on field laws that Weyl termed "gauge invariance", the demand not only that the field laws remain invariant under arbitrary transformations of the space-time coordinates ("general covariance") as in general relativity, but also that fundamental field quantities be invariant under arbitrary local transformations of "gauge".[2] As I will show in detail below, already by the end of 1919 Weyl had surrendered his belief that atomic phenomena could be accounted for in a unified field theory, based on classical conceptions of causality. Yet long after he himself recast his "gauge principle" in the context of quantum theory in 1928 as pertaining not to a non-integrable factor of scale but to a non-integrable phase factor of the wave function of the electron, he continued to regard his original "systematic" approach as a superior epistemological alternative to Einstein's use of rigid rods and clocks in measuring the interval ds^2, for reasons considered more closely in chapters 5 and 6.[3] Although Weyl's term "gauge" is no longer really appropriate, the idea of local symmetries survived; revived in somewhat different form in the 1950s by Yang and Mills, and by Shaw, it has since the 1970s been a key framework principle of the quantum field theories comprising the Standard Model of fundamental interactions.[4] In any case, Weyl's theory met with stiff resistance from Einstein and others and did not survive. But it spurred related attempts by A. S. Eddington, T. Kaluza, and also Einstein, the latter thereby embarking on a search that, in tandem with his nonacceptance of the fundamentally probabilistic character of quantum mechanics, led him into what nearly all of the rest of the theoretical physics community viewed as a scientific wilderness from which he would not again emerge.

Weyl's proposed "broadening" of the "geometrization" of general relativity was a particularly prominent target of Hans Reichenbach's "neo-conventionalist" and empiricist account of the metric of space-time, the first notable, even defining, result of logical empiricist philosophy of science.[5] In the general theory of relativity, physics and geometry are "entangled" in a way in that geometric conventionalism had not previously envisaged: the metric of space-time is no longer accounted as a globally rigid structure, fixed for all time, but as dynamically dependent in a given region, according to the Einstein field equations, upon surrounding matter and energy distributions.[6] But the philosophical questions prompted by the dynamical nature of the geometry of space-time in general relativity are rendered even more salient in a theoretical context that aspired to bring electromagnetism (and so, it was thought, the structure of matter) together with gravitation into a single unifying geometrical representation of space-time. Despite the speculative (not to mention, premature) character of these efforts, they gave rise to a fundamental epistemological disagreement regarding the permissible character of the physical objects or processes that could be employed to connect the geometrical-physical theory to measurement and observation. Ostensibly a dispute about the appropriateness in general relativity about the rigid rods and perfect clocks sanctioned by empiricism, the far deeper issue concerned the significance of mathematical representation in constituting the very notion of physical objectivity.

In its own way, the disagreement reflects two alternative perspectives on general relativity that could, and still can, be adopted. From one perspective, emphasis is placed on the principle of equivalence, affirming the local validity of the laws of special relativity in the variably curved space-times of general relativity. Then there can be no objection to carrying over to the general theory the rigid rods and regular clocks (measuring "local time") that are the accepted measuring implements of the special theory of relativity. As discussed in chapter 3, this view of measurement in general relativity, corresponding to the "practical geometry" of Einstein's essay "Geometry and Experience" (1921), has an obvious attraction to an empiricism seeking the most direct links of the mathematical constructs of physical theory to observation and measurement. But from the vantage point of unification, the most fundamental thing about general relativity is the dynamical character of the space-time metric and the general covariance of the theory's field equations. In this standpoint, corresponding to what Einstein termed (see chapter 2, §2.2.3), that of a "completed physics", the behavior of all fundamental physical objects is in principle to be accounted for by nonlinear field equations of dynamically interacting matter fields and space-time geometry. Reichenbach and Weyl are exemplars of these quite distinct and rival appraisals of general relativity. What lends heightened interest to the confrontation is that Einstein himself begins the decade of the 1920s as the leading proponent of the former perspective and ends it as virtually the only practitioner of the latter. Thus it is that Reichenbach and Weyl can both appear as champions of Einstein, but indeed of different Einsteins.

Weyl and Reichenbach accordingly stand on opposite sides in an epistemo-logical debate that ostensibly turns on whether measuring rods and clocks do or should play an epistemologically fundamental role in the new theory. According to Weyl's "broadened relativity theory" (*erweiterte Relativitätstheorie*), the behavior of such complicated material structures as rods and clocks can only be data to be explained, that is, to be derived from the equations of the "total field", and not stipulated as independent primitive "facts" licensed in the physical definition of metrical notions. On the other hand, Reichenbach, established early on as a philosophical *savant* regarding the theory of relativity, took up the task of episte-mologically combating the Weyl heresy concerning rods and clocks, under the banner of the new empiricism concocted by Schlick, rooted in the idea of physical cognition as mere coordination of mathematical representations to concrete phys-ical objects. In doing so, Reichenbach followed Schlick in adhering to Einstein's view of "practical geometry" that rests upon a stipulation regarding rigid mea-suring rods and uniform clocks. Hence, for Reichenbach, the metrical notion of congruence is physically definable (via a "coordinative definition") by the stipu-lated coincidence of end points of rigid rods, both before and after an intervening spatial separation. The collision of Reichenbach and Weyl occurs precisely here over the meaning of *congruence* in the new context of the general theory of rela-tivity and its possible generalization. Einstein, however, occupied a somewhat am-biguous position throughout this controversy, defending (although increasingly pragmatically) the use of rods and clocks as legitimate indicators of the ds^2 while, in his own odyssey of unification, serendipitously proceeding epistemologically unhindered, even to the extent of suggesting a fundamentally nonmetrical theory as the basis of the unified field.

In the following, a presentation of the relevant parts of Weyl's theory is given in §4.2; chapter 5 details the epistemological context of Husserlian transcendental phenomenology within which it was conceived. The response to Weyl's challenge evident in the writings of Reichenbach on relativity theory in the mid-1920s is given in §4.4. Particular attention is directed to a little-studied work of Reichenbach, the *Axiomatization of the Theory of Relativity* (*Axiomatik der relativistischen Raum-Zeit-Lehre*, 1924), an expressly "epistemological" axiomatization of relativity theory. Here, in treating rods and clocks as empirical postulates of the theory, Reichenbach directly opposed Weyl's proposal for metrical determinations of space-time without them. The central component of Reichenbach's "mature" neo-conventionalist treatment of the metric in general relativity in the more familiar *Philosophy of Space and Time* (*Philosophie der Raum-Zeit-Lehre* 1928), an *agreement* regarding the nonexistence of "universal forces" relies heavily on the analysis conducted in this earlier work. Interspersed between, in §4.3, is a discussion of Einstein's ambivalent attitude regarding the two perspectives on relativity theory.

4.2 Weyl

4.2.1 A Hidden History

With an enthusiasm that could scarcely be disguised, Hermann Weyl wrote from Zürich to Einstein in Berlin on 1 March 1918: "This day, as I believe, I have succeeded in deducing electricity and gravitation from a common source". The ingenious character of Weyl's achievement was unmistakable and did not fail to make the corresponding impression on Einstein, who replied, on the 6th of April, that Weyl's theory was "a stroke of genius of the highest magnitude". Yet within a few more days, on 15 April, Einstein was to object that despite its seductive mathematical elegance, Weyl's theory was empirically untenable, leading to consequences that did not seem to be in accord with implacable observational fact, the constancy of atomic spectral lines.[7] The Einstein objection was given a rigorous formulation by the 20-year-old Wolfgang Pauli, Jr., in his canonical survey of the theory of relativity published in 1921. And, so the usual story goes, the objection sealed the fate of Weyl's theory, even before the advent of quantum mechanics in the mid-1920s changed forever the classical framework of field–matter interactions. As a result, Weyl's theory is remembered today only as the locus of the now not quite appropriately named notion of "gauge invariance", which, taken as involving a complex factor of phase of the wave function of an elementary particle, and not of scale as Weyl originally thought, has become a major theoretical construct in the physics of elementary interactions.[8]

Like many other chronicles of the history of modern science, this capsule summary of the "received view" of the fate of Weyl's theory achieves its streamlined finality at the rather considerable cost of a good bit of inattention to historical and scientific detail. For one thing, Weyl countered the Einstein–Pauli objection with an elaborate reply designed to show how, in his theory, material bodies, in particular rods and clocks, "adjust" to the field strengths *where they are*, hence exhibit the congruence behavior that we familiarly attribute to them. For another, this sanitized narrative completely ignores the disquieting fact that the Einstein–Pauli objection—ostensibly concerning a rather straightforward empirical matter—did not succeed in convincing Weyl even long after he had abandoned his theory on other grounds. Moreover, as evidenced by the direction of his own initial work on a "unified theory of fields" in the next six or so years (see §4.3), Einstein did not, apparently, consider that he had delivered a knockdown argument against Weyl's criticism of the use of rods and clocks as direct metrical indicators in general relativity. For such appliances cannot be simply assumed in a fundamentally nonmetrical theory such as Einstein, following Eddington, adopted at various periods, beginning in 1923. Finally, it becomes apparent that Pauli's objections to Weyl turn not so much upon the apparent empirical disconfirmation pointed out by Einstein, but rather upon the failure of Weyl's attempted unified theory (and also the later ones of Eddington *and* Einstein) to abide by a rather severe positivist covenant regarding what can be said to be physically meaningful. But this is a stricture with which Einstein, in his pursuit of a unified theory of fields, was hardly in agreement. Thus, when considering Einstein's seemingly empiricist plumping on behalf of practically rigid rods and ideal clocks, we have to take to heart the injunction of his 1933 Herbert Spencer lecture—to heed what scientists do, and not what they say about what they do.[9]

4.2.2 "Pure Infinitesimal Geometry"

Weyl's theory of "gravitation and electricity" is perhaps best seen from the conditions of its genesis: as a bold, but ultimately unsuccessful attempt to remove a "blemish" ("*Schönheitsfehler*") from Riemannian geometry,[10] the mathematical framework in whose terms Einstein's gravitational theory is cast. In Riemannian geometry, the magnitude or length and the direction or orientation of a vector are treated asymmetrically. As would only become clear with Levi-Civita's "discovery", in 1917, of the concept of infinitesimal parallel transport of a vector (see chapter 6, §6.3.1), in a Riemannian manifold the infinitesimal parallel transport of a vector around a closed curve, in general, changes the orientation but not the length of the vector on return to its initial point. For this reason it is, in general, meaningful to speak of an immediate comparison of lengths but not of directions separated by an arbitrary distance. Even as the first edition of his classic *Raum-Zeit-Materie* (1918) was going to press, Weyl conceived a "pure infinitesimal geometry" (*Reine Infinitesimalgeometrie*) that removed, as he put it, this last *ferngeometrisch* Euclidean remnant from Riemann's theory of manifolds. According to Weyl, "a genuine local geometry [*wahrhafte Nahegeometrie*] can only be acquainted with a principle of transport of length from one point to another infinitesimally adjacent to it"; from such a perspective, Riemann's geometry is only "a half and inconsistent local geometry" ("*eine halbe und inkonsequente Nahegeometrie*") since it assumes the meaningfulness of a direct (path-independent) distant comparison of lengths.[11] In Weyl's new geometry, comparisons of both direction (like Riemann) and length (unlike Riemann) are not direct, but depend on parallel transportation of a comparison vector in infinitesimal increments along a given path between vectors at two finitely separated points P_1 and P_2. In removing this asymmetrical treatment of length and direction, Weyl created a purely infinitesimal "world geometry" for field physics, fully *d'accord* with the Leibniz–Riemann principle that genuine understanding of nature only results from formulating its behavior in the infinitesimally small.[12]

Chapter 5 will show that Weyl's "purely infinitesimal" understanding of nature is an epistemological concomitant to the constitutive transcendental-phenomenological idealism of Husserl while chapter 6 provides a much more explicit account of the mathematical constructions of Weyl's theory set against this epistemological context. For present purposes, the following schematic outline will suffice.

Once the Riemannian assumption of length comparison "at a distance" is abandoned, then, as Weyl showed, one begins with a weaker geometry of the Riemannian manifold that has only a conformal (angle-preserving) structure. Hence, the length of a vector at a given point P is arbitrary up to a choice of scale ("gauge") at P. Thus, it is initially possible to compare only the *relative* lengths of two vectors at P or to determine the angle between them; the *absolute* length of a vector at P has no meaning. A metrical structure of this kind is only invariant under the *conformal transformation*,

$$g_{\mu\nu} \Rightarrow g'_{\mu\nu}(x) = \lambda g_{\mu\nu}(x),   \text{(1)}$$

where the conformal factor λ (>0) is an arbitrary function of position, and only the ratios, and not the absolute values, of the ten independent $g_{\mu\nu}$ have a determinate value. Weyl showed that this conformal geometry corresponds to the light cone structure of space-time in that any two (nonisometric) space-times that are conformally equivalent will have the same light cone structure, noting that Kretschmann (1917) had independently arrived at the same result.[13]

At this point, there are two decisive considerations regarding the conformally invariant metric. First, such a metric determines only an equivalence class of symmetric (torsion-free) affine connections and not a unique connection as does, according to earlier results of Christoffel and Levi-Civita, the metric of Riemannian geometry. This is unsatisfactory because now there is no unique notion of parallel displacement of a vector. Additional structure must be sought to ensure unique compatibility with a metric so that the class of affine geodesics and that of metrical geodesics will coincide. Weyl regards this postulate as fundamental; it is "the basic fact [*Grundtatsache*] of infinitesimal geometry".[14] Then a "metric connection" or "length connection" must be added so that it is possible to speak of a vector at P and one at P' in the "infinitesimal" region (tangent space) at P, possessing the *same* length. That is, although length comparisons "at a distance" have been forsaken, it still must be possible to unambiguously "metrically connect" P with all points lying "infinitely near" P.

To restore the condition of unique metric determination of an affine connection, Weyl showed that a single connection could be determined, coupled to given choice of a metric tensor, by incorporating a pseudovector φ within the definition of the standard Riemannian ("Levi-Civita") connection (see the more detailed discussion in chapter 6, §6.3.1). Given such a "Weyl connection", it is possible to speak of "a manifold with an affine connection" where, as in the Riemannian case, there is a unique determination of parallel displacement of a vector at every point.[15]

The next step is to show that there is a metrical concept of infinitesimal "congruent displacement" that already carries with it a compatible concept of infinitesimal parallel displacement, hence that the parallel displacement of a vector leaves vector magnitude unchanged. This is a requirement that the affinely connected manifold must additionally become a *manifold with a metric connection*; that is, not only must there be a metric defined at every point, but also that each point P must be *metrically connected* to the points in the infinitesimal region around it, comparing vector magnitudes at P with those at the various points P' in the tangent space of P. A particular gauge may be chosen so that congruent displacement in this region does not change magnitude (a "geodetic gauge"). But in the general case, choice of gauge is allowed to vary continuously from point to point. Then, where l is the length of an arbitrary vector at P, $l + dl$ is the length of the vector arising through displacement to a neighboring point P'. This change is defined

$$dl: = -ld\varphi, \tag{2}$$

where the factor $d\varphi$ is independent of the transported length. Changing the gauge at P so that $l' = \lambda l$ (where λ is the conformal factor noted above), yields

$$dl' = -l'd\varphi'$$

where

$$d\varphi': = d\varphi - \frac{d\lambda}{\lambda}. \tag{3}$$

Weyl showed that a necessary and sufficient condition for dl to vanish at P (as desired) is that φ is just a linear differential form, $\varphi_\mu dx^\mu$ (Einstein summation convention). A Weyl metric then consists of two "fundamental forms", the quadratic form of Riemannian geometry $ds^2 = g_{\mu\nu}dx^\mu dx^\nu$, and the linear form just defined. These are defined up to the "gauge transformations",

$$g_{\mu\nu} \Rightarrow g'_{\mu\nu}(x) = \lambda g_{\mu\nu}(x), \tag{4}$$

and

$$\varphi \Rightarrow \varphi' = \varphi - \frac{d\lambda}{\lambda}. \tag{5}$$

Physically, these would appear to have the effect of slightly changing, for example, the lengths of measuring rods and the rates of clocks, at each successive point P. Taking the "curl" (or "*rot*") of the linear differential form, Weyl defined a co-variant second rank antisymmetric tensor

$$F_{\mu\nu}: = \frac{\partial \varphi_\nu}{\partial x^\mu} - \frac{\partial \varphi_\mu}{\partial x^\nu}, \tag{6}$$

giving the curvature of his length connection (*Streckenkrümmung*), showing that by "gauge invariance" that it satisfies the condition,

$$\frac{\partial F_{\nu\sigma}}{\partial x^\mu} + \frac{\partial F_{\sigma\mu}}{\partial x^\nu} + \frac{\partial F_{\mu\nu}}{\partial x^\sigma} = 0. \tag{7}$$

Although momentarily speaking of rods and clocks, up to this point, we have not left the purely mathematical domain of *Reine Infinitesimalgeometrie*. Weyl then naturally identified this antisymmetrical second rank tensor with the electromagnetic field tensor (the so-called Faraday tensor) and thus to the first system of

Maxwell's equations (in their space-time formulation), by setting φ_μ identical to the space-time four electromagnetic potential A_μ. As purely mathematical consequences of his geometry, equations (1) and (2) are held to express "the essence of electricity"; they are an "essential law" (*Wesensgesetze*) whose validity is regarded completely independent of the actual laws of nature.[16] Furthermore, Weyl could show that a vector density and contravariant second rank tensor density follow from the *general* form of a hypothetical action function invariant under local changes of gauge $\lambda = 1 + \pi$, where π is an arbitrarily specified infinitesimal scalar field. These are respectively identified with the four current density $\mathbf{j}^\mu$ and the electromagnetic field density $\mathbf{h}^{\mu\nu}$, through the relation

$$\frac{\partial \mathbf{h}^{\mu\nu}}{\partial x^\nu} = \mathbf{j}^\mu, \tag{8}$$

that is, the second ("inhomogeneous") system of Maxwell equations. Thus, Weyl claimed that, without having to specify a particular action function, "the entire structure of the Maxwell theory could be read off of gauge invariance".[17] Again, using only the general form of such a function, he demonstrated that conservation of energy-momentum and of charge follow from the field laws in two *distinct* ways.[18] Accordingly, he asserted that, just as the Einstein theory had shown that the agreement of inertial and gravitational mass was "essentially necessary" (*wesensnotwendig*), his theory did so in regard to the facts finding expression in the structure of the Maxwell equations, and in the conservation laws, "an extraordinarily strong support" for the "hypothesis of the essence of electricity" (*Wesen der Elektrizität*).[19] The domain of validity of Einstein's theory of gravitation, with its assumption of a global unit of scale, was originally held to correspond to $F_{\mu\nu} = 0$, the vanishing of the electromagnetic field tensor. By 1919, Weyl had fashioned a sketch of a "dynamical" account of the origin of this global unit, "the natural gauge of the world", as discussed in chapter 6, §6.4.2.1.

In setting out to remove a glaring blemish in Riemann's theory of manifolds, Weyl had produced a geometrical theory containing invariant expressions for all the fundamental quantities of gravitation and electromagnetism. Since these were the only physical forces recognized in 1918, Weyl was led to triumphantly proclaim the unity of geometry and physics:

> Everything real [*Wirkliche*] that transpires in the world is a manifestation of the world-metric: Physical concepts are none other than those of geometry.[20]

For such declarations Weyl was taken to task by Einstein and Reichenbach among others; the latter already in his first book, *The Theory of Relativity and A Priori Knowledge* (1920). There, Reichenbach wrote, expressing what was probably the consensus of the physics community:

> Purely mathematical axiomatization never leads to principles of an *empirical* theory [*Theorie der* **Naturerkenntnis**].... So it is entirely false if one would conclude, as, e.g., Weyl and also Haas, that mathematics and physics are fused into a single discipline.[21]

(Inspired by Hilbert's "axiomatic method" (see §4.4.2) and by Weyl's theory, the Leipzig physicist Arthur Haas had written in 1920 a provocative paper entitled

"Physics as Geometrical Necessity"). The young philosopher's accusation of such an absurd confusion elicited an indignant response from the prominent mathematician: all that Weyl had claimed was that "the concepts of geometry and field physics had come together".[22] Nonetheless, the speculative mathematical road that led Weyl to his theory would continue to enlist the prejudice of most physicists. As I will show in chapters 5 and 6, in holding that it is the task of geometry to investigate the "essence of metric concepts", whereas physics is concerned to determine "the law" (i.e., the integral invariant formulated in an action principle) according to which the actual world is singled out from all other possible four-dimensional metric spaces,[23] Weyl gave a distinctive transcendental constitutive significance to the relation of mathematics to physics of which physicists and indeed most philosophers could not have been aware. In this way, Weyl's concept of a "world geometry" engendered a radical, if largely unheeded, extension of the meaning of *Weltgeometrie* in the sense, familiar since Minkowski, of the geometry of actual space-time.[24]

4.2.3 Empirical Determination of the Space-Time Metric Without Rods and Clocks

We need now sketch Weyl's constructive procedure for empirical determination of the metric of space-time without the use of Einstein's "practical geometry" of rigid rods and uniform clocks. Since for Weyl, the essence of the metric lies in the "purely infinitesimal" concept of congruence,[25] the metric is to be viewed solely as a structure of the continuous field whereupon the concepts of vector and "tract" (*Strecke*) employed for its characterization have, in themselves, "nothing to do" with material measuring rods and clocks.[26] Moreover, because of the gauge invariance in Weyl's theory, it is not the $g_{\mu\nu}$ themselves but only their ratios that have an empirically determinable meaning. From this perspective, it is no longer permissible to stipulate that chronogeometrical notions are "coordinated" to rigid measuring rods and uniform clocks, a situation that, to Einstein, robbed the line element ds of its empirical foundation (see §4.3). Instead, Weyl provided a so-called geodetic method involving two "directly observable" physical processes that are geodesics in space-time: the propagation of light rays (null geodesics) and the freely falling trajectories of "test particles" or mass points (timelike geodesics).[27] The association of these physical objects with the metric of space-time comes via a theorem according to which the conformal and the projective properties of a metric space (in Weyl's sense) univocally determine its metric up to a factor of scale ("gauge").[28]

As Weyl first observed, the conformal and projective structures of a metrical manifold have intuitively clear physical counterparts in the theory of relativity.[29] The conformal properties could be identified by the paths light rays that also fix the causal structure of space-time; hence, by observing the arrival of light at points in the immediate neighborhood of a point O, the *ratios* of the quantities $g_{\mu\nu}$ at O may be determined. But, as shown above, the propagation of light determines the space-time interval only up to a conformal factor while leaving the linear form φ unrestricted. This latter may be fixed by considering that the projective properties may be taken to be physically instantiated by the trajectories of freely falling point masses (the so-called geodetic hypothesis). One may then assume that the proper time s may be read off from the motion of these unaccelerated "ideal" particles, and hence that these projective geodesics carry affine parameters and so are affine geodesics. There is a compatibility requirement for the conformal and projective structures such that the conformal (null) geodesics $ds^2 = 0$ are included within the class of projective geodesics. Then, through a comparison of two such point masses passing through O in different directions, a unit of measure at O may be uniquely determined. (One shows that the difference in directional derivatives at O is proportional to the gauge at O.) In effect, the affine structure forces the conformal factor to be a constant (at each point).[30] The physical significance of the theorem is then that the space-time metric can be empirically determined *without reliance on measuring rods and clocks*, measuring appliances of vastly more complicated structure. Weyl's constructive method was revived by Ehlers et al. (1972); more recently, Ehlers (1988b) has adduced a microsymmetry criterion providing a nonmetrical means for picking out geodesic paths, thus redeeming the "geodesic hypothesis" against conventionalist leveling arguments.[31]

4.2.4 The Einstein "Prehistory" Objection and Weyl's Rejoinder

Weyl's unified theory, put forward in *two* explanatory versions, received an almost uniformly unfavorable reception from the theoretical physics community, beginning with Einstein himself. Einstein first expressed his disagreement privately in a letter of 15 April 1918 to Weyl and then publicly, in a note appended to Weyl's Prussian Academy paper (1918b) announcing his theory; as shown in chapter 2, the objection recurs in indirect form in his widely read "popular" lecture to the Prussian Academy of January 1921, "Geometry and Experience". The obstacle appeared to be that Weyl's theory could not be in agreement with observation. In fact, Weyl's theory, in its original form (i.e., without sketching the origin of the "natural gauge of the world"), predicted minuscule "second clock" effects, which, as Eddington showed, were far below the threshold of observation[32] and, as it later turned out, were within the limits of quantum mechanical tolerance.[33] And as Pauli himself had helped to show, a solution of Einstein's field equations for the case of a static, spherically symmetric field surrounding a "material particle" (e.g., the solar gravitational field) is also a solution of Weyl's theory, even though the latter generally has field equations of the fourth (and not second) order. As this case was decisive for all the known empirical predictions of Einstein's theory—the perihelion precession of Mercury and the bending of light rays in the solar gravitational field—Weyl's theory as well as Einstein's was capable of explaining these observed phenomena.[34]

What, then, was Einstein's objection? We do observe that measuring rods retain their length under transport in electromagnetic fields; *prima facie*, this is evidence that Riemann's geometry, not Weyl's, is the geometry of space-time. This is because, Einstein argued, if Weyl's theory is correct, then the spectral lines emitted by atoms would not be the very sharp and well-defined frequencies that in fact they are observed to be. For if two atoms, say of hydrogen, are together at an initial time in one space-time region and then transported, via different paths to another region of space-time where they are brought together again, then, according to

Weyl's theory, we should expect to observe a difference in their spectral lines corresponding to their past histories, that is, to the differing values of the electromagnetic fields in the space-time regions they passed through in the interim. In fact, according to Pauli's calculation, no matter how small the initial difference in the spectral lines of the two atoms posited by Weyl's theory, this difference would "increase indefinitely in the course of time".[35] But astronomical observation tells us that hydrogen atoms everywhere in the heavens exhibit the same spectral signature. So Weyl's theory did not, apparently, correspond to the facts of observation.

Despite his enormous respect for Einstein, Weyl was not persuaded by the objection at the time, or even three decades later, long after surrendering the details of his theory on grounds of the new quantum mechanics. Instead, he adopted a two-pronged argumentative strategy to counter it, producing in effect a second explanatory version of his theory.[36] On the one hand, he adamantly maintained that the behavior of physical objects such as rods and clocks or, for that matter, atoms, has "as such nothing to do" with the ideal metric notions defined by vector transport:

> The functioning of these instruments of measurement is however a physical occurrence whose course is determined through laws of nature and which has as such nothing to do with the ideal process of congruent displacement of world tracts [*Verpflanzung von Weltstrecken*].[37]

To critics like Pauli, this meant that although there was no longer a "direct contradiction with experience", there also no longer existed an "immediate connection" between electromagnetic phenomena and the behavior of measuring rods and clocks"; consequently, the connection between electromagnetism and the world metric posited in Weyl's theory is only "purely formal". Eddington similarly (although more sympathetically; see chapter 8) interpreted Weyl as giving up any claim to characterize the geometry of the real world and instead as providing only a "graphical representation", that is, a kind of conventional representation, of "world geometry".[38] There objections might be summarized: if geometrical relations are not concerned with measuring rods and clocks, with what *are* they concerned?[39]

Weyl's retort, however, was that this response dodges an explanatory burden that cannot be shirked from the viewpoint of either a pure field theory of matter or by a systematic account of field-matter interactions.[40] For it is, as Weyl expressly stated, "perverse" (*verkehrt*) to use physical bodies such a rods and clocks that are *indicators* of the gravitational field, as at the same time instruments to *stipulate* metric relations. To do so is just to treat as a definition ("rigid rod", "clock") regular behavior should be explained, that is, *should be derived* from the field equations of a systematic theory.[41] For such a systematic theory, "Einstein's *definition* of measure determinations in the metrical field with the help of measuring rods and clocks has validity only as a preliminary connection to experience just as does the definition of electrical field strengths as the ponderomotive force on a unit charge". In order, in Weyl's terms "to close the circle" it is

> necessary, once a suitable action law has been set up, to *prove* that here, the charged body under the influence of the electromagnetic field, there, the measuring rod

under the influence of the metrical field, exhibit, as consequences of the action-laws, that behavior we had originally utilized for the physical definition of the field magnitudes.[42]

The general theory of relativity, where admittedly "the conceptual foundations of the theory have no relations with the electromagnetic field",[43] is certainly not systematic in this sense. Such a theory, in Einstein's estimation, lay in the future, and so for the time being it is sufficient for general relativity to rely, as it does, on the notion of "practically rigid rod" that corresponds to "congruence at a distance" (i.e., path independent transport of length). But if it is supposed that the fundamental metric concept, congruence, is only to be properly conceived as a "purely infinitesimal" concept, as Weyl did on the epistemological grounds of Husserlian phenomenology (see chapter 5), the "practically rigid rod", rather lamely defended by Einstein, becomes an unprincipled and gratuitous assumption.

Certainly Weyl had to account for why we do observe the congruence-preserving behaviors of rods and clocks that we do, as well as for the constancy of spectral lines of atoms. He did so (and this is the other component of Weyl's explanatory strategy) by invoking a dualism regarding the manner in which physical magnitudes are determined, a distinction that he sees as reprieving his theory from the empirical refutation sketched by Einstein and elaborated by Pauli. Physical quantities are fixed either by a body's following a "tendency of persistence" (*Beharrungtendenz*) or by its "adjustment" (*Einstellung*) to the field strengths where it is, a distinction made concrete by appealing to the different physical behaviors of a spinning top and the magnetic needle of a compass.[44] Whatever its initial orientation, the axial direction of a spinning top is transferred from instant to instant by a tendency of persistence; that is to say, it is governed by the inertial or "guiding" field (*Führungsfeld*). On the other hand, as Weyl still maintained three decades later, the magnetic needle of a compass *adjusts* to the value of the magnetic field wherever the compass is carried; "adjustment...enforces a definite value that is independent of past history and hence reasserts itself after any disturbances and any lapse of time as soon as the old conditions are restored".[45]

Accordingly, Weyl objects that the Einstein–Pauli "prehistory" criticism of his theory unjustifiably presupposes that measuring rod lengths and clock periods are altered (or not) through a time-dependent process of *persistence*, whereby the magnitude in question at a given instant is some function of its magnitude at a previous instant. But given the distinction above, this is not at all a *necessary* presupposition, in which case the explanatory burden runs in the other direction. If, for instance, a measuring rod is moved around within a physical field assumed to be *inhomogeneous*, that is, where the field strengths have different magnitudes at different points, an account is surely required as to why we do detect no noncorrectable variable behavior in our measuring rods. One *prima facie* reason may be that such "deforming" forces at each space-time point are universally present but are counteracted by electromagnetic forces within the atom, producing a state of force equilibrium. This would mean that a massive object, such as a measuring rod, carries with it a determinate magnitude representing the interaction of the gravitational forces of the field and the electromagnetic forces obtaining between the rod's constituent molecules and atoms. If this is so, then a measuring rod may be held to "adjust" to the field strengths where it is *now*; it does not exhibit what was for Weyl a qualitatively distinct, but miniscule (i.e., far below the threshold of observation) quantitative difference in pattern of behavior, the tendency for its length to persist, that is, to remain the length it was a moment ago at another point of space-time. The commonsense objection that of course the length of a measuring rod persists in moving it from one end of a room to another can, presumably, be countered by taking into account the crudity of our everyday experience with middle-sized objects, which are not presumed to interact with the "empty space" in which they (and we!) are located.[46]

What is this field strength? Weyl appears to have been motivated by Einstein's 1919 "scalar-free field equations" (which were intended to provide a more principled footing for the cosmological constant, introduced in 1917), to take the relevant field strength here to be the equivalent in his theory to the Riemann scalar of curvature in Einstein's, that has now become a constant, playing the role of keeping electrodynamical forces within electrical "corpuscles" in equilibrium.[47] Analogously to Einstein, Weyl can appeal to inertial-gravitational forces as responsible for maintaining an equilibrium of intra-atomic electrical forces (see chapter 6, §6.4.2.1). Hence, we *observe* that measuring rods display the behavior regarded as "the natural gauge of the world"; their length—due to "adjustment", that is, constant force equilibrium—is unaltered under transport in (weak) gravitational fields.[48] Weyl's response to Einstein and Pauli is, then, to say that the alleged "empirical refutation" (prehistory objection) does not touch his theory, since the constancy of "atomic clocks", as also the congruence behaviors of measuring rods, is to be accounted for as arising through *Einstellung* (as indeed it must in a principled "systematic" theory), not through *Beharrung*, as the "prehistory objection" wrongly presupposes.[49]

4.3 Einstein

As noted above, since it was first made the Einstein-Pauli "prehistory" (constancy of spectral lines) objection has been viewed as an authoritatively convincing rejoinder to Weyl's highly speculative theory by a majority of the community of physicists, although perhaps, even given what was known at the time, it should not have been.[50] But if we turn to consider a wider range of Einstein's writings and activities in the period 1918–1925, we find that, for Einstein, the purported empirical disconfirmation of Weyl's theory was not at all the end of the matter. Taking these into account, it can be seen that Weyl's criticisms of Einstein's tie of the empirical basis of general relativity to the assumption of rigid rods and uniform clocks left their mark. To these criticisms is probably accountable Einstein's shift from defending the behavior of rods and clocks as evidential requirements of general relativity to a more tempered *pro tem* justification. Besides, the speculative schemes for unification initiated by Weyl, and then modified by Eddington (1921a), served as Einstein's own point of departure for his first attempts to formulate a unified field theory. The conclusion then emerges that, far from regarding the "prehistory" objection as decisively undermining Weyl's entire approach, Einstein himself was to become and remain the leading proponent of mathematically speculative schemes for unifying physics, long after Hilbert and Weyl had quit the field. It remains to explore this development, and its grounds, a bit further.

Already at the Bad Nauheim meeting of the German Society of natural scientists in late September 1920, a widely publicized confrontation of Einstein with his antirelativity critics, especially Phillip Lenard, Einstein admitted that at the current stage of the theory's development, it was a "logical weakness" (*Schwache*) of the theory that "it must separately introduce rods and clocks instead of being able to construe them as solutions of the differential equations [i.e., of the field equations]".[51] Turning to Einstein's several more explicitly "philosophical" discussions of the status of rods and clocks as legitimate concepts in the 1920s, we find this ambivalence in abundance. It is famously voiced in perhaps the most widely read text of Einstein within subsequent logical empiricist philosophy of science, the lecture of January 1921, entitled "Geometry and Experience". As previously discussed, Einstein's thoroughly pragmatic justification of "practical geometry" is somewhat deviously coupled in this lecture with a fundamental criticism based upon the in principle difficulties attending the concept of a rigid body in the theory of relativity. Recalling the discussion in chapter 3 (§§3.3–3.4), Einstein first made a case for the validity of the supposition of the existence of Helmholtzian "practically rigid bodies", that is, bodies upon which "two 'tracts' [*Strecke*] found to be equal once and anywhere are equal always and everywhere". This is just to assume, *contra* Weyl, that congruence relations are path independent, an assumption for which, alluding to his prehistory objection, "the existence of sharp spectral lines is a compelling empirical proof".[52] Nonetheless, Einstein went on to lodge a criticism of this point of view which he attributes to Poincaré. There are no actual rigid bodies that correspond to the ideal rigid body of geometry; it is not possible to thus disentangle geometry from physics in this manner. Accordingly, only the whole comprising G(eometry) + P(hysics) is empirically testable; although G and parts of P may be chosen arbitrarily, all that matters is that the whole not conflict with experience. This latter viewpoint is correct *sub specie aeterni*, Einstein admitted, for reasons quite analogous to, if not identical with, whose given by Weyl. Yet even so, for the time being, the former position is to be preferred:

> The concept of the measuring rod and the concept of a clock coordinated with it in the theory of relativity do not find their exact correspondence in the real world. It is also clear that the solid body and the clock do not play the role of irreducible elements in the conceptual structure of physics, but the role of composite structures, which should not play an independent role in the construction of theoretical physics. However, it is my conviction that these concepts, at the present stage of development, still must be introduced as independent concepts.[53]

Einstein again adopted a cautious "on the one hand, on the other" mode of presentation in reiterating this conclusion some four years later in the essay on non-Euclidean geometry cited in chapter 3. Only now, in addition, there is a warning that in adopting "the standpoint of the practical physicist",

> [w]e must however be continually conscious of the fact that the idealization which lies in the fiction of rigid [measuring] bodies [*Körper*] as objects of nature

might one day prove unjustified or justified only with respect to certain phenomena of nature.[54]

Noting that Riemann had already anticipated this possibility, both in regions that are "not astronomically small" and also in the microworld of "electrical elementary quanta" (*elektrische Elementarquanta*), Einstein observed that Riemann had had the "audacious idea" that the geometric behavior of bodies might be conditioned by physical realities or forces. This brief essay concluded with mention of the attempts of Weyl and Eddington to generalize Riemannian geometry so as to "find a place for the laws of electromagnetism in the accordingly expanded conceptual system", endeavors about which Einstein at this time reserved judgment.[55] In later years Einstein continued to work both sides of the aisle, sometimes characterizing geometry, without qualification, as "the study of the possible positions [and displacements] of rigid bodies"[56] while, on the other, holding that a "*complete* theory of physics" has "no room for the supposition" of rods and clocks.[57]

We might now inquire into the reasons underlying the provisional character of Einstein's choice of rods and clocks as the physical correlates of chronogeometrical notions in the context of general relativity where, as he repeatedly points out, coordinate differences do not have an immediate metrical significance in terms of unit rods and clocks.[58] One likely reason seems to be that the assumption of standard measuring instruments played a vital heuristic role in the key thought-experiment of the rigidly rotating disk, concluding that the space-time geometry of gravitational fields was non-Euclidean.[59] Essential to this conclusion is that rigid and periodically regular measuring instruments exist with which the geometry of the disk may be determined through actual measurements. Without such intuitive means of disentangling physics from geometry, the thought-experiment would not work. So there was a heuristic and motivational reason for Einstein's retention of rods and clocks in the context of general relativity.

There is a related epistemological attachment. Already in the letter to Weyl in April 1918 first stating his "prehistory" objection, Einstein affirmed that the "empirical basis" of relativity theory lay in "the connection of the (line element) *ds* with rod and clock measurements" and that, if this connection is severed, the theory would lose its basis in experience.[60] Elsewhere Einstein elaborated upon the "prehistory" objection by highlighting the obvious evidential virtues in retaining in general relativity a path-independent concept of congruence corresponding to "measuring rod geometry". This is a consideration not to be taken lightly, as Einstein wrote to his friend Besso in Zürich in July 1920. Besso, it seems, had been entertaining favorable opinions of Weyl's theory. Einstein responds:

> You think: there is no need to find the invariability of relative extension of a body in the foundation of the theory, that it would be more beautiful if this resulted as a consequence or more acceptable if it had a place in theory as a special hypothesis. However, don't forget that the theory is based on measuring rod geometry [*Maßstabgeometrie*]. Then one accepts that the relative length of measuring rods is a function of its prehistory. It follows that one should find that *actual* measuring rods are relatively invariant. This is why the measuring rods employed in the foundation of [Weyl's] theory are only *imaginary* [*gedachte*] measuring rods which behave otherwise than actual ones. That is detestable [*abscheulich*].[61]

A more complete articulation of this line of thought was publicly made in responding to Weyl's presentation at the aforementioned 86th *Deutsche Naturforscherversammlung* at Bad Nauheim in September 1920:

> In the arrangement of my conceptual system, for me it has become decisive [*massgebend*] to bring elementary experiences into the language of signs [*Zeichensprache*]. Temporal-spatial intervals are physically defined with the help of measuring rods and clocks. If I consider two (such) structures, then their equality is empirically independent of their prehistory. Upon this rests the possibility of coordinating [*zuzuordnen*] a number *ds* to two neighboring world points. Insofar as the Weyl theory renounces this empirically-grounded coordination [*Zuordnung*], it robs the theory (general relativity) of its most solid empirical support and possibilities of confirmation.[62]

Such declarations enamoured Einstein to the logical empiricists. Employing the language of coordination (*Zuordnung*) that lay at the center of the account of the nature of cognition in Schlick's epistemology book (1918), Einstein here stated in no uncertain terms that he considered "norming" the *ds* to ("infinitely small") unit rods and clocks to be essential to the empirical interpretation of his theory. This thought is again in evidence when, a few months later, in a talk recorded in Vienna in January 1921, he warned (with presumably Weyl's theory in mind) that unless the line element *ds* is connected with "the observable facts", a "reality-alien" (*wirklichkeitsfremde*) theory is the result.[63]

Nonetheless, in several letters to Weyl during 1918 and 1919, Einstein privately revealed that his opinion of the latter's theory is far less one-sided than this public posture would suggest.[64] More important, as evidenced in his scientific work of the period, the initial attempts to construct his own unified theory of gravitation and electricity, it is quite clear that Einstein is not at all constrained by these empiricist attachments to the suitability of rods and clocks as linking mathematical theory with experience. Already on 3 March 1921, he submitted a paper to the Prussian Academy hypothetically considering a relativity theory in which, as in Weyl, the *ds* is only conformally invariant (i.e., only $ds^2 = 0$ is invariant), thus "without making use of the concepts of measuring rods and clocks".[65] The next step was yet more radical. By January 1923, in a contribution sent to the Berlin Academy from the ship *Haruna Maru* en route to Japan, Einstein was now prepared to follow Eddington in jettisoning the metrical basis of a combined gravitational and electromagnetic theory altogether. Weyl, Einstein noted, had not accorded an invariant meaning to the magnitude of a line element (or vector), but only to the relation of two such magnitudes, a theory that could be designated "half metrical". But following Eddington (1921a), Einstein's new perspective was *nonmetrical*; it now appeared, "from a purely logical starting point", that "it is much more satisfactory to adopt as the basis of such a theory only [an affine connection] while letting the invariant ($ds^2 = g_{\mu\nu}dx^\mu dx^\nu$) fall".[66] Writing to Bohr from the same ship "near Singapore" on 11 January, Einstein expresses more than hypothetical interest in this option:

> I believe that I have finally understood the connection between electricity and gravitation. Eddington has come closer to the truth than Weyl.[67]

All thought of unduly jeopardizing empirical basis of general relativity by refusing to allow rods and clocks to be physical counterparts of the metric interval ds^2 is surely abandoned in pursuit of this new nonmetrical route to a unified theory. Having embarked upon such a course, Einstein can evidently no longer consider it "decisive" to link the fundamental concept of spatiotemporal intervals with "elementary experiences". To the contrary, his method of mathematical speculation in striving to find a unified field theory occasionally even led to apparent expressions of Platonism, as in his Herbert Spencer Lecture at Oxford in 1933.[68] In fact, what Einstein was obliquely referring to, in claiming there that "the genuinely creative principle resides in mathematics" was his own methodological transformation from basing the search for new physics on an "intuitive physical principle" (such as the Principle of Equivalence) to relying almost exclusively on mathematical (differential geometric) speculation. This shift is fully evident in September 1923, in a report on his new affine theory of the field. Upon choosing a nonmetrical basis for the theory, Einstein noted that the following methodological shift ensued:

> The search for the mathematical laws which shall correspond to the laws of nature then resolves itself into the solution of the question: What are the formally most natural conditions that can be imposed upon an affine relation?[69]

As Weyl would point out much later in 1952, there is a certain personal irony in this transformation, for whereas Einstein in 1918–1919 had upbraided Weyl for following so purely a speculative approach to physics without "a guiding intuitive physical principle", their roles were soon thereafter reversed. According to Weyl, Einstein came to believe that

> the chasm between ideas and experience is so large that only the path of mathematical speculation, whose consequences must naturally be developed and confronted with the facts, has a prospect of success...

whereas for Weyl, chastened by recognition of the premature character of geometrical unification schemes in the present state of physics,

> my confidence in pure speculation has sunk and a closer connection with quantum mechanical experience seems necessary...[70]

Ironies aside, there can be little question that Einstein was far from convinced of the total invalidity of the speculative approach of Weyl's theory as the lore of his "prehistory" objection might suggest. On the contrary, he thought enough of Weyl's theory and of the related generalization offered by Eddington to adopt them as starting points for constructing a unified theory of fields.[71] And once underway, even as he repeatedly complained that this route led only to exasperating dead ends, he continued to explore his own variants of the Weyl/Eddington unification schemes, exhausting all the possibilities he deemed reasonable within what he called the "Weyl-Eddington complex of ideas".[72]

In this regard Einstein's attitude is instructively contrasted with the untrammeled positivism of Pauli. Writing to Eddington just after the publication of Einstein's pure affine theory in September 1923, Pauli stated that he considered such nonmetrical endeavors to be "physically meaningless":

The most beautiful achievement of the theory of relativity was certainly to have brought the metrical results of measuring rods and clocks, the paths of freely falling mass particles and those of light rays into a determinate inner bond (*Verbindung*). . . . [However,] the magnitudes $\Gamma^{\mu}_{\nu\alpha}$ [of the affine connection] cannot be directly measured, rather they must be obtained from the directly measured magnitudes first through complicated calculations. No one can empirically determine an affine connection between two vectors in neighboring points if he has not already ascertained the line element. For this reason, I maintain, in opposition to you and Einstein, . . . that to attempt to base a geometry upon an affine connection without a line element is above all meaningless for physics . . . [for] in that case, we not only have no "natural geometry" but also no "natural theory".[73]

Pauli's positivist strictures on physical meaning were already leveled in 1919 against Weyl's theory. There, Pauli pointed out, one continually operated with a "meaningless fiction" in supposing a determinate value can be given to field strengths in the interior of an electron, a procedure violating the rule that legitimate quantities in physics must be "observable in principle".[74] In a letter to Born in January 1920, Einstein underscored his disagreement with such strict constraints on physical meaningfulness by pointing out that Pauli's criticism extended to continuum theories in general.

Pauli's objection is directed not only against Weyl's, but also against any other continuum theory.[75]

Indeed, Einstein's anticipation was fully borne out in the penultimate paragraph of Pauli's renowned monograph on relativity theory, appearing just a year later:

Finally, a conceptual doubt should be mentioned. The continuum theories make direct use of the ordinary concept of electric field strength, even for the fields in the interior of the electron. This field strength is however defined as the force acting on a test particle, and since there are no test particles smaller than an electron or a hydrogen nucleus, the field strength at a given point in the interior of such a particle would seem to be unobservable, by definition, and thus be fictitious and without physical meaning.[76]

As Pauli's previous discussion makes evident, Einstein is also the target of these remarks. But confessed epistemological opportunist that he was,[77] Einstein would not elevate the twinges of epistemological conscience visible in his objection to Weyl's theory, to the status of positivist strictures on the practice of theoretical physics. In point of fact, within a few years, Einstein would come into open combat with positivism, the "epistemologically-soaked orgy" of the new quantum theorists.[78]

4.4 Reichenbach

4.4.1 The Birth of a Method

Discussions of Reichenbach's epistemology of geometry have understandably focussed on the "mature" presentation of his views in *Philosophy of Space and Time* (*Philosophie der Raum-Zeit-Lehre*, 1928; English translation, 1958). To be sure, some recognition has been given to the circumstance that the neo-conventionalism there in force differs considerably from the position adopted in his first philosophical monograph on relativity theory published in 1920, indeed at variance with his "mature" position. As shown in chapter 2, in 1920 Reichenbach maintained that the metric of space-time is not at all conventionally fixed but "an objective property of the world". However, the metric does not initially appear to be such an objective property because it only arises through constitutive "principles of coordination", first enabling the coordination of abstract mathematical representations to physical reality that

> define the individual elements of reality and...are *constitutive* of [*sind* **konstitutiv** *für*] the real [*wirklichen*] object; in Kant's words: "because in general only by their means can any object of experience be thought".[79]

Now it would be of enduring significance for logical empiricism that here Reichenbach first outlined a new method for epistemology of science replacing Kant's "analysis of Reason", termed *die wissenschaftsanalytische Methode*, to become the logical empiricist "method of logical analysis of science". The idea is that one can sharply distinguish the "subjective contribution of Reason" in physical theories from the "objective" contribution provided by the world, thereby identifying the necessary subjective factors (the coordination principles) from the "axioms of connection", that is, empirically attested physical laws describing relations between physical state variables, notably the metric of space-time. As shown in chapter 2, such neo-Kantian liberalization of the meaning of synthetic *a priori* principles was expressly, and successfully, challenged by Schlick.

Reichenbach's 1920 neo-Kantian analysis of cognition is directed, in the first instance, against unnamed empiricist views that countenance no role for the object-constituting role of mathematical/conceptual elements. But already in 1922 the essential piece of the "mature" conventionalist view of the metric of space-time fell into place: the fundamental geometrical concept—congruence—requires, as Reichenbach deemed Helmholtz in particular to have shown,[80] a *stipulation* governing transported rigid bodies (measuring rods). The stipulation states, of course, that our measuring instruments suffer no nondetectable, hence noncorrectable, deforming forces under transport (variously termed, in chronological order, "*force d'espèce X*",[81] "metrical forces",[82] "universal forces"[83]). Once such a stipulation is made, the fiduciary measuring instruments ("normed" as the unit of line element *ds*) can be employed for an empirical determination of the geometry of space-time. In further accord with the shift to empiricist neo-conventionalism, by 1924 the category of constitutive "coordinative principles" of 1920 is transformed into that of "coordinative definitions" that link purely formal concepts with alleged facts concerning empirically given objects.[84]

A noteworthy aspect of Reichenbach's method of analysis that proposes to cleave a physical theory into its empirical and its nonempirical parts (to be designated, after the "linguistic turn" prefigured in Schlick (1925), its synthetic and its analytic statements) concerns its implied opposition to various contemporary holist views of the relation of physics and geometry in general relativity. For in the contemporary writings of Schlick (1917, 1918), Carnap (1922), Einstein (1921a), Eddington (1923a), and Weyl (1918a), are credible (if *sub specie aeterni*, as Einstein has it) arguments to the effect that only the whole of geometry and physics can be brought into connection with experience, with the implication that such a winnowing of subjective or conventional chaff from the kernels of pure empirical content such as Reichenbach envisaged is "an epistemological chimera".[85] However, in Reichenbach's hands, the analysis of cognition as coordination gives rise to what is, to all appearances, a strict "logical analysis" of scientific theories according to which a theory's empirical content is identified by the requirement that its fundamental conceptual elements be directly connected (via "coordinative definitions") with physical objects posited as independent of the theory but used in observational tests of the theory. On the other hand, while it would be futile to deny the apparent similarity of result of viewing scientific theories through the interpretive lenses of Reichenbach's *wissenschaftsanalytische Methode* and as given in, say, the analysis of Bridgman (1927),[86] Reichenbach's philosophical motivations were, by genesis and intent, sharply different from strict positivism or operationalism wherein concepts are just shorthand for observations or pencil-and-paper operations. Reichenbach, for example, is more than willing to accord theory-ladenness to the "elementary facts" expressed by his axioms (as long as *relativity* theory is not involved (as will be seen, this scruple is moot). Moreover, he firmly opposed positivist readings (by Frank and Petzoldt) of the curious remarks of Einstein concerning "point coincidences" in the canonical exposition of general relativity of 1916. Such coincidences, Reichenbach correctly observed (following Schlick)[87], are just the intersections of world lines; as such the meeting of two elementary particles surely counts as a legitimate "coincidence" in Einstein's sense.[88]

4.4.2 A Constructive Axiomatization

The "method of logical analysis", schematically introduced in 1920, received its first detailed application in Reichenbach's *Axiomatization of the Theory of Relativity* (*Axiomatik der relativistischen Raum-Zeit-Lehre*), published in 1924. This is a work with several levels of interest. It has immediate significance for the history of logical empiricism in that it is the first sustained attempt to give what would become known as a "rational reconstruction" of a physical theory,[89] that is, an *exposé* of the "logical structure" of a scientific theory wherein empirical and definitional components are clearly distinguished. The proclaimed character of the work as an "epistemological-logical investigation" ("*erkenntnistheoretisch-logischen Untersuchung*")[90] distinguishes it from axiomatizations of physical theories as usually conceived. For as method of analysis of science, "philosophy is only interested in the logical separation of empirical and logical components" of a theory and for this purpose, "the value of the axiomatic method" is that "it directly reveals the places where definitions are present; it separates the conceptual components of the theory from experimental content and shows where the discernable problems of physics first begin".[91]

Already in 1921, a preliminary version of his *Axiomatik* had appeared in which Reichenbach announced the epistemological-logical goal of sharply distinguishing the basic empirical assumptions of the theory of relativity, expressed in two groups of axioms governing 1) light signaling and 2) rods and clocks (*Materialaxiome*).

These are verifiable by observation and, as such, distinct from the theory's "conceptual ingredients" (*begriffliche Gehalt*), which are definitional, mere matters of convention. Together, the basic empirical propositions and the conceptual components comprise the theory's "logical structure" upon which depends all of the remaining propositions of the theory of relativity. In the initially published version, a brief report made to the Congress of German Physicists at Jena in 1921, Reichenbach asserted that the *Materialaxiome* simply affirm the complete identity of measurement results made with rods and clocks with those attainable from the "*Lichtgeometrie*" set out in the five axioms of the first group, and that the latter alone suffice for construction of a complete "*Raum-Zeit Lehre*", his most important result.[92] On the basis of the light axioms alone, it can be demonstrated that the theory of relativity is "a valid and complete physical theory".[93] And although no treatment of the metric in general relativity is given, the striking (but false) claim is made that this procedure also fixes the metric of general relativity, a result he deemed essential.[94]

Similar claims are later made on behalf of the 1924 *Axiomatik*'s "physical significance". Now there is actually a purported demonstration of how, in the flat Minkoswski space-time of special relativity, a metrical determination can be made using only light signaling (i.e., the *Lichtgeometrie*).[95] Later on, this claim will be somewhat qualified, only to be subsequently withdrawn. Furthermore, Reichenbach contended that what is "physically new" in Einstein's assertion (*Behauptung*) about the metric in special relativity can be summarized in saying that rods and clocks adjust (*einstellen*) not to the classical but to relativistic light geometry.[96] The agreement of metrical determinations made by the light geometry and as made with rods and clocks, the core of his *Axiomatik*, as also taken as the centerpiece of special relativity. Moreover, Reichenbach alleged that "on the basis of our axiomatic representation we can finally pick out *what is affirmed concerning reality (Wirklichkeit) by the relativistic doctrine of space and time.*[97] He will thus respond to the *Axiomatik*'s critics that "Einstein's theory stands or falls with my Axiomatik",[98] a contentious remark, especially in view of the fact that the posited agreement breaks down in gravitational fields. However, the agreement (which is affirmed by Axiom VIII; see below) "holds only in infinitesimal regions for neighboring points", as, not surprisingly, we learn much later.[99]

Although the *Axiomatik* combines a logical-epistemological orientation with a technical discussion requiring familiarity with the calculus of tensors, it does not appear to have been successful in bridging the disparate communities of relativity physicists and scientifically minded philosophers eager to draw out the philosophical significance of the theory of relativity. Due to its explicitly epistemological and "constructive" character, the work did feature a certain novelty. But it was, in Reichenbach's own admission, widely "misunderstood".[100] Most noticeably, in its "constructive" concern to partition the theory into an empirical content as distinct from the conceptual structure of the theory, it completely departed from Hilbert's axiomatic treatment of general relativity (1915), whose starting point is a Hamiltonian (i.e., variational) principle, the empirical confirmation or disconfirmation of which, Reichenbach duly noted (in accord with Pauli), is rather far removed from actual experiment.[101] By pointedly proclaiming a different and "constructive" goal for axiomatization, the Reichenbach *Axiomatik* deliberately ventured into epistemological *terra incognita* so far as the relativity community was concerned, apparently not making much of an impression among theoretical physicists and mathematicians, perhaps the readership Reichenbach most wanted to take notice. But also because of its considerable use of the tensor calculus, the Reichenbach *Axiomatik* similarly received little attention from philosophers. One can speculate that the far more straightforwardly philosophical treatment given the subject in his *Philosophie der Raum-Zeit-Lehre* of 1928 was, at least in part, a response to the relative neglect of his earlier book. This, at any rate, was the judgment of Schlick (1929).

4.4.3 "Elementary Facts", "Metrical" and "Physical Forces"

Like its predecessor, the 1924 *Axiomatik* contains two central classes of axioms and, corresponding to these, two classes of definitions pertaining to the behavior of light (*Lichtaxiome, Lichtdefinitionen*) and pertaining to the behavior of material bodies (*Körperaxiome, Körperdefinitionen*). The axioms express "elementary facts" (*elementare Tatsachen*), observable facts on which the theory of relativity is based and which are independent of the relativity theory, although not of all theory. Each axiom "signifies [*bedeutet*] an intuitively presentable fact in which nothing further remains that is mysterious or unrepresentable".[102] The *Körperaxiome*, rendered in the English translation as "matter axioms", implement the notions of rigid rods and natural clocks, which are "closed systems", that is, systems that may be considered isolated from any "physical forces", in that these effects may be correctable or considered negligibly small; whereas "metrical forces" are disregarded. The axiomatic standing of the congruence behavior of measuring rods and clocks as "elementary facts" rests upon a distinction (Definition 21) between "physical" and "metrical" forces by virtue of which the "rigid rods" and "natural clocks" figuring in these axioms are *defined*.[103] The difference recurs in his 1928 book in the more familiar guise of a distinction between "differential" and "universal" forces. But only by making rods and clocks independent—by stipulation—of "metrical forces", can Reichenbach preserve their standing as "elementary facts", basic empirical postulates of the theory. For the existence of metrical forces ("which depend on the choice of metric") is not independent of general relativity; indeed, that theory asserts that lengths of rods and periods of clocks are dependent on the surrounding gravitational field strengths. So here the separation between "physical" and "metrical" forces is a consequence of the *erkenntnislogische* character of the axioms as "elementary facts". Despite their appellation, these "facts" are really idealizations; for instance, it is necessary to specify that rods and clocks are only "infinitesimally closed systems".[104]

The axiomatization takes as its basic concepts the notions of *real point* (points at which physical objects may be considered to be at rest), *signal* (a physical process propagating between real points), and *simultaneity at a real point*. *Earlier* and *later* at a real point are then defined by reference to the departure, and return, of a signal traversing a closed circuit. The great bulk of the book concerns the special theory of relativity. In motivating this approach, Reichenbach provided an imaginative picture of the world as a space filled with mass points similar to the molecules of a gas; these are to be real points at which observers are located who can signal to one another using light rays. The space-time metric within such reference systems ("rigid systems") can be determined solely by the light geometry corresponding to the "light axioms" grouped as I–V. Axioms I and II are so-called topological axioms; in fact, they are statements regarding causal chains, the spatial and temporal order of sequences of events issuing from or at a given real point, and it must be observed that this sense of "topological" has nothing to do with the characteristic notions of neighborhood, convergence, and continuity of topology. The six axioms grouped under II provide the means for making time comparisons at different points, the most important affirming the existence of "first signals", those signals traversing a closed circuit in least time. Axiom III is a Fermat axiom identifying first signals as directly emitted light signals. The two axioms of Group IV, introducing the concepts of "stationary" and "static" systems,[105] have the effect of making the simultaneity relation both symmetric and transitive. The standard $\varepsilon = 1/2$ ("Einstein") simultaneity relation is then defined (Definition 8). After giving a coordinative definition identifying straight lines as light rays (Definition 9), Reichenbach proceeded to define "spatial straight" and "spatial length" via light rays and return time of light signals; hence, he claimed, "congruence" is defined by using merely the "light geometry". The final light Axiom V affirms that the light geometry in static systems is Euclidean, whereas inertial systems are defined as stationary spatial coordinate systems conforming to the light Axioms I–V.[106] From three further definitions (15–17) concerning comparison of units in stems moving with respect to one another and two auxiliary theorems, Reichenbach easily derived the result that the Lorentz transformations are mappings preserving the Euclidean character of static systems.

An ensuing claim that the thus developed light geometry suffices for the determination of the metric in the flat space-times of special relativity requires some qualification. Reichenbach had shown how observers in reference systems situated within a finite distance may use light signals alone to distinguish systems at rest relative to one another. Then he showed that within such "rigid systems" it is possible to define a metric (up to a linear factor) using only light signals. But the extension of this method of metrical determination to the entire space(time) presents a difficulty, for the most general transformations that carry light cones into light cones are not the linear Lorentz transformations but spherical transformations that carry spheres into spheres, the group of so-called Möbius transformations. But in full three-dimensional Euclidean space, or four-dimensional Minkowskian space-time, these contain singularities wherein a point is carried into infinity.[107] So not all real points can be reached with light signals, and so if light geometry alone is held to be sufficient for the determination of the class of inertial systems, the following problem arises. If an observer who believes that he is in an inertial system after light signaling in his own region of space-time, then signals arbitrarily far outside of this region, he will encounter a singularity, leading to the conclusion that his is not an inertial system. For the light axioms hold without singularities only in inertial systems.

Two possible solutions to this difficulty are offered. To uphold the sufficiency of the light axioms for the determination of the metric, a procedure is sketched wherein an observer, assuming his own system S is inertial, can, by constructing another reference system S' relative to his own, calculate the limit at which singularities should occur in S' if it is not an inertial frame. Then, if these singularities are in fact found in S', the observer may conclude that his system S is an inertial system. If the calculated singularities in S' are not found, then S' is inertial and S noninertial, and the observer may then calculate the location of singularities in his own system S.[108] A second route is to adopt the criterion that the metrical determination must be without a singularity at any space-time point. This entails giving up the sufficiency claim, making the reasonable assumption that observational determinations occur anyway only in finite regions. But then it is necessary to invoke "material structures" to determine the class of inertial systems: points at rest relative to one another (as determined by light signaling) and only these may be connected by rigid rods (*Körperaxiom* VII; see immediately below). Hence, the problem of singularities is circumvented by a restriction of metrical determinations to within finite regions and by relying upon the posited existence of rigid bodies.[109]

At this point (§19) come six *Körperaxiome*, grouped as VI–X. The two Axioms VI assert path-independent congruence of lengths (as measured by "rigid rods") and intervals (as measured by "natural clocks"); they express "old presuppositions of measuring with rods and clocks".[110] Axioms VII–X then assert the identity of the geometry of rigid rods and clocks with the light geometry, an identity, per the considerations above, initially claimed only for inertial systems.[111] Axiom VII affirms that only points at rest with respect to one another can be connected by rigid rods. Axioms VIII–X state the identity of lengths and unit length and time intervals with those of the light geometry. There will be an attempt to extend this identity to noninertial frames with the addition of two general relativistic Axioms XI (see §4.4.5).

4.4.4 *Countering Weyl*

Critical engagement with Weyl in the *Axiomatik* appears prominently in two strategic places, both involving a general defense of the *Körperaxiome*. The first concerns Weyl's proposal to use the trajectories of force-free mass points in the construction of the metric. Here Reichenbach adamantly insisted that such a method offers no epistemological advantages to his own proposals involving rods and clocks. This response occurs twice: in the context of the construction of the metric of special relativity, where Weyl would use free mass points to specify the class of inertial systems, and in the context of the construction of the metric of general relativity, where Weyl would use them to provide a determination of the values of the $g_{\mu\nu}$ (up to a factor of scale) that have been conformally fixed only as ratios by the use of light signals in arbitrary gravitational fields.[112] In notes in two separate places, the first in the section introducing the *Körperaxiome*, the second in a section on light geometry in a gravitational field,[113] Reichenbach observed that one can, as did Weyl, employ force-free point masses, rather than rods and clocks, as the needed material structures, referring to the treatment in the first appendix to the fourth (1921) edition of *Raum-Zeit-Materie*.[114] But Weyl's approach, it is claimed, faces epistemologically the same problems as does the coordinative definition of the metric employing rods and clocks. For just as a rigid body must be *defined* to be one free from the effects of "metrical forces", so can a mass point be said to travel in a straight line (a geodesic) only if it is *defined* as a body upon which no net forces act. Weyl's use of mass points offers no epistemological advantages; it is on a par with rods and clocks. That is, it is open to someone to argue that a putatively force-free particle in fact is not traversing a geodesic but describes a trajectory of a particle with a (nondetectable, "metrical") force acting upon it.[115] This little argument appears to be a specific application of what will become known as "the method of equivalent descriptions", a conventionalist leveling of the epistemological playing field achieved by pointing out the role of definitional elements in different proposals for tying theory to observation. The key move lies in generating a class of "equivalent descriptions": rival but empirically equivalent characterizations of "the facts" of a physical situation, here pertaining to the metric of space-time, the choice among them to be made on the grounds of "descriptive simplicity". Subsequently, Reichenbach located "the philosophical achievement" (*Leistung*) of the theory of relativity precisely in that it enables one to see metrical coordinative definitions are really called for where previously empirical cognitions (*Erkenntnisse*) had been sought.[116] In this specific instance, one may object that the epistemological comparison is framed too narrowly, in that it is a *law* (admittedly, a law with, in all probability, a vacuous antecedent) that bodies on which no forces are acting travel in uniform rectilinear motion whereas we do not speak of laws of rigid bodies. But there is a substantially more trenchant objection. The general relativistic analogue to the Galilean law of inertia in classical mechanics is a specific structure of the space-time manifold, termed by Weyl the "guiding field" (*Führungsfeld*), the combined gravitational-inertial field mathematically represented as a manifold endowed with an affine connection.[117] As the affine connection (representing the potential of the gravitational field) is not a tensor, it cannot be split into gravitational and inertial parts in a nonarbitrary way, yet it is not a merely conventional mode of mathematical description of freely falling bodies. In fact, that the paths of test particles (neutral, spinless, and "small" enough so as to be negligible sources of the gravitational field) are geodesics of the space-time metric may be derived from the Einstein field equations.[118] In view of this *constitutive* role, invocation of the behavior of bodies under the influence of the "guiding field" is epistemologically on quite a different footing than an appeal to the approximately rigid behavior of rods and periodic behavior of clocks. In particular, these are complicated material structures whose exact behavior remains a task for physical explanation (say, in a many-body quantum theory) that will require assumptions far outstripping the "guiding field" postulate.[119]

In a second rejoinder to Weyl, Reichenbach observed that one does speak of the "adjustment" (*Einstellung*) of rods and clocks—as seen above, Reichenbach himself referred to the "adjustment" of rods and clocks to the "light geometry". Noting that Weyl had first used the term "adjustment" in this connection, Reichenbach states that this characterization must be taken with a grain of salt since it only provides (as Weyl surely would agree) "a statement of the problem". As a merely verbal characterization of the behavior of material structures, it cannot be taken as explanatory (but, Weyl might counter, it can be taken as having heuristic value). Of course, if the term is understood literally, it is incompatible with the property of rigidity stipulated in the definition of congruence. Accordingly, Reichenbach maintained that the "situation" regarding the admitted "adjustment" of measuring rods and clocks to the fields in which they are embedded is "formulated rigorously by the matter axioms" without use of the term "adjustment". After all, Weyl's term merely named, but does not solve, a problem to be resolved by a future theory of matter of which there was, in 1924, scarcely an inkling.[120]

4.4.5 Critique

Reichenbach, in defense of his claim that the physical content of the Einstein theory is expressed by the claimed agreement between light and matter axioms, has sought to defend the use of rods and clocks by countering Weyl's proposals to do without them. But what justification supports Reichenbach's defense of rods and clocks? Reichenbach's epistemological analysis of the theory of relativity has been previously criticized by Torretti as "putting the cart before the horse" as giving pride of place to special relativity.[121] The difficulty, of course, is that while the global validity of general relativity entails the local validity of special relativity, the converse, of course, is not the case, since the global validity of special relativity entails that general relativity is false. Nowhere are the grounds for his preposterous attempt to derive general relativity over finite regions from the infinitesimal validity of special relativity more clearly displayed than in the *Axiomatik*. In particular, Reichenbach's guiding strategy is to attempt to accommodate, to the greatest extent possible, the rods and clocks that are supposedly licensed in the inertial frames supported in the special theory within the context of gravity (that is, general relativity). To accomplish this, he exploited the limiting process that is at work in the admissible principle of equivalence, bootstrapping from the "infinitesimal" validity of the special theory into the more general finite setting of gravitational fields. But Reichenbach characteristically viewed this process through, as it were, an inverted lens, seeing metrical determinations in the general case as mere extrapolations from what can be established (and justified) infinitesimally. In a certain respect, this perspective on general relativity has endured among elementary particle physicists, for whom "the geometrical approach" to the space-time manifold "has driven a wedge between general relativity and the theory of elementary particles".[122] But, of course, Reichenbach did not have the unsuitability of variably curved space-times for quantum field theory in mind in 1924; rather, his was the epistemological project of providing a suitably empiricist basis for the general theory of relativity.

This inverted perspective governs the path taken by Reichenbach in extending his axiomatization of the special theory (Axioms I–X, Definitions 1–21) of part I of the book to the general theory in part II. For the general theory of relativity, two new "differential" *Körperaxiome* are required (Axiom XI, 1, 2; Definitions 23–25, §33). From the first general relativistic axiom (XI, 1) asserting the infinitesimal validity of special relativity in every frame ("coordinate system of real points"), Reichenbach took it to follow that "around every world point a finite region can be defined in which a ("spatial") coordinate system exists"; that is, "a rigid reference system" is both "everywhere present" and "infinitesimally stationary" at every point;[123] thus, we are to generally view measurements made *in finite regions* as if they are in agreement with metrical determinations made *at each point* in such a special coordinate system. Indeed, this is just what is stipulated in Definition 24: that in any such coordinate system, the metrical determination of the world is to be made in such a way that it will become identical at every point with the metrical determination that is locally prescribed by axioms I to X and definitions 1–21 (of the special theory of relativity) while the same measuring rods and clocks are used everywhere.

There is a rather glaring difficulty with this proposal that metrical determinations are to generally be conceived as occurring within such "spatial coordinate systems" (frames of reference) wherein the validity of the special theory of relativity is maintained.[124] For consider, in geometrical formulation, the claim that special relativity is locally valid in the sense that the Riemann curvature tensor vanishes everywhere ($R_{\mu\nu\sigma\tau} = 0$). Of course, the same criterion for the local validity of special relativity cannot apply identically to every point of space-time on penalty of passing from curved to the globally flat space-times of nongravitational physics. Reichenbach's account of space-time measurement in general relativity presupposes a space-time pieced together of local bits at each of whose points the Riemann curvature tensor vanishes, a condition that certainly precludes every interesting space-time of general relativity (for the only solution of the Einstein field equations in which the Riemann tensor is identically 0 is Minkowski space-time). Reichenbach's proposal is an illegitimate extrapolation from the principle of minimal gravitational coupling, that is, that the theory of special relativity can be expected to hold in a sufficiently small neighborhood of a point P in which there is an inertial frame. A related difficulty obtains with his second general relativistic axiom (XI, 2). It asserts that accelerating rods and clocks can be considered as "differentially at rest" and hence give the same measurement results as rods and clocks "permanently at rest".

> Accordingly, axiom XI, 2 asserts that every rigid rod l...behaves in the same way as a rigid rod l_o...that is permanently at rest in an inertial system.[125]

Observing that "this result is by no means obvious", Reichenbach nonetheless stated that it "is an assumption [*Annahme*] of the general theory of relativity" from which "follow Einstein's well-known statements concerning the behavior of rods and clocks in a static gravitational field".[126] Considered in itself, Reichenbach's axiom (XI, 2), termed by Torretti the "rod hypothesis", is no less problematic. The grounds of the difficulty lie with the notion "differentially at rest" that is, "the momentary inertial rest frame of an accelerating body". This concept cannot be made exact, that is, for an extended body, there is no such thing as *the* momentary inertial rest frame; to the contrary, its various point constituents at any single moment will be "co-moving with different inertial frames".[127] Reichenbach seemed aware of the difficulties consequent upon his "rod hypothesis", but his means of overcoming them involves another problematic application of the inverted limiting principle noted above. For he noted that

> the rigid rod and the natural clock are defined as closed systems; but closed systems exist only in inertial systems.[128]

Note that this last sentence is still not quite adequate if we take into account the problem of elastic forces, hence that of rigid bodies, already in the special theory.[129] But Reichenbach believed he could counter the problem of elastic forces (whereas "metrical forces are to be ignored") by stipulating a limiting process in terms of which these forces can be made to vanish. This is done in the definition of an "infinitesimally closed" material system that is

> sufficiently small relative to space-time changes of the gravitational field strength
> and if the quotient of external physical supporting forces and internal physical
> forces approaches o with the shrinkage (*Verkleinerung*) of the system.[130]

Once again, the attempt is made to extrapolate the legitimacy of physical concepts that have a certain restricted domain of validity to more general domains through the inverted application results obtained from a limiting process. This mode of procedure is in full force in the very brief treatment given "the construction of the metric in the general case" where Reichenbach maintained that units of $ds^2 = \pm 1$ can be introduced in any given coordinate system by assuming rigid rods and clocks at rest at each point, thereby (with additional use of light signals) determining the $g_{\mu\nu}$ in this system of coordinates.[131] However, as Weyl would point out in a critical review, this "fibrillation" of the world has nothing to do with the nature of the metric field.[132]

Independently of these conceptual problems, what remains of the central claim that the geometry of light rays and rods and clocks will always agree? Certainly Reichenbach cannot provide a general demonstration of this claim for in a generic general-relativistic space-time, there can be no rigid congruences representing rigid bodies.[133] Although he was able, in a section entitled "Light Geometry in a Gravitational Field", to exploit a mathematical result showing that there will be agreement in the special case of the field equations of the vacuum;[134] in general, there is agreement only in the infinitesimal regions compatible with the validity of the special theory. For other gravitational fields, Reichenbach proceeded to show that the metrical determinations of the *Körperaxiome* will *not* agree with light geometrical ones; in a step-by-step discussion, the failure is tracked by considering more and more general types of gravitational fields. Already for static gravitational fields, the metrical determinations of the light geometry and the *Körpergeometrie* no longer coincide (so *Körperaxiom* VIII falls), whereas for stationary fields, such as the rigidly rotating disk (the simplest case; see §4.2) the round-trip light axiom (IV, 2) fails; hence, measurement of spatial lengths involves an additional complication. For in order to say that rods everywhere on the disk have the same unit length, a correction factor (corresponding to Lorentz contraction) is required for rods lying tangentially to the motion of the disk. Thus, in order to preserve the customary definition of congruence (i.e., "no metrical forces") for the full four-dimensional manifold of space-time, it is necessary to invoke "metrical forces" in the definition of congruence for three dimensional rigid rods.[135] With respect to even more general "real systems" (restricted only by a coordinate condition that it is impossible to transform any two distinct point-events on the same timelike worldline onto the same plane of simultaneity), only the so-called topological Axioms I and II are held to be valid.[136] For such systems, "the failure of axioms III, IV, and V means that metrical particularities [*Besonderheiten*] no longer exist".[137] Finally, in the most fully general case, where the $g_{\mu\nu}$ are fully variable in position and time, restricted only by the requirement that the ds^2 be of "inertial index 1" (of signature [+,+,+,−]; or vice versa, with one plus index), the restricted claim is made that the topological order of time, as given by Axioms I and II, obtains just in "cut-out" finite domains, while global order properties are wisely left in abeyance. As the result of these considerations, which show that "topological properties turn out to be more constant than metrical ones", the startling admission is that in such general gravitational fields, where there are no rigid congruences, there can no longer be a metric (hence, chronogeometry). The rather stunning conclusion of the *Axiomatik* is that if rigid rods and clocks, as the empiricist correlates of metrical notions, are no longer physically possible in such fields, it is necessary to *renounce* the metrical properties of space-time altogether!

> [T]he step from the special theory to the general one represents merely a re-nunciation of metrical particularities [*ein Verzicht auf metrischen Besonderheiten*], while the fundamental topological character of space and time remains the same.[138]

Recalling Reichenbach's special sense of "topological", this result is, moreover, held to be in harmony with his central contention that the so-called topological structure, in particular, the "topological distinction" between space and time is, far more than the metrical structure, an object of visualization (*Anschaulichkeit*) in the general theory of relativity. So in the *Axiomatik*, and in the more widely read treatment of Reichenbach (1928), metrical properties of space-time are deemed less fundamental than "topological" ones, while the latter are derived from empirical facts about time order.[139] But time order in turn is reduced to facts about causal order, and so the whole edifice of structures of space-time is considered episte-mologically derivative, resting ultimately upon basic empirical facts about the causal ordering of events, and, it must be added, a lingering *a priori* prohibition against "action-at-a-distance". The end point of Reichenbach's epistemological analysis of space-time is then his "causal theory of time", a relational theory of time presupposing (until sometime in the 1930s) the (seemingly *a priori*) validity of a principle of local causal action (*Nahewirkungsprinzip*). Of course, just what sense can be made of such a principle in the absence of a metric, or at the very least a topology (in the usual, not Reichenbach's, acceptance), remains to be seen.[140]

At this juncture, where the metric is subordinated to "the causal order of time", we have reached essentially the same end point attained in *Philosophy of Space and Time* (1928), where also the causal order is regarded as "the physical structure into which space-time order can be embedded even when all of the metrical properties of the space-time continuum are destroyed by gravitational fields".[141] This is, it must be said, an astonishing finale. The talk of "renouncing metrical particu-larities" and of "destroying" metrical properties cannot be taken literally, unless metrical properties are *necessarily* associated with rods and clocks, an association that, after all, was inaugurated as a convention! But if the association is a nec-essary one, then Reichenbach has unwittingly provided a telling (and damning) illustration of *the* fallacy of positivist metascience: an epistemological tail wags the physical dog.

Ironically, in Reichenbach's reference to topological order as "an ultimate fact of nature", we can see just how far his empiricist analysis of the metric of space-time has philosophically strayed from Einstein's. For, on the one hand, it was shown in §4.3, that Einstein was content to wield an empiricism about rigid rods and clocks chiefly as a cudgel against Weyl's theory, while proceeding epistemologically unencumbered on the winding paths of unified field theory. Then again, we can recall the lesson of the *Lochbetrachtung*, the "Hole Argument". As discussed in chapter 2, §2.2.2, the key step taken by Einstein in arriving at his generally covariant field equations consisted in recognizing that *nothing*, including the "points", remained of the space-time manifold in the absence of the metrical field. Without an "individuating field" (as the metric tensor field), there is no physical way, Einstein reasoned, to accord physical meaning to the points of the space-time manifold, a conclusion underlying his elliptical, and in itself quite puzzling, statement of that the "requirement of general covariance . . . takes away from space and time the last remnant of physical objectivity". As seen there, in the clipped formulation of the later Einstein, without the metric field, there is not only no residual "empty space", but "absolutely *nothing* and also no 'topological space' ". The implication, surely, is that one cannot conceptually bind the metric field of gravitation to a conventional stipulation regarding rods and clocks. Reichenbach's ostensibly conventional, but in fact empiricist-mandated, tie of metrical notions to rigid rods and clocks by a physical "coordinative definition" issues in the bizarre conclusion that where such physical structures are no longer possible, metrical characteristics are to be "renounced" in favor of "facts" ultimately concerning causal order. As subsequent developments have shown, this is a highly unsatisfactory and even preposterous epistemological analysis of the theory of general relativity.

4.5 Conclusion

From a contemporary perspective, an amalgam of interrelated philosophical issues concentrate along the Weyl–Einstein–Reichenbach axis. For one thing, subsequent philosophical thinking about how the geometry of physical space has been transformed by the general theory of relativity has largely crystallized along this axis. Attention to its several way stations underscores the difficulty of accommodating the epistemologically revolutionary message of this theory within positions antecedently understood as "empiricism", or "conventionalism" or even "holism".

There is the still broader consideration, that a crucial juncture in the philosophy of physical science lies along this axis. For inasmuch as logical empiricism, through the interpretive works of Schlick and Reichenbach, was able to claim the philosophical mantle of Einstein regarding the theory of relativity, at least the crucial formative years of the 1920s and early 1930s, a decisive turn was taken for subsequent philosophy of science. As I have shown, just at the time of appropriation Einstein happened to be in the process of readjusting his philosophical attire. And it subsequently mattered little that significant philosophical differences would emerge between Einstein and the positions of logical empiricism in the early 1930s. What did matter was the supposed *imprimatur* of Einstein to logical empiricist orthodoxy concerning the structure of scientific theories and the relation of theory to observation that has been duly transmitted to subsequent generations of philosophers of science. In no small measure, this desirable genealogy enabled an adherence to empiricist epistemological imperatives, institutionalized as the method of "coordinative definitions", to become the cornerstone of the "received view" of scientific theories. In its train came a phalanx of attendant philosophical problems and *insolubilia* surrounding the issue of "empirically equivalent descriptions", an unduly exulted role accorded to "purely conventional" elements in scientific theories and, perhaps most important, a willful inattentiveness to the constitutive role of mathematical structures in physical theory. However this may be, our reconsideration of the scientific context of Reichenbach's account of the "philosophical significance" of general relativity may serve to remind that philosophy is part and parcel of a wider intellectual culture, and that even epistemological analyses of fundamental physical theories do not virginally spring Minerva-like from the brow of Jove, but are dialectically forged within the contingent circumstances of that culture.

5

TRANSCENDENTAL-
PHENOMENOLOGICAL IDEALISM
Husserl and Weyl

When it is actually natural science that speaks, we listen gladly and
as disciples. But it is not always natural science that speaks when
natural scientists are speaking. . . .

Husserl (1913, 38)[1]

5.1 *Annus Mirabilis*

From the relative comfort of neutral Zürich during the first six months of the war
year of 1918, Hermann Weyl published three works that left indelible marks on
20th century physics and foundations of mathematics. First to appear, just after
the beginning of the year, was a controversial monograph on the continuum. Judging that "the house of analysis has, to an essential degree, been built on sand",
Weyl coupled a critique of the set-theoretical foundations of mathematics with the
outline of a purely predicative alternative, recently judged to have largely realized
its goal of being the basis for scientifically applicable mathematics.[2] The month of
May brought publication of *Raum-Zeit-Materie*, the first comprehensive treatise on
general relativity, extolled by Einstein as a "symphonic masterpiece" wherein
"every word has its relation to the whole".[3] By 1923, the work had already cycled
through a fifth edition, chronicling, among other changes, the waxing and waning

of Weyl's hopes for a "pure field theory of matter". Eight decades later, the book continues to exercise an appeal. Until recently, a seventh German edition was in print, with appended notes by the noted relativist Jürgen Ehlers, while the 1922 English translation of the fourth (1921) edition remains available in a paperback edition. Finally, on 1 March, in the same note informing Einstein to expect the printer's proof sheets of *Raum-Zeit-Materie*, from the publisher (Julius Springer, in Berlin), Weyl announced that "this day, as I believe, I have succeeded in deriving gravitation and electromagnetism from a common source".[4] In distinct contrast to the dim view he accorded to Hilbert's 1915 schematic unification of gravitation and electromagnetism via "the axiomatic method", Einstein immediately hailed Weyl's theory as "a stroke of genius of the first rank". However, within a few weeks, he raised the fundamental objection that, as shown in chapter 4, effectively sealed its fate. Despite this, the paper was submitted as Weyl had wished, under Einstein's sponsorship, to the Proceedings of the Berlin Academy; it appeared there late in May with Einstein's objection appended, at the demand of Walter Nernst, Secretary of the Academy, and, at Einstein's insistence, with Weyl's response.[5]

Despite Weyl's considerable, and somewhat subtle, efforts to elaborate that response over the next five years, Einstein's criticism, together with a widespread perception that the theory provided merely a formal unification, were taken as decisive by the physics community, an opinion that has not wavered. Even so, Weyl's theory of gravitation and electricity was enormously influential, launching the first phase of the geometrical "unified field theory" program that Einstein continued up until his death in 1955. More recently, it has been remembered for introducing the requirement of "gauge invariance", as Weyl called his demand of local scale symmetry, of invariance of field laws with respect to arbitrary dilatations at each space-time point. As has been well documented in recent years, the modern concept of gauge (phase) invariance emerged, with Weyl's assistance, some time later, and in several stages, from its swaddling clothes of a local scale invariance in space-time geometry.[6]

Such a richly creative and diverse output within the space of six months has but few equals in the annals of modern science. What makes Weyl's achievement all the more remarkable is that each of these contributions bears the decisive imprint of Husserlian transcendental-phenomenological idealism. Thus, Weyl's critique of the impredicative methods of classical analysis in *Das Kontinuum* stemmed from his opposition to the conception of mathematical objects as abstract entities existing independently of consciousness that had resulted in the set theoretic paradoxes. Instead, he adopted phenomenology's fundamental epistemological principle of *Evidenz*, what is "given" in the insight of immediate, nonsensuous, intuition rather than formal proof, as the ultimate source of cognitive authority in mathematics. And although Weyl envisaged the rational justification of his predicative alternative to classical mathematics as lying in its provision of sufficient mathematics for physics, phenomenological reflection on such applications uncovers the ineliminable transcendental subjectivity framing the "geometrical-physical" world manifested in the necessary posit of a coordinate system. Next there is *Raum-Zeit-Materie*'s densely philosophical introduction. Largely unheeded, and nearly unintelligible in the flawed English translation, it contains a highly condensed recapitulation of the argument for the thesis of transcendental-phenomenological idealism of Husserl's *Ideen I* (1913). In Weyl's rendering, this thesis states that

> the actual [*wirkliche*] world, all of its constituent parts, and all their determinations, are and can only be given as intentional objects of conscious acts.

Such is the *Ansatz* for Weyl's epistemological elucidation of Einstein's new theory, a task deemed essential lest "knowledge be turned into a meaningless chaos". The accompanying phenomenological method of "eidetic analysis" underlies Weyl's theory of "gravitation and electromagnetism". Broadening the Riemannian geometry of Einstein's theory to satisfy what Weyl termed "the epistemological principle of relativity of magnitude", the resulting "pure infinitesimal geometry" is proposed as the "world geometry" for field physics, enabling a rational understanding of the world solely from its behavior in the infinitely small. Resting upon the fundamental posit of the infinitesimal "congruent displacement" of a vector, a notion having, notoriously, "nothing to do" with the actual behavior of rods and clocks, the intent of Weyl's ideal world geometry was almost uniformly misunderstood. Still more, both Weyl's epistemological method and his theory violated central tenets of the largely positivist metascience to which many physicists, at least nominally, adhered. Yet without reference to these philosophical motivations, neither the origin nor implementation nor Weyl's tenacious defense of his "epistemological principle of the relativity of magnitude", the precursor of the modern gauge principle, is fully understandable.

It is the burden of this chapter to identify, and form a coherent picture of, the various currents of Husserlian transcendental-phenomenological idealism within Weyl's writings pertaining to these achievements of 1918. Then, in chapter 6, I will show how, in Weyl's hands, they comprise an essential part of the "context of discovery" of gauge principle, one of the most productive ideas of 20th century theoretical physics. In what follows, I largely presuppose the details of Weyl's theory and its historical reception laid out in chapter 4. §5.2 presents information regarding what is known of the personal contacts between Weyl and Husserl, since these are not irrelevant to an appreciation of the extent of Weyl's embrace of phenomenology in the years roughly extending from 1917 to 1926. Illuminating the transcendental-phenomenological context of Weyl's "broadening" (*Erweiterung*) of relativity theory is a more exacting labor, and the next several sections are considerably more ambitious. Leading off with several striking passages in the introduction to Weyl's classic *Raum-Zeit-Materie*, §5.3 undertakes to characterize the principal motivations for, and central thesis of, Husserl's transcendental-phenomenological idealism. Through juxtaposition with similar expressions in Weyl, these themes are seen to be salient in his "broadening" of relativity theory. In particular, it is shown how "transcendental subjectivity", in Husserl's sense of the "absolute being" of "pure consciousness" surviving the phenomenological reduction, plays the fundamental role in Weyl's understanding of the constitution of objectivity in physical theory. In §5.4 I provide further exposition of those parts of phenomenological method that are seen to be at work in Weyl's "pure infinitesimal geometry" that is intended as a "world geometry" in which field physics is to be constituted. Only then, in chapter 6, can a convincing case be made, showing how, in particular, Husserl's phenomenological method of eidetic analysis guided Weyl in arriving at, and in mathematically formulating, his "pure infinitesimal geometry". I will further argue there that Weyl's conception of a "world geometry", the ideal frame for field physics, was conceived along the lines of a Husserlian "material regional ontology" and so was not put forward simply as a physical hypothesis about the geometry of space-time.

5.2 Weyl–Husserl Personal Contacts

In the years 1904–1913, Edmund Husserl (1859–1938) and Hermann Weyl (1885–1955) overlapped at the University of Göttingen. Husserl had come to Göttingen as *extraordinarius* professor of philosophy in 1901, at least partly at the instigation of David Hilbert.[7] Promoted to a personal chair in 1906, he nonetheless accepted a call to Freiburg in 1916; by the early 1920s, he was the leading philosopher in Germany.[8] Weyl entered the University of Göttingen in 1904 to be a student of David Hilbert, who directed Weyl's Ph.D. dissertation on integral equations in 1908. Habilitating in 1910, Weyl taught, as *Privatdocent*, mathematics in Göttingen until 1913, when he accepted an appointment in Zürich as professor of mathematics at the Federal Institute of Technology (ETH). Hilbert and Husserl shared a number of students, including Kurt Grelling and Kasimierz Adjukiewicz, and it is known that a number of Hilbert's students, including Max Born, Ernst Hellinger, and Rudolf König, attended Husserl's seminar on the philosophy of mathematics in the summer of 1905. In the *curriculum vita* appended to his Göttingen dissertation, Weyl reported attending lectures of Husserl, who substituted for Hilbert and chaired Weyl's oral *Prüfung* on 12 November 1908, a courtesy to Hilbert in view of the latter's dislike of the formality of such affairs.[9]

Despite these contacts, Weyl's passionate interest in phenomenology was first kindled by the woman he was to marry before leaving for Zürich in 1913, Friederike Bertha Helene Joseph (1893–1948). In a philosophical reminiscence written the year he died, Weyl wrote that by marrying this student of Husserl in Göttingen, "it thus came to be Husserl who lead me out of positivism . . . to a freer outlook upon the world".[10] Hella, as she was known, had come to Göttingen in 1911 for the express purpose of studying with Husserl. In a memorial tribute written in June 1948, just after her death, Weyl reported that the first article of Husserl she ever read, as a student in Rostock, "hit her like a lightening bolt [*wie ein Blitz*]" and that "phenomenology always remained for her the foundation of her philosophical thinking".[11] Considering Husserl's dearth of publications between 1901 and 1913, this is quite probably a reference to "Philosophy as Rigorous Science" in the first volume of the journal *Logos* in 1910–1911.[12] As discussed below, that paper was the first published articulation of the new transcendental idealist direction Husserl had given phenomenology. The Weyls were acquainted with the Husserl family from the period in Göttingen, becoming close friends with Husserl's youngest son, Gerhard, who as a refugee from Hitler in the 1930s stayed for some time with the Weyls in Princeton.[13] Hella Weyl would later be the translator of works of Ortegay Gasset into German and English, and works of Eddington[14] and James Jeans into German. But in view of what is to follow, of particular relevance are the comments of Richard Courant concerning her influence on Weyl's book *Raum-Zeit-Materie* (1918). Speaking at her memorial service, on 5 September 1948, Courant recalled how, in rekindling his interest in mathematics after three years' service in the German army in World War I, Weyl's book "obviously would not have been written in its inspiring form without the influence of Hella's personality".[15]

"Inspiring form" is presumably an allusion to the book's profuse intertwining of phenomenological themes into the mathematics and physics of the new theory. All the German editions of *Raum-Zeit-Materie* published in Weyl's lifetime carry the dedication "*Meiner Frau gewidmet*".

5.2.1 Weyl–Husserl Correspondence

Four letters between Husserl and Weyl survive and have recently been published.[16] For present purposes, several passages provide further documentation of the close affiliation of Weyl with phenomenology in the years 1917–1926 that will be established in the rest of this chapter. Their transcendental-phenomenological background is developed in subsequent sections.

Husserl to Weyl (10 April 1918) on receipt of Weyl's gift of a copy of *Das Kontinuum* (1918):

> Finally a mathematician shows appreciation for the necessity of phenomenological modes of treatment in all questions of clarification of fundamental concepts, and hence returns to the original soil [*Urboden*] of logical-mathematical intuition, on which alone a really authoritative foundation of mathematics and an insight into the sense of mathematical achievement is possible ... I see, in all you have written, what I have sought in a similar inclination, a greater, wider perspective: of a philosophically based *mathesis universalis* and this again linked to a new formal metaphysics (of the *a priori* and general doctrine of individuation)— on which I have worked for years and continue to do so.

Husserl to Weyl (5 June 1920) on receipt of Weyl's gift of a copy of the third edition (1919) of *Raum-Zeit-Materie*, the first to contain Weyl's "pure infinitesimal geometry" and "theory of gravitation and electromagnetism":

> For a whole free afternoon I remained seated over and reading your work, which flowed with increasing delight. How near this work is to my ideal of a physics permeated by a *philosophical spirit*. What joy it is that our time has brought about such a universal knowledge of the mathematical form of the world, guided by the highest ideas, and that I may yet experience it! How much your own most characteristically deep cognitions concerning the Riemannian space form, concerning the distinction of 4 dimensionality, etc. has impressed me. Without closely reading the mathematical parts, still I have, as *Exmathematicus*,[17] presumed understanding of the *sense* of such deductions and, from the side of *my* studies, I am above all moved here by the transcendental significance, which points to similar, correlative, problems and thus anticipates such theories as yours.

Weyl to Husserl (26 March 1921) on receipt of Husserl's gift of the second edition of volume two, part two of the *Logische Untersuchungen* (the group-theoretical investigation referred to is discussed in chapter 6, §6.3.2):

Despite all the faults you attribute to the *Logical Investigations* from your present standpoint, I find the conclusive results of this work, which has rendered such an enormous service to the spirit of pure objectivity [*reiner Sachlichkeit*] in epistemology—the decisive insights on evidence and truth, the recognition that "intuition" ["*Anschauung*"] extends far beyond sensuous intuition—established with great clarity and concision. . . . Recently, I have occupied myself with grasping the essence of space [*das Wesen des Raumes*] upon the ultimate grounds susceptible to mathematical analysis. The problem accordingly concerns a similar group theoretical investigation, as carried out by Helmholtz in his time. . . . However, today the situation is altered through the theory of relativity, which also enables a notable deepening of the foundations [*Tieferlegung der Fundamente*]. . . .

It may be noted that the foreword to this Husserl volume contains a remarkably sharp rejoinder to critics who have misunderstood the phenomenological method and, in particular, the fundamental distinction between sensible and "categorial" intuition (see §5.3). One of them, Moritz Schlick, is singled out for Husserl's special wrath:

How readily many authors employ critical rejections, with what conscientiousness they read my writings, what nonsense they have the audacity to attribute to me and to phenomenology, are shown in the *Allgemeine Erkenntnislehre* of Moritz Schlick.[18]

Schlick's animadversions against phenomenology completely dismissed the fundamental concept of *Evidenz*, levying the charge that phenomenology claimed to employ a peculiar intuition that is "not a psychically real act" ("*kein psychischer realer Akt*"). Freely lifting quotations from Husserl, Schlick wrote that, according to phenomenology, anyone who failed to find such an "experience", even though outside "the domain of psychology", had not understood properly the phenomenological doctrine. Attaining such understanding, and so achieving the "correct adjustment of experience and thought", Schlick sarcastically continued, apparently required "peculiar and rigorous studies".[19]

In what may well have been a calculated response to Schlick's contemptuous dismissal of phenomenology, Weyl wrote a belated, but highly critical, review of Schlick's book in a mathematical yearbook dated 1923.[20] Published in this venue, there is every reason to suppose that Weyl's criticism was addressed to a scientific, not a philosophical, audience. An additional motivation is perhaps expressed in the one surviving letter from Weyl to Husserl, of 26–27 March 1921. In it, Weyl ruefully observed that Schlick's book had found "great resonance among the leading theoretical physicists". It is quite possible that Weyl had Einstein in mind, for Einstein had praised Schlick's "epistemology book" in both public and private.[21] In fact, the centerpiece of Schlick's book, the "purely semiotic" account of scientific cognition, dismissing any grounding in the intuition of "the given", is precisely the target of Weyl's attack.

In Schlick's opinion, the essence [*Wesen*] of the process of cognition is exhausted by [the semiotic character of cognition]. To the reviewer, it is incomprehensible how anyone, who has ever striven for insight [*Einsicht*], can be satisfied with this. To be sure, Schlick also speaks of "acquaintance" ["*Kennen*", in opposition to cognizing, *Erkennen*] as the mere intuitive grasping of the given; but he says nothing of its structure, also nothing of the grounding connections between the given and the meanings giving it expression. To the extent that he ignores intuition, in so far as it ranges beyond the mere modalities of sense experience, he outrightly rejects self-evidence [*die Evidenz*] which is still the sole source of all insight [*Einsicht*].

The polemical passages against phenomenology that so offended Husserl were removed from the second (1925) edition of Schlick's book. Referring to their omission, and to "Husserl's very sharp comments directed against me", Schlick noted that this was in accord with the policy of the new edition, announced in the preface, of "eliminating all nonessential polemical excursions".[22] Nonetheless, Schlick couldn't resist insertion of one last dig against the "*Evidenztheoretiker*": How do we know self-evidence (*Evidenz*) obtains? Is it itself "self-evident"?[23] Ironically, this criticism perfectly parallels that lodged by Otto Neurath against Schlick's own quasi-experiential notion of "*Konstatierungen*" ("affirmations") as the foundation of knowledge later on during the so-called Protocol Sentence Debate of the early 1930s in the Vienna Circle. At that time, Schlick, freely "utiliz[ing] stretches of the Cartesian road" (as did Husserl), held that judgments of perception that univocally correspond to facts are accompanied by "a sense of *fulfillment*, a wholly characteristic satisfaction; we are *content*". For Neurath, such private and inexpressible experiences were entirely "alien to science".[24]

Weyl's affiliation with Husserlian phenomenology is readily apparent in his fundamental works on relativity theory in the period 1917–1923, and also in *Das Kontinuum* (1918), his predicative alternative to classical analysis. Yet it is well known that Weyl gave up his predicative theory in late 1919, in favor of the intuitionism of the Dutch mathematician Brouwer. In fact, Weyl did not so much "join" intuitionism as interpret it in his own distinctive way.[25] In ways that are still not completely transparent, this involved amalgamating phenomenological intuition, an "originary giving" intentional act, with Brouwer's notion of "primal intuition". Then, in the mid-1920s, Weyl turned away from intuitionism to a more favorable view of Hilbert's finitism, ostensibly on the pragmatic grounds that the constructive methods of intuitionism were too restrictive, but also underscoring the epistemological and semantic virtues of holism available within finitism. Noting Hilbert's reference of the similarity of the epistemological situation regarding his ideal transfinite elements in mathematics to that in physics, where evidence and meaning accrue not to particular statements individually but only to "the theoretical system as a whole", Weyl judged in 1927 that Hilbert's finitism appeared triumphant.

> If Hilbert's view prevails over intuitionism, as appears to be the case, *then I see in this a decisive defeat of the philosophical attitude of pure phenomenology*, which thus proves to be insufficient for the understanding of creative science even in the area of cognition that is most primal and most readily open to evidence—mathematics.[26]

One may question whether Weyl's subsequent turning away from intuitionism need have implicated, as he indicated here, "pure phenomenology", since there are considerable differences in the two approaches in their respective accounts of intuition as a source or ground of mathematical knowledge. But this was not

Weyl's last word on foundations, or on phenomenology. Although in later years he advanced a method termed "symbolic construction", its guiding idea, the projection of *being* onto an *a priori* background of constructed symbolic *possibilities*,[27] expresses in its own fashion the central motif of the constitutive procedure of Husserlian transcendental phenomenology (see further below). Weyl consistently understood the theme of "potentiality" in phenomenological terms; after Husserl's death in 1938, this was the topic of an essay contributed to a memorial volume to Husserl.[28] In the last year of his life, as the late reminiscence mentioned above demonstrates, Weyl was still grappling with the principal themes of Husserl's phenomenology, and to this extent, his engagement with Husserl never ended. Over the years, writings of other philosophers—all idealists of one stripe or another—were particularly mentioned, Leibniz, Kant, and even Fichte.[29] While, by his own admission, he was "all-too-prone to mix up his mathematics with physical and philosophical speculations",[30] those seeking a definitive statement of his philosophical position will find that Weyl, in the end, did not settle into any readily definable or antecedently recognized category. Instead, there are muted expressions of a sense that he had not reached any definite conclusions, an outcome certainly abetted by the possession of

an epistemological conscience [*Erkenntnisgewissen*], sharpened by work in the exact sciences, [that] does not make it easy for the likes of us to find the courage for philosophical statement. One cannot get by entirely without compromise.[31]

5.3 Transcendental-Phenomenological Idealism

5.3.1 The "Introduction" to Raum-Zeit-Materie

The first nine pages of *Raum-Zeit-Materie* surely must come as a shock to readers who believe they hold in their hands what the journal *Nature* (in the person of A. S. Eddington) described, on its appearance in English, as "the standard treatise on the general theory of relativity".[32] Few readers prepared to take on a work of its level of mathematical sophistication could be possibly expected also to possess the background for successfully grappling with the "several philosophical discussions" these pages contain. Within them are found the following passages, whose meanings, even if but dimly perceived, seem quite remote indeed from the kind of "philosophy" customarily found in technical monographs.[33]

The real world [*wirkliche Welt*], each of its pieces and all their determinations, are, and can only be, given as intentional objects of conscious acts. Absolutely given are conscious experiences that I have—just as I have them. It certainly is in no way the case, as positivists often state, that these experiences consist in the mere stuff of sensations. Rather in a perception, for example, there is indeed an object standing there incarnately [*leibhaft*] before me to which that experience relates in a wholly characteristic, but not further describable, manner known to everyone. Following Brentano, it shall be designated through the expression "intentional object".

The immanent is absolute, that is, it is exactly what it is as I have it and I can eventually bring this, its essence [*Wesen*] to givenness [*Gegebenheit*] before me in acts of reflection.

The given to consciousness [*Bewußtseins-Gegebene*] is the starting point in which we must place ourselves in order to comprehend the sense and the justification of the positing of actuality [*Wirklichkeitsetzung*].

"Pure consciousness" is the seat of the philosophical *a priori*.

To the *cognoscenti*, Weyl's orientation to Husserlian phenomenology is immediately apparent in the language he used; to others, it is announced in the book's first footnote:

The precise wording of these thoughts is closely modeled upon Husserl, *Ideen zu einer reinen Phänomenologischen Philosophie* (*Jahrbuch f*[*ür*] *Philos*[*ophie*] *u*[*nd*] *phänomenol*[*ogische*] *Forschung*, Bd. I, Halle, 1913).[34]

For those who know the book only in the 1922 English translation by the Australian physicist H. L. Brose, the difficulty of understanding these mysterious incantations is considerably compounded. With excruciating consistency, Weyl's philosophical statements are so garbled that either no comprehensible meaning can be assigned to them at all, or their intended meaning is completely subverted, with the ensuing obscurities and incoherence widely, and unjustly, attributed to Weyl. Yet judging by the extant literature on Weyl, even those who read him in the original German have found these "discussions" too murky or idiosyncratic to merit serious attention and further investigation. This is unfortunate since the expressed philosophical alignment is not a momentary infatuation. Despite many changes in the various editions of other parts of the book, these passages remain unmodified, while additional evidence of Weyl's phenomenological inclination accumulates through the successive editions. One such instance appears midway through the third and all later editions, just following the new sections on "pure infinitesimal geometry". There readers are again reminded that "the true problems occupying us" are those of

attaining insight into the essence [*Wesen*] of space, time and matter, in so far as these participate in the construction [*Aufbau*] of objective reality [*Wirklichkeit*].[35]

No doubt correctly, Weyl went on to observe that this goal is all too readily obscured by "the Flood [*Sintflut*] of formulas and indices deluging the guiding ideas of infinitesimal geometry". Despite such declarations, it would appear that the vast majority of the book's readers and users have considered Weyl's philosophical remarks wholly irrelevant, unnecessary to understanding the book and its exposition of relativity theory. Admittedly, the mathematical and physical content can be extracted and independently studied, a fact explaining the longevity of the book's publishing history. But such selective attention ignores Weyl's declaration that "philosophical clarification" of science, although a task of a completely different kind than that of the individual sciences, "remains a great responsibility". For, he noted, "as things stand today", there is no alternative to allowing the individual sciences to proceed unhindered along the fresh paths opened up by revision of fundamental principles and newly emerged ideas. In this they are guided by the "reasonable motive" of "good faith" in the domain of competence of their particular methods, but, by the same token, they "proceed in this sense, dogmatically" ("*in diesem Sinne dogmatisch zu verfahren*"). It is precisely for this reason that philosophical elucidation is needed, while not obstructing the forward steps of the special sciences and respecting the difficulty of the problems they face.[36] Hence, the incursion of philosophy into this technical treatise is a necessity, not mere window dressing. An eloquent testimony to this effect concludes the introduction:

> All beginnings are obscure ..., from time to time the mathematician, above all, must be reminded that the origins [*die Ursprüngen*] lie in depths darker than he is capable of grasping with his methods. Beyond all the individual sciences, the task of comprehending [*zu begreifen*] remains. In spite of philosophy's endless swinging from system to system, to and fro, we may not altogether renounce it, or else knowledge is transformed into a meaningless chaos.[37]

Much of *Raum-Zeit-Materie*, together with Weyl's related papers on his theory of "gravitation and electricity, on "pure infinitesimal geometry", and "the new problem of space", manifest this striving to attain philosophical comprehension of the general theory of relativity. On recognition of the many clearly visible intimations of transcendental-phenomenological idealism, these texts evince a remarkably sustained attempt to probe the "darker depths" of the "origins" of the objective physical world portrayed in relativity theory through mathematical construction guided by the phenomenological method of "essential analysis". Weyl himself judged his treatment of the "epistemological questions" raised by the new theory as preliminary and tentative, lamenting that he had not been able to provide such answers as would salve his scientific conscience.[38] Even so, there can be little doubt that his attempt to cast illumination in these murky regions was carried out in agreement with the fundamental thesis of transcendental-phenomenological idealism, as stated at the end of §49 of Husserl's *Ideen I* (1913; emphasis in original).

> [T]he whole *spatiotemporal world*, which includes man himself and the human Ego as subordinate single realities is, *according to its sense, a merely intentional being*, thus one having the merely secondary, relative sense of a being *for* a consciousness.[39] It is a being which consciousness posits in its experiences and which, in principle, can be determined and intuited only as what is identical in concordantly motivated manifolds of experience. *Beyond that* it is nothing.

Understood as guided by this beacon, the intent of Weyl's setting of general relativity within the new framework of a "pure infinitesimal geometry" is an exhibition that the mathematically constructed objective world of relativistic field physics in 1918 has the sense of an intentional being, of "a being for a consciousness". His new "world geometrical" framework for that physics is the fruit of a second-order "eidetic analysis", carried out by phenomenological reflection upon the differential geometrical framework of general relativity. Its "purely infinitesimal" basic relations are explicitly anchored in "the given to consciousness", the starting point from which "the sense and the justification" of general relativity's conception of physical reality, according to Weyl, is to be understood.

> The given to consciousness [*Bewußtseins-Gegebene*] is the starting point in which we must place ourselves in order to comprehend the sense and the justification of the posit of actuality [*Wirklichkeitsetzung*].[40]

As I will show in chapter 6 (§6.3), that relativity theory's conception of the physical world admits of a completely geometrical representation is crucial to

Weyl's endeavor. For a "geometrization of physics" will enable him to portray the actual world of physics, the "objective physical world", against the background of a world geometrical domain of possibilities opened up by an eidetic analysis of the concepts of space and of congruence. Weyl's "pure infinitesimal geometry" and so his theory of "gravitation and electricity" are accordingly mathematical-physical constructions primarily motivated by an attempt to understand Einstein's revolution in physical cognition from the standpoint of transcendental-phenomenological idealism.

However, without at least a nodding acquaintance of the animus and central themes of Husserl's phenomenology (ca. 1918), it is virtually impossible to recognize this principal intent of Weyl's recasting of classical field physics in the mold of "pure infinitesimal geometry". To provide this minimal assistance, several preparatory, and largely uncritical, expository steps will first be taken. It first is useful in §5.3.2 to briefly recapitulate Husserl's renewed attack on "philosophical naturalism" that played a central role in transforming the earlier "categorial phenomenology" of the *Logical Investigations* into a transcendental-phenomenological idealism. The resulting criticism of "philosophical naturalism" (or in contemporary terms, scientific realism) provides the background for understanding Weyl's comment, quoted above, about the individual sciences "proceeding dogmatically". Next, Husserl's argument for transcendental-phenomenological idealism, chiefly in §§33–55 of the *Ideen I* (1913), are rehearsed in §5.3.3. Until the late 1920s, Husserl's *Ideen I* is the principal published source of transcendental-phenomenological idealism; it is cited by Weyl not only in *Raum-Zeit-Materie* but also in *Das Kontinuum* and in "*Philosophie der Mathematik und Naturwissenschaft*", all written between 1917 and 1926. Our exposition will provide the requisite context for identifying the transcendental idealist position clearly assumed by the above-quoted remarks from the introduction to *Raum-Zeit-Materie*.

By Husserl's own admission, the language of the *Ideen* was "so difficult, even for Germans";[41] translation inevitably compounds the problem. It must be emphasized at the outset that Husserl's phenomenology is not a completed philosophical system but an imperfectly worked out variety of approaches to the investigation of the phenomena of consciousness, much of the published work consisting of reformulations of an "introduction" to phenomenology. At his death in 1938, Husserl's "constitutive" phenomenology was still very much a work in progress. It is so even today. For these reasons alone, the question of Weyl's faithfulness to Husserlian orthodoxy can scarcely arise. Moreover, as already indicated in Husserl's several letters to Weyl concerning the latter's writings on the continuum and on "pure infinitesimal geometry", Weyl was no mere acolyte but forged his own method, coupling phenomenological eidetic investigation with specific mathematical constructions in framing a "world geometry" for field physics. Husserl himself, under the rubric of "regional ontologies" had only considered the possibility of particular applications in highly abstract general terms. But whether in the original German or in translation, Husserl's philosophy is scarcely, if at all, known to most philosophers and others primarily interested in the mathematical and physical sciences. It is for this reason that the following expository sections are offered. They are intended only for the express purpose of facilitating identification of, and underscoring as vividly as possible, the currents of transcendental-phenomenological idealism salient to Weyl's "epistemological principle of relativity of magnitude", the precursor of the modern "gauge principle". In no way do they pretend to "introduce" or "critique" phenomenology, nor to give anything beyond a selective and even superficial exposure to the guiding ideas of transcendental-phenomenological idealism. But with these pieces in place, I go on to consider in §5.3.4 the transcendental-phenomenological meaning of Weyl's otherwise puzzling reference to coordinate systems as "the unavoidable residuum of the ego's annihilation". This designation gives explicit recognition to the thesis of "transcendental subjectivity", the "purified consciousness" that is the residue of the "phenomenological reduction" and the ground from which all objectivity is "constituted".

5.3.2 Husserl's Transcendental Turn

Even before receiving his Ph.D. on the calculus of variations at Vienna in 1883, Husserl's interests had shifted to philosophy, especially to problems of intentionality as presented in the writings of Franz Brentano. Yet these mathematical beginnings left a deep imprint on the terminology and method of phenomenology, seen particularly in Husserl's guiding ideal of a "philosophy as a rigorous science" and, above all, in his conception of phenomenology as an "eidetic science". The rigorously scientific character of Husserl's philosophical temperament was recognized and praised by thinkers as diverse as Ernst Cassirer and David Hilbert. As late as 1925, Cassirer lauded Husserl as "the leading representative of scientific philosophy in Germany".[42] A quarter of a century earlier, Hilbert, who highly valued Husserl as a "*Naturwissenschaftler*" concerned with "philosophical problems", had been instrumental in bringing him to Göttingen from Halle in 1901, the year of publication of the last volume of his *Logische Untersuchungen*.[43] Primarily because of its strident attack on psychologism in logic, and so its close association with central themes of Frege's anti-psychologism, this book, in J. N. Findlay's 1970 translation of the second (1913) edition, is the only major text of Husserl generally encountered in the education of Anglo-American "analytic" philosophers. In it, phenomenology is presented as a method of "descriptive psychology", a purely descriptive, internal analysis of consciousness and its varieties. In particular, the investigations concern those conscious acts whose intended objects are meanings or "semantic categories", "objects" in the broad sense of predicative discourse, having "categorial structures" cognized through immediate apprehension in a non-sensuous or "categorial" intuition, "given" just as they are, independently of all genetic and other theories.[44] Susceptible to an unintended "platonizing" reading, the "categorial phenomenology" of the *Logical Investigations* predates Husserl's transcendental idealism. The "transcendental turn" occurred in 1906 while Husserl was *extraordinarious* professor of philosophy at Göttingen.[45] The new direction of phenomenology was first publicly announced in a set of five lectures delivered in Göttingen in April and May 1907, entitled *Die Idee der Phänomenologie*; these remained unpublished until 1950, thirteen years after Husserl's death in 1938.[46] While Weyl was in Göttingen in 1907, it is not known whether he attended any of these lectures. If he did, presumably little impact was made for, as noted above, Weyl first came to an appreciation of Husserl's philosophy only after 1911, through his wife Hella.

In these lectures, as well as in a paper of 1910–1911, "*Philosophy as Rigorous Science*",[47] the only published indication of the new orientation of phenomenology prior to the *Ideen I* of 1913, Husserl made especially clear the intended critical role of the new conception of phenomenology. Its "first and principal part" is that of *Erkenntniskritik*, the epistemological clarification and critique of cognition. This task itself has two moments. Positively, it is to "explicate the essential meaning of being an object of cognition", which is to say, in accordance with the thesis transcendental idealism (see further below), of "being an object at all".[48] As critique of cognition, it is to engage in "*the critique of natural cognition* in all natural sciences". This is not a matter of adopting a skeptical stance regarding the "objective truth" of theories of natural science and mathematics, in the commonly and uncritically understood sense of truths about an "objective reality". Any manner of global skepticism of the exact sciences is regarded as simply unreasonable.

> No reasonable person will doubt the objective truth or the objectively grounded probability of the wonderful theories of mathematics and the natural sciences. Here there is, by and large, no room for private "opinions", "notions", or "points of view".[49]

To clearly identify the target of phenomenology's *Erkenntniskritik*, it is necessary to distinguish between what Husserl termed the naive (or "pre-epistemological") "natural attitude" of "*natural thinking* in science and everyday life untroubled by the difficulties concerning the possibilities of cognition", and "philosophical naturalism".[50] The former is a mode of thinking that is ingrained in both the conduct of science and of worldly affairs; explicitly or (usually) not, it assumes a pregiven world of material objects, located in space and time, having properties that are entirely independent of human perception and conception. Within natural science this is the view that

> [t]he nature it will investigate is for it simply *there* . . . as things at rest, in motion, changing in unlimited space, and temporal things in unlimited time.[51]

Certainly it is "the aim of natural science to know these unquestioned data in an objectively valid, strictly scientific matter". In this, recalling Weyl's remark in the "Introduction" to *Raum-Zeit-Materie*, the particular sciences are guided by the "reasonable motive" of "good faith" in the domain of competence of their particular methods. But while the natural attitude is an adequate, and perhaps even necessary, working mode of thought, in unselfconsciously adopting it, the natural sciences proceed, as Weyl noted, "dogmatically". By accepting nature as simply given, the natural sciences display, to use Husserl's term, an "immortal naiveté", exhibited each time recourse is made to experience. This is not to say that natural science does not have, in its own way, a "very critical" account of experience, as little stock is placed in "isolated experience", but a great deal in the methodological arrangement and connection of experiences, according to the interplay of observation and theory. However, natural science does not put experience itself in question; it does not consider "how experience as consciousness can give or contact an object",[52] a specifically epistemological issue whose adequate treatment is accordingly and anti-naturalistically regarded as lying beyond the competency of the methods of natural science. In virtue of this necessary omission, "all natural science is naive in regard to its point of departure".[53] To the extent that natural scientists engage at all in epistemological reflection upon their theories and results, the default and uncritical supposition is that their methods are capable of establishing truths about nature-in-itself, securing knowledge of objects completely transcendent to consciousness.

Still, phenomenology's critical animus is directed not to the "pre-epistemological" realism of the "natural attitude" but rather to its hypostatization in the philosophical naturalism that "dominates the age" posing "a growing danger for our culture".[54] On account of its "*philosophical* absolutizing" of the natural world, an interpretation far outstripping any conclusions following from the experimental methods of the natural sciences, it is deemed "completely alien" to the natural attitude.[55] In maintaining that a "genuinely scientific philosophy" must itself be based upon physical natural science and its methods, this hypostatization of scientific method has both epistemological and metaphysical components. Husserl considered each not only wrong, but "counter-sensical"—absurd. Epistemologically, naturalistic philosophy affirms the *ur*-thesis of scientific realism, that cognition of "nature in itself" is indeed possible through the methods of the natural sciences, and even, in certain areas, wholly or partly achieved. To Husserl, such a claim is literally "counter-sensical" since the very sense of objectivity, of what it means to be an object of a rational proposition, either of prescientific or scientific cognition, can only be made evident or understandable within consciousness itself.[56] On the other hand as a metaphysics, philosophical naturalism "sees only nature, and primarily, physical nature". Thus, it maintains that "whatever is, belongs to psychophysical nature", namely, is either physical, belonging to the unified totality of physical nature, or else is derivatively mental, but a variable depending on the physical. The mental is, then, "at best, a secondary "parallel accompaniment", to the physical and nature is regarded "as a unity of spatio-temporal being subject to exact laws".[57]

Consciousness, values, reason and logic are thus completely naturalized; these are social, psychological, and biological phenomena to be accounted for in empirically attested psychophysical causal laws. As he had already argued at length in the *Logical Investigations*, Husserl regarded naturalism or psychologism as self-refuting; in particular, it led to absurdity in its supposition that that "exemplary index of ideality, formal logic" is rooted in "natural laws of thinking".[58]

Hence, in its critical mode, phenomenology is explicitly antinaturalistic, undertaking to "engage in a radical criticism of naturalistic philosophy".[59] From this vantage point, Husserl's turn to transcendental idealism appears to be a further step necessary to confront the naturalism that, as a philosophical world-view modeled on natural science, is kith and kin to the naturalism and psychologism in logic that was the critical target of the *Logical Investigations*. It is worth noting, however, that Husserl did not consider his antinaturalist transcendentalism as anti-empiricism but rather, in recognizing the origin of cognition in phenomena immediately "given" to consciousness, as the "fulfillment" of the "intentions" of English empiricist philosophy, as he stated in an unpublished lecture at Göttingen in 1908:

My transcendental method is transcendental-phenomenological. It is the ultimate fulfillment of old intentions, especially those of English empiricist philosophy, to investigate the transcendental-phenomenological "origins" . . . the origins of objectivity in transcendental subjectivity, the origin of the relative being of objects in the absolute being of consciousness.[60]

Husserl's transformation of phenomenology from an unfortunately named "descriptive psychology" to transcendental idealism thus extended the earlier critique of naturalism and psychologism in logic to philosophical naturalism generally. The crucial move in this transition is the methodological procedure of the phenomenological reduction, the suspending or "bracketing" or "putting out of action" all of the existential posits of the natural attitude. Considered as a *"transcendental ἐποχή"*,[61] this operation first opens up the "absolute being of pure consciousness", the "residuum of the world's annihilation" (*Residuum der Weltvernichtung*).[62] With it, phenomenology necessarily becomes transcendental inasmuch as phenomenological investigation is concerned to give an exhaustive description of this revealed region of "transcendental subjectivity" together with its structures of intentionality. Consequent to the phenomenological reduction, all reality (*Realität*), ideal as well as actual, is exhibited as having being in virtue of "sense-bestowal" (*Sinngebung*), and indeed, the notion of an "absolute reality" independent of consciousness is as nonsensical as that of a "round square". By the same token, "pure consciousness", the ultimate origin of all "sense-bestowal", "exists absolutely and not by virtue of another (act of) sense-bestowal".[63] It is the ultimate conferee of sense or meaning, the source of all representations, and so of all objectivity. To the argument for these conclusions I now turn.

5.3.3 The Argument for Transcendental Idealism
in Ideen I

On its inception in 1906–1907, Husserl's transcendental-phenomenological idealism was, in the first instance, explicitly an *Erkenntniskritik* concerned to exhibit "the origins of objectivity in transcendental subjectivity". Not altogether surprisingly, a number of students and followers, nurtured on the *Logical Investigations* and its guiding conception of phenomenology as a rigorous method for objective description of "the facts" of the intentional structures of meaning and judgment, refused to follow Husserl's "turn toward subjectivity as the basic phenomenological stratum".[64] Modern critics continue to argue that Husserl's transcendental idealist critique of cognition presupposes a nonnaturalistic Cartesian dualism.[65] While such criticisms can be satisfactorily rebutted,[66] they need not be considered here since our interest is limited to reconstructing those aspects of Husserl's position salient to Weyl's "epistemological principle of relativity of magnitude". As far as can be determined from the relevant texts, these aspects are exclusively epistemological. Moreover, we have it on Weyl's own authority that only later on, after the *Ideen I*, did phenomenology "shyly grope towards the metaphysical idealism that receives its most candid and forceful expression" in Fichte.[67] These later developments have no bearing on Weyl's understanding of phenomenology in 1917–1926. Hence, it is as an epistemology of science opposed to philosophical naturalism, specifically concerned to draw out "the origins of objectivity in transcendental subjectivity", that transcendental-phenomenological idealism played a guiding role in Weyl's geometric reconstruction of relativity theory and our comments are restricted to this conception.

Husserl's basic argument for transcendental idealism, presented in §§33–55 of the *Ideen I*, is neither identified as such nor presented systematically.[68] But as will be shown in §5.4.2, it is essentially recapitulated by Weyl in the introduction to *Raum-Zeit-Materie*, even down to striking similarities of wording. The argument appears in the guise of a detailed meditation on the essential difference between inner and outer perception. The declared point is to motivate and justify the crucial methodological step of the "phenomenological reduction" introduced previously in §32; there, the positing of the reality of the whole natural world, including ourselves and our bodies, is suspended or "put out of action". What remains after the phenomenological reduction, as the "phenomenological residuum" (§33), is pure consciousness, which has an "absolute mode of being" untouched by the phenomenological exclusion. This reduction is the *sine qua non* of phenomenological method, for it is necessary to uncover the phenomena of pure consciousness, the nested structures of intentional acts and their objects that comprise the exclusive field of phenomenological inquiry. But how is it possible to carry through such a reduction? Surely the idea that one could suspend belief in an external world is ridiculous, mere philosophical conceit? In attempting to respond to these objections Husserl, in the considerations that follow, claimed to have "at last done justice to a core of Descartes's *Meditations* (which were directed to entirely different ends)".[69]

In accordance with the theoretical attitude, and its emphasis on the cognizing subject, that dominates the *Ideen I*, Husserl assumed that continuous perception is the usual manner of consciousness, the "normal, wakeful Ego-life"; in later writings this cognitivist assumption is considerably liberalized.[70] Perception is either inner ("immanent", "nonactional") or outer ("transcendent", "actional") (§39). The problem of "pure consciousness" is just the problem of separating out "consciousness itself" ("as a concrete being in itself") from that which is intended in it, from the "*perceived being*". The manner in which this is done shows that there is a fundamental difference in inner and outer perception or, rather, of two modes of "givenness" to consciousness: one immanent, one transcendent.

Briefly, the argument from perception first establishes that the *senses* of being of a physical thing and that of a mental process are completely different since a physical thing is transcendent to the perception of it. This contrast immediately calls for a "deeper inquiry" (§43) into the *sense* of a being transcendent to perception or consciousness, an inquiry that will require distinguishing between the physical thing as "object of the sensuous *imaginatio simpliciter*" and as "object of the physicist's *intellectio*". Since this "deeper inquiry" is not concluded until later (see below), its consideration here may be deferred. But the aforementioned difference in being stems from the fact that a mental process is perceptually given as something absolute, whereas a physical thing is given in perception only "one-sidedly", imperfectly, through various "adumbrations" that can always be extended but never completed. For example, it is impossible to see all sides of a cube at once from a given perspective. In the perception of a physical thing there always remains "a horizon of determinable indeterminateness".

Nonetheless, there is an analogy between inner and outer perception that makes the objects of both capable of being objects of knowledge. Namely, just as with ordinary perception, there is a kind of "seeing" of mental processes, *reflection*, that has the "remarkable property" that what it grasps perceptually is characterized as not only present but also "as something which *already existed before* this regard was turned to it" (§45). Thus, mental processes are a background "ready to be perceived", analogously to unnoticed physical things in the external field of perception. Even so, the analogy is not exact. For the *sense* of perception of a hitherto unheeded mental process is that only a "regard of simple heeding", or reflection, is required. Whereas the *sense* of saying that an unperceived physical thing "is there" is that it is possible to perceive it through concatenations of perceptions, the completion of "continuously harmoniously motivated perceptual sequences", each with its own unheeded background. Moreover, this holds for a plurality of Egos, instead of, as with inner perception, for a single one. Finally (§46), the perception of something immanent to consciousness guarantees the existence of its object, its absolute factual being, a claim obviously not obtaining for objects of outer perception. The *sense* of a "positing" of the world (of non-immanent objects) is accordingly that the positing is contingent, whereas that of the positing of being of the pure consciousness and its stream of mental experiences is not contingent but "absolute".

These analogies and differences between inner and outer perception motivate the crucial methodological step, already introduced, of the phenomenological reduction. The phenomenological reduction, in turn, is necessary to reveal the proper domain of phenomenological investigation, that of the "pure Ego", of transcendental subjectivity, in whose terms the positive account of the "constitution" of physical (and all other) objectivity must be given (see below). But even without the details of the positive account, only sketchily outlined in the last part of *Ideen I*, the argument from the difference in inner and outer perception supports the conclusion of transcendental-phenomenological idealism in the following way. In virtue of the contingency of the positing of the world of physical things, including other humans, there are no limits on "the process of conceiving the destruction of objectivity" of any nonimmanent object of consciousness in the phenomenological reduction. It should be emphasized that all that Husserl seeks to show is that it is *possible* to entertain this conception. But if this "destruction" *can* be conceived, whereas that of the being of consciousness *cannot*, on pain of absurdity, then the *sense* of the being of the "whole spatiotemporal world" is not only entirely different, it is that of a "merely intentional being", a being relative to the "absolute being" of consciousness. Nature, in the sense of the entire world of spatiotemporal transcendent things that are objects or possible objects of empirical scientific cognition, is the intentional correlate of consciousness, a conclusion anathema to philosophical naturalism.

> The existence of a Nature *cannot* be the condition for the existence of consciousness, since Nature itself turns out to be a correlate of consciousness: Nature *is* only as being constituted in regular concatenations of consciousness.[71]

Perhaps it is a contingent fact of human mental development that inquiry seeks to go beyond perceptually presented physical things in order to explain experience through the "truth of physics". But this does not change the circumstance that all truths about physical things, whether perceptually presented or not, *can* pertain only to objects as intentionally thought, conceived, or imagined, objects that can be subject of rational propositions and so about which definite statements can be made. To be an object at all, in the *sense* of a possible object of knowledge or conception, is to be an object of a possible intentional act of consciousness. Accordingly, the truths of physics pertain only to things that are *experienceable physical things*, in the broadest sense of conscious experience: objects or potential objects of cognitive acts within the "determinable horizon" of a subject's "prevailing experiential actuality".[72]

> In the nature of the thing, whatever really is [*was auch immer realiter ist*] but is not yet actually experienced can come to givenness, and this then affirms that it belongs to the undetermined but *determinable* horizon of my prevailing experiential actuality.[73]

Phenomenologically considered, the thing of physics *is* an "intentional unity of sense, continuously persisting in the manifolds of experience", experience taken again in the broadest sense of what is actually or potentially present to consciousness. Since the meaning of the statements of physics is composed, if at all, within that experience, the thing of physics is misleadingly deemed "an object existing in itself"; misleadingly, for according to its *sense*, it is not an object with which consciousness "has nothing to do".

> An object existing in itself is never such that it has nothing to do with consciousness and the Ego of consciousness.[74]

Such affirmations will remind of positivist or verificationist meaning principles. The similarity is superficial on account of the vastly broader phenomenological conception of experience, of what is "given" to consciousness. Whereas positivism restricts experience, and so the "given", to objects of perception and/or observation (either actual or also potential), phenomenology recognizes a wide variety of modes of "givenness" to consciousness, of which "outer" perception is but one. In addition to sensory givenness, phenomenology extends experience, in the form of the "given", to intuition, to the nonsensuous, to "experience in fantasy" or in imagination and, in so doing, locates the source or origin of all cognition in what is given "originarily" in intuition. This had been affirmed in §24 in what Husserl called "*the principle of all principles*":

> That *every intuition originally giving* (something) [*originär gebende Anschauung*] *should be a validating source* [*Rechtsquelle*] *of cognition*, that everything *offered in* "*intuition*" *to us originarily* (so to speak, "in the flesh" [*sozusagen in seiner leibhaften Wirklichkeit*]) *should be simply taken for that as which it presents itself* [*es sich gibt*], but also *only within the bounds in which it presents itself here.*

On account of this extension of intuitive knowing as the basis of all cognition, comprehending also the nonsensuous domain of "pure essences" and so concepts of mathematics, Husserl would proudly proclaim that "we [phenomenologists] are the true positivists".[75] On the other hand, the phenomenological account of experience is immensely richer than that of empiricism or positivism in still another way. As further discussed in §5.4.2, while an object may be "given" in intuition just as it is, "but also *only within the boundaries in which it presents itself here*", it is "given" together with a fluid "halo" of possible relations and connections to other objects of reflective acts, and so possible meanings. In consequence, its characterization relates to this changing but always co-posited "horizon" against which the positing of the object receives its sense. This places a requirement of consistency upon the concatenation of connections of meaning belonging to a physical object. Such an object, as one "transcendent" to consciousness, must necessarily be an object of possible experience not as a mere logical possibility but as a demonstrable unity of sense in the concatenations of experience, and this, not only for a single consciousness, but for any actual Ego.[76]

At the same time, this requirement of concordance does not lead to realism. The harmony in the course of mental experiences of several cognizing subjects cannot be explained by the hypothetical assumption of an "intrinsically foreign" hidden cause of this course of experience, "only indirectly and analogically characterized by mathematical concepts". Because the very sense of a "thing" to the physicist is that of a sign for the collected phenomenal dependencies corresponding to its wealth of causal properties, it is "countersensical" to construe it as signifying something "reaching out beyond the world" constituted by "every ego acting as a cognizing subject".

> The thing, appearing with such and such sensory determinations under given phenomenal circumstances, is *for the physicist* (who *has already attained physical determination in general* for such things in connected appearances of the sort in question), a sign [*Anzeichen*] for the abundance of causal properties of this particular thing, showing themselves in the controlled manner of well-known phenomenal dependencies. What evidently reveals itself here—plainly revealed in intentional unities of conscious experiences—is in principle transcendent. According to all this it is clear that *even the higher transcendence of physical things does not* signify *for consciousness a reaching out beyond the world*, respectively, beyond every ego acting (singly or in an empathetic context) as cognizing subject.[77]

This is not to say that it must be possible to "definitely demonstrate" the existence of physical things in any human experience, for there are factual limits to such experience.

The transcendental idealism consequent upon the phenomenological reduction, is not a "re-interpretation" or a denial of "reality" or "actuality", but the removal of "a countersensical interpretation" of it, that is, an interpretation contradicting the very sense of these terms as that sense is clarified within phenomenological insight. There is nothing wrong with the ordinary, pre-philosophical use of these terms within "the natural attitude". But naturalism as a philosophical attitude, and so distinct from the "natural attitude", "absolutizes" these ordinary meanings so that they purport to designate an "absolute being" of which, naturalism claims, scientific inquiry can and does provide cognition. The countersensical interpretation is accordingly the result of a philosophical absolutizing of "Nature", "Reality", and "World" by a philosophical naturalism or, more familiarly, scientific realism. Realism attributes

> a mythical absolute reality to the being determined by physics, while completely failing to see what is truly absolute: pure consciousness as pure consciousness in its purity".[78]

Thus, "causality, which belongs essentially to the context of the constituted intentional world and has a sense only within that world", is made into "a mythical bond between the 'objective' being determined in physics and the merely 'subjective' being appearing in immediate experience". In "absolutizing Nature as conceived by physics", realism has absolutized the concept of reality, an "intentional correlate of logically determinative thinking". This is an absurdity: "*An absolute reality is just as valid as a round square*". Whereas, according to transcendental-phenomenological idealism, "*all real unities [realen Einheiten] are 'unities of sense' [Einheiten des Sinnes] presupposing . . . [a] sense-bestowing consciousness*", and the terms

> "reality" [*Realität*] and "world" are here simply names [*eben Titel*] for certain valid *unities of sense*, namely, unities of "sense" according to their essence, exactly so and not otherwise, related to certain sense-bestowing connections, exhibiting sense-validity, of absolute, pure consciousness [*sinngebende und Sinnesgültigkeit ausweisende Zusammenhänge des absoluten, reinen Buwußtseins*]".[79]

In affirming that reality or the world is not an "absolute being" but an intentional unity or correlate of pure consciousness, that all objectivity is "constituted" within the "absolute being" of "transcendental subjectivity", phenomenology is obviously a species of idealism. Casual appearances to the contrary, it is not, however, a subjective idealism *à la* Berkeley; anyone who reaches this assessment has not, Husserl claimed, "understood the *sense* of my statements".[80] The difference with subjective idealism is brought out clearly in the difference and the relation in meanings of the terms *transcendental* and *transcendent* and in the "enigmatic" nature of the "constitution' of the latter from the former. In Kant's sense of the term, essentially adopted by Husserl, an inquiry or discipline or body of knowledge is *transcendental* inasmuch as it is concerned "not so much with objects but rather with our mode of cognition of objects insofar as this is to be possible *a priori*" (B25). Phenomenological inquiry into the intentional structures of "pure consciousness" is *transcendental* in precisely this sense. On the other hand, the actual world of physical objects, as regarded within the natural attitude and by physical science, is *transcendent* to consciousness; one may even say that "the physical thing as determined by physics" is "an utterly transcendent thing".[81] Yet this does not alter in the slightest the *transcendental* finding that, considered as objects of mathematical natural science, these objects are constituted in, and tied to, consciousness:

> [T]he transcendence of the physical thing is, in principle, the transcendence of a being bound to consciousness, constituting itself in consciousness. . . .[82]

The "riddle" or "mystery" of cognition concerns precisely the relation of the domain of *transcendental subjectivity* (as revealed in the phenomenological reduction) to that of objects *transcendent* to all consciousness, *rightly and unproblematically assumed within the natural attitude*. Much of the difficulty encountered in reading Husserl (and Weyl) stems from a failure to keep these two perspectives, which in a certain sense are complementary to one another, carefully distinct. As will be seen, the failure to comprehend Weyl's claim that the fundamental posits of his purely infinitesimal "world geometry" have an ideal meaning, only indirectly physical, occasioned much of the misunderstanding of Weyl's theory, and (in chapter 8) directly led to Eddington's "affine field theory". Keeping this in mind, the "mystery" can be approached through the thesis of intentionality, for cognitive mental processes have an intended object ("an *intentio*") to which they refer or are otherwise related, and this activity of intending inherently belongs to the processes whether or not the object does.[83]

> Consciousness (experience) and real being [*reales Sein*] are anything but coordinate modes of being, which dwell peaceably side by side, and occasionally become "related to" or "connected with" one another.... Between consciousness and reality [*Realität*] there yawns a true abyss of sense.[84]

On account of this abyss, "the puzzles, the mysteries, the problems concerning the ultimate meaning of the objectivity of cognition", of "how experience as consciousness can give or contact an object",[85] will always remain to one extent or another. The phenomenological account of the "constitution" of objectivity in transcendental subjectivity is not intended to solve these puzzles or to dissolve the problem, but to ascertain just what can be said about them. It takes away nothing of the "fully valid being of the world" but rather aims to clarify the sense of this being. "The point", Husserl wrote almost at the end of his life, "is not to secure objectivity but to understand it".[86]

5.3.4 Space and Time Coordinates: "Residuum of the Annihilation of the Ego"

In 1922, Oscar Becker completed his *Habilitatationsschrift* in Freiburg im Breisgau on the phenomenological foundations of geometry and its application to physics.[87] A student of mathematics who, after service in World War I, turned to philosophy, especially philosophy of mathematics, Becker would become Husserl's *Assistent* in Freiburg in 1923. A subsequent work, *Mathematische Existenz* (1927), followed the existentialist and "anthropological" direction of phenomenology initiated by Heidegger and further assisted him in attaining a professorship in Bonn in 1931. Like Heidegger, Becker became a Nazi sympathesizer after 1933, while openly opposing Heidegger's "nihilism" and writing articles on such elevated topics as "Nordic Metaphysics". "De-nazified" after World War II, he returned to teaching in Bonn, remaining there till his death in 1964, publishing only articles on the history and philosophy of mathematics.[88] However unsavory his later career, Becker's thesis in 1922 made a distinctive and highly favorable impression on Husserl. According to the latter, informing Weyl of the work in a letter of 9 April 1922, Becker had written "nothing less than a synthesis of Einstein's and your discoveries with my phenomenological investigations of nature".[89] A year later, when his *Habilitatationsschrift* appeared in print (in Husserl's *Jahrbuch*), Becker naturally sent a copy to Weyl. In the introductory remarks, Weyl's influence, as second only to Husserl's, is accorded laudatory recognition. An accompanying letter, dated 12 April 1923, similarly warmly conveyed "deeply felt thanks for the *decisive* scientific stimuli that I have received from your writings on the foundations of mathematics and on relativity theory and the theory of space". Indeed, it was Weyl's work that first "made possible a complete phenomenological foundation for geometry (in the sense of 'world geometry')".[90] Becker goes on to express his conviction that "the same basic idealistic conception" governing his work "also constitutes the background of your theory of the continuum and of your 'pure infinitesimal geometry'". This common "philosophical starting point" is identified as "the principle of transcendental idealism ... out of which the fundamental problem of the phenomenological constitution of nature arises". In evidence of this fundamental agreement, Becker pointed to Weyl's enigmatic designation of coordinate systems as "the unavoidable residuum of the ego's annihilation" (*das unvermeidliche Residuum der Ich-Vernichtung*). That elusive phrase, recurring in insignificantly different variants in several key texts, from 1918 through 1926, does indeed signal Weyl's broad concord with the fundamental thesis of transcendental-phenomenological idealism.

Before 1921, it appeared only in §6 of *Das Kontinuum* (1918), entitled "Intuitive and Mathematical Continuum",[91] an overtly philosophical reflection on the epistemological problem presented by the "deep chasm" between the intuitive and the mathematical continuum. Recall that in Weyl's book, the following position is adopted regarding the mathematical continuum. The set theoretic continuum, and with it its epistemology of Platonism (or, as Weyl later termed it, "naive realism") is rejected. In its place, a new theory of the mathematical continuum is proposed. This is the theory of a purely predicative analysis, whose basic object is the sequence of the natural numbers, while sets and functions of these are constructed through explicit or recursive definitions. Underlying the basic category is a single primitive relation (*Urbeziehung*) $S(x, y)$ whose meaning, *that y is the immediate successor of x*, is immediately exhibited in "pure intuition".[92] In rare agreement with Poincaré ("whose philosophical position I share in so few other respects"), Weyl maintained that "*the idea [Vorstellung] of iteration, that is, of the sequence of the natural numbers, is an ultimate foundation of mathematical thinking*".[93] From the natural numbers, and from certain immediately experienced individual properties and relations of them, predicative principles (prohibiting a *circulus vitiosus*) are laid down for the construction of the rationals, and subsequently for "real numbers", as arithmetically definable lower Dedekind sections in the set of rational numbers; moreover, as admissible "functions" of these are considered only such as can be arithmetically defined according to the construction principles specified at the outset.[94] In this way, a weaker surrogate for the set theoretic real number continuum is constructed on the basis of the intuitively evidenced succession of natural numbers. The resulting theory of the mathematical continuum admittedly involves surrender of treasured parts of classical analysis, such as the principle that every bounded set of real numbers has a least upper bound. Yet such a drastic step can be, according to Weyl, "in very essential measure rationally justified" to the extent it provides mathematics enough for physics. In particular, the arithmetically constructed real numbers and permitted functions of them must be found sufficient for giving an "exact account of what 'motion' means in the world of physical objectivity".[95]

With that task in mind, Weyl turned his attention to the epistemological issues raised by the envisaged rational justification of *Das Kontinuum*'s predicative theory. Yet consideration of the application of the theorems of analysis (such as may be developed in Weyl's arithmetical theory of the continuum) to physics has made glaringly apparent just how wide is the "chasm" between the intuitive continua of space and time, and the mathematical continuum. For these applications ordain a "transfer principle", whereby the theorems of analysis may be taken over into the geometry of physical space and time. This obliges a one-to-one correspondence between real numbers (whether in the sense of the set theoretic continuum or in Weyl's sense) and points of physical space and time. For definiteness, the discussion is initially restricted to "*time*, as the most fundamental continuum" and so to the "chasm" between immediately given phenomenal time, the direct experience of temporal continuity, and objective measurable quantitative time represented by the one-dimensional continuum of real numbers. If one has the epistemological conscience to recognize the primal givenness of phenomenal time, the "chasm" is apparent for there is nothing in the continuum of phenomenal time that can serve as an object of immediate awareness on which to base the construction of the continuum of physical time. In particular, phenomenal time contains nothing corresponding to the immediately evidenced relation between a discretely given number x and its unique successor that underlies the construction of the purely predicative mathematical continuum. Any attempt to directly construct a quantitative continuum of time—physical time—from phenomenally experienced time encounters a fundamental obstacle right at the outset in that the latter possesses an "essential and undeniable inexactness". In it a "point of time" can be approximately, never exactly, determined (in support of this claim, Weyl cites §§81–82 of the *Ideen I*). It would appear that the sole way to overcome this hurdle is to stipulate a counterfactual idealization that points in phenomenal time *can be* displayed in intuition:

> In order to have some hope of connecting phenomenal time with the world of mathematical concepts, let us grant the ideal possibility that a rigidly punctual "now" can be placed within this species of time and that time-points can be exhibited.[96]

If this be allowed, and if it is granted that "earlier", as a relation between two time points, and "equality", as a relation between two time intervals, are both immediately evidenced relations (the latter, Weyl noted, certainly seems questionable), then quantitative time might be constructed from what is directly exhibited in intuition. The evidential grounds of such a construction rest on the basic category of the immediately exhibited "time-points" A, B, A′, B′, and the immediately experienced relations, "A is earlier than B", and "AB is equal to A′B′". But, Weyl conceded, the idealization is not at all admissible. For phenomenal time is presented in consciousness as "an enduring and changing being-now" ("*Jetzt-sein*"). Only in reflection can this "constant now" seem a "flow" in which "points" can be placed. Even so, the notion of an intuitive "flow" of time, on examination, collapses from internal inconsistencies. Indeed, Weyl maintained that the direct experience of phenomenal time, due to its genuine primitiveness, cannot even be described.[97]

The conclusion, therefore, is that exact time points are *not* given in intuition in any manner, and so are not "absolute", but are *concepts* ("the product of reason"), attaining "full definiteness" only in the purely formal "arithmetico-analytic concept of the real number". The same holds for the points of physical space *vis-à-vis* the intuitive continuum of space; in sum, the intuitive continua have been replaced by the exact concept of the real number. Thus, it might seem that the "transfer principle" taking theorems of analysis into physical geometry circumvents entirely the intuitive continua of space and time. That this is not the case is seen in the necessary reference to a coordinate system. While the points of physical space and time can only be defined relative to a coordinate system, Weyl noted elsewhere (see immediately below) that a coordinate system is "exhibited only through an individual act"; it does not, in any sense, "follow" from the objective geometrical axioms. The "exact, conceptual" definition of points of time as real numbers presupposes a one-dimensional directed axis whose orientation is not conceptually, but intuitively, determined. Similarly, only on the basis of three spatially directed axes issuing from an arbitrarily exhibited origin can the totality of points in a given region of space be defined or constructed as a three-dimensional set of these numbers. Of course, the exact definitions of geometrical concepts such as "curve" or "surface" presuppose the exhibition of a coordinate system, as does indeed differential calculus on a manifold. For this reason, the discussion in *Das Kontinuum* concluded, the objective "geometrico-physical world", that mathematically constructed world free of all sensory qualities, still bears unmistakable imprint of its origin in what is given in intuition, in transcendental subjectivity.

> The coordinate system is the unavoidable residuum of the ego's annihilation [*das unvermeidliche Residuum der Ich-Vernichtung*] in that geometrico-physical world which reason sifts from the given under the norm of "objectivity"—a final scanty token in this objective sphere that existence [*Dasein*] is only given and *can* only be given as the intentional content of the conscious experience of a pure, sense-giving ego.[98]

The language is telling. Recall from the preceding section that in transcendental phenomenology, "pure consciousness" is the "phenomenological residuum" following upon the reduction that "brackets" the reality of objects uncritically posited in the natural attitude. In the similar terms of *Ideen I*, it is the "absolute being" that "'constitutes' within itself, all worldly transcendences".

> Instead, then, of living naively in experience and theoretically exploring what is experienced, transcendent Nature, we effect the "phenomenological reduction"... we put all these positings "out of action", we do not "participate in them"; we direct our grasping, theoretically inquiring regard to *pure consciousness in its own absolute being*. That, then, is what is left as the sought-for *"phenomenological residuum [Residuum]"*, although we have "excluded" the whole world with all physical things, living beings, and humans, ourselves included. Strictly speaking, we have not lost anything but rather have gained the whole of absolute being which, rightly understood, contains within itself, "constitutes" within itself, all worldly transcendences.[99]

Just as "pure consciousness" remains when the natural attitude's posit of the reality of a "transcendent Nature" is "put out of action" through the phenomenological reduction, so also *within* that posited reality, namely, the "geometrico-physical world" of physical science, and so beginning *within the natural attitude*, reference to a coordinate system is a reminder that this "worldly transcendency" has been constituted from "pure consciousness".

The considerations on the "transfer" of the theorems of analysis to physical geometry required for a rational justification of Weyl's predicative theory of analysis have shown that physical, quantitatively measurable, space and time have no evidential mooring in the immediately given phenomenal continua of space and time. The points of physical space and time are nothing more than conceptual structures of real numbers or sets of real numbers. But they are only constituted as such on the exhibition in intuition of a coordinate system. The requirement of a coordinate system needed for the application of analysis to geometry is the *residuum* of the "pure, sense-giving ego" and its "immediate life of intuition" in the otherwise completely "geometrico-physical" world of relativity theory. It bears the ineradicable trace of transcendental subjectivity that "'constitutes' within itself" the *sense* of this objective, purely conceptual, world. Of course, according to the principle of general covariance, the choice of coordinate system *is* essentially arbitrary since the laws of nature are to be formulated in tensor form, valid for *all* coordinate systems. Again, for Weyl, choice of a coordinate system, arbitrarily exhibited by an act of the constituting ego, implies as well a local choice of unit length or gauge. For, as I show in chapter 6, §6.2, only the immediate spatial-temporal neighborhood surrounding the "ego center" ("*Ich-Zentrum*") has a "directly clear meaning exhibited in intuition", the relation between the intuitive space and physical space "becoming the vaguer the farther one departs from the ego center".[100]

The envisaged rational justification of his purely predicative analysis thus led Weyl back to the "origin" of transcendental subjectivity. Scrutiny of the required applications of analysis in physics shows that objectivity in physical theory, or at least of all quantities into which dimensions of space and time enter, cannot be obtained without, as Weyl would later put it, "taking subjectivity into the bargain". Lest it be thought that physical quantities, mathematically represented as functions of the space and time (or, space-time) coordinates are without any taint of subjectivity, pertaining to "existences in themselves", Weyl's discussion reminds that these concepts invoke necessary and ineliminable reference to the arbitrary introduction of a coordinate system. Although the choice of a particular coordinate system may in principle be regarded as arbitrary, the dependence of the concept of physical quantity on the notion of a coordinate system is not, at least if one heeds the "deep chasm" between intuitive and the quantitative, mathematical continuum. The notion of a coordinate system bears "the unavoidable residue of the ego's annihilation" in the objective world of physical theory; it is a reminder that "existence is only given and *can* only be given as the intentional content of the conscious experience of a pure, sense-giving ego". This, indeed, an affirmation of transcendental idealism in Husserl's sense that all objectivities are constituted from pure consciousness, the residuum that survives the "phenomenological reduction".

The salient epistemological issue looming here, to which Weyl repeatedly referred as "the problem of relativity" (see further below), concerns precisely the fundamental opposition between the subjective/intuitive and the objective/conceptual, the "abyss" separating intuitive exhibition and precise mathematical concepts. The general significance of "the problem of relativity" is that objectivity in physics, more specifically, the purely symbolic world of the tensor fields of classical relativistic physics, is constituted or constructed only via subjectivity and is not understandable as pertaining to objects of completely mind-independent reality, transcendent to consciousness.

This message is elaborated again in a second occurrence of the *Ich Vernichtung* passage that first appears in the introduction to the fourth edition of *Raum-Zeit-Materie* (1921), in a parenthetical sentence amidst remarks appended in that edition to a discussion of "the essence of measurement" (*das Wesen des Messens*), undertaken in order to make understandable "how mathematics comes to play its role in exact natural science". Crucially, this role is only enabled through the introduction of a coordinate system. In *Das Kontinuum*, as just shown, Weyl had been led to consider how mathematics "plays its role" by attending to what is necessarily presupposed in the attempt to rationally justify his predicative theory of the continuum. At the outset of that inquiry he encountered the "deep chasm" dividing the intuitive and the quantitative continua of space and time. At roughly the same time, Weyl noted in the introduction to the first edition of *Raum-Zeit-Materie*, essentially the same conclusion had been reached, within physics, only on the basis of the theory of relativity:

> Within physics, it has perhaps only become wholly clear through the theory of relativity, that nothing of the essence [*Wesen*] of space and time, given to us in intuition, enters into the mathematically constructed physical world.[101]

What is meant is spelled out later on in the book:

> One may even maintain that only this standpoint [of Riemann and Einstein] gives justice to the circumstance that space and time are opposed to the material content of the world as *forms* of phenomena [**Formen** *der Erscheinungen*]: only the physical state magnitudes can be measured, ... but not the four world co-ordinates that rather are assigned in an *a priori* arbitrary manner to the world points in order to represent the state magnitudes extending in the world by mathematical functions (of four independent variables).[102]

That is to say, because of the general covariance of Einstein's theory (and Riemann's implicit unrestricted use of coordinate transformations in specifying the metric fundamental form), a justification is found for regarding space and time as *forms* of phenomena. From chapter 2 we learned that Cassirer similarly regarded the requirement of general covariance as a decisive "clarification" of the doctrine of the Transcendental Aesthetic. However, Weyl's allegiance to the transcendental-phenomenological account of the origin of objectivity in "the given to consciousness" ("*Bewußtseins-Gegebene*") pulls in a different direction from Cassirer's Marburgian assessment of this requirement as an "form of thought" (*Denkform*) mandating the ordering of phenomena according to coexistence and succession, a "synthetic principle of unity". Pure intuitions of space and time are not altogether jettisoned, but are reinterpreted phenomenologically. In compliance with the starting point of "the given to consciousness", time is recognized as "the primitive form of the stream of consciousness" ("*die Urform des Buwußtseinstromes*") while space is "only a form of our intuition". However, as forms of intuition, both time and space are completely homogeneous. Only through the introduction of real numbers (here taken as unproblematic) can time points and space points be discriminated so that field magnitudes can be represented as mathematical functions of space and time.

As in *Das Kontinuum*, Weyl's illustration here again considered how this mathematical representation is attained, beginning with the intuitive continuum of time, based upon given primitively experienced relations of "earlier than" and "equality" of time intervals, now taken as evidentially unproblematic. For measurement, a unit for time must be chosen, a choice that presupposes, on conceptual grounds, three distinct time points standing in the intuitively experienced relation of "earlier than" (or, "later than") to a given "*being now*" ("*jetzt-seiend*"). Taking as unit of time the "distance" between two of them, the third is then obtained as some real number multiple of this unit. This assignment of numbers enables the construction of points of time generally; it is possible to conceptually fix any point as a time coordinate along an axis based upon the unit. In this way, time made amenable, as a magnitude, for the construction of the objective world of physics. On the other hand, this arithmetization of time is achieved only upon a directed axis "exhibited only through an individual act" of the cognizing, experiencing ego. This is indicative: the exact determination of the concepts of physics obtained through symbolization cannot be accomplished without the introduction of a coordinate system. It must not be thought, however, that such an "objectification", relative to a coordinate system, is absolute:

> But this objectification [*Objektivierung*] through exclusion of the ego and its immediate life of intuition, is not attained without remainder; the coordinate system, exhibited only through an individual act (and only approximately) remains as the necessary residue of this annihilation of the ego [*das notwendige Residuum dieser Ich Vernichtung*].[103]

Elaborating, Weyl emphasized that it is *measurement* that mandates this necessary reference to a coordinate system, and that consequently measurement always gives rise to a "theory of relativity":

> *For measurement the distinction is essential between the 'giving'* [*dem 'Geben'*] *of an object through individual exhibition* [*individuelle Aufweisung*] *on the one side, in conceptual ways on the other.* The latter is only possible relative to objects that must be immediately exhibited [*unmittelbar aufgewiesen*]. That is why a *theory of relativity* is perforce always involved in measurement.[104]

For measurement, and thus for linking the mathematics of physical theory to observation and experiment, a coordinate system must be immediately exhibited with respect to which any given individual object O can be singled out and referred to, with arbitrary exactness, from a continuously extended object domain. A "theory of relativity" then establishes what "lawful connection" exists between the coordinates of one and the same arbitrary object O in two different coordinate systems. It is also in this sense of a "theory of relativity" that Weyl referred to his theory of gravitation and electromagnetism, with its requirement of "gauge invariance", as a "broadening" (*Erweiterung*) of relativity theory. For Weyl, these presuppositions of measurement disclose the subjective ground of physical objectivity according to which, as we have seen, "existence [*Dasein*] is only given and *can* only be given as the intentional content of the conscious experience of a pure, sense-giving ego". The very sense of space and time magnitudes (and so all physical magnitudes that are functions the space and time coordinates) cannot entirely exclude reference to the constituting acts of pure consciousness. Necessary reliance upon a coordinate system, posited by the willful act of a cognizing consciousness, belies the dogmatic assumption that the objective physical world portrayed in general relativity, a world in which "the physically real is conceptually fixed, in all its determinations, through numbers",[105] lacks any trace of (transcendental) subjectivity.

Subsequently, in his *Philosophie der Mathematik und Naturwissenschaft* (1926), Weyl returns several times to "the problem of relativity", in one instance repeating the *Ich Vernichtung* passage of *Raum-Zeit-Materie* almost *verbatim*.

> On the basis of objective geometrical relations, with which the axioms are concerned, it is not possible to determine a point absolutely, but conceptually only *relative* to a coordinate system, through numbers. For understanding the application of mathematics to reality the distinction between the "giving" [*dem "Geben"*] of an object through individual exhibition on the one side and, on the other in conceptual ways, is fundamental. The objectification through exclusion of the ego and its immediate life of intuition [*Objektivierung durch Ausschaltung des Ich und seines unmittelbaren Lebens der Anschauung*] is not attained without remainder. The coordinate system, exhibited only through an individual act (and only approximately), remains as the necessary residuum of this annihilation of the ego [*das notwendige Residuum dieser Ich-Vernichtung*].[106]

In still another, Weyl made especially vivid his principal claim that the "objective" world portrayed in theoretical physics always carries with it a necessary "relative" amendment, whereas the "absolute" is the sole possession of "subjective" immediate experience.

> Immediate experience is *subjective* and *absolute*; even as hazy as it may be, it is given in its very haziness as it is and not otherwise. On the other hand, the objective world . . . which natural science attempts to crystallize out of our practical lives—through methods that are the consistent development of those criteria according to which we construe reality in the natural attitude of daily life—this objective world is necessarily *relative*; it is only representable in a determinate manner (through numbers or other symbols) after a coordinate system is arbitrarily introduced into the world. This oppositional pair: *subjective-absolute* and *objective-relative* seems to me to contain one of the most fundamental epistemological insights that can be extracted from natural science.[107]

The epistemological thesis that cognition in the exact natural sciences is the joint ("relative") product of objective characterization in precise mathematical concepts, *and* the subjective "immediate life of intuition", belongs to transcendental-phenomenological idealism. In its terms, the very *sense* of objectivity is constituted within "transcendental subjectivity", and accordingly, there must indeed be a vestige within the "objective world" represented by the mathematical/conceptual theories of physics of its "origin" in the "absolute" being of the given-to-consciousness. Through reflection on the applications of mathematics in physics, Weyl recognized an ineliminable trace of the subjective source of all objectivity in the arbitrary fixing of a local coordinate system. In the light of general covariance and "the epistemological principle of relativity of length", this posit is, in itself, completely without any objective physical significance. As is apparent from the 1949 edition of *Philosophy of Mathematics and Natural Science*, and other late texts, Weyl ever afterward located the central problem of epistemology in "the problem of relativity", of presentation in intuition as against conceptual determination. In maintaining that the objective relies upon the subjective, whether upon acts of intuitive "seeing" or of willing, or free constructive operations, Weyl remained continually critical of any realism laying claim to cognition of an objective world "which has nothing to do with consciousness".

A more refined treatment of "the problem of relativity", of intuitive givenness and conceptual determination, appears in chapter 6, §6.3.2, in considering Weyl's group-theoretic justification of the infinitesimal "Pythagorean" metric of any manifold suitable to represent space-time. There I will show that, whereas in Kant transcendental idealism rested on the central claim of the Transcendental Aesthetic, that space and time are *a priori* forms of intuition, necessary conditions of the possibility of experience of objects of outer sense, Weyl truncated this claim in two ways. Space and time are indeed forms of intuition, but intuition's "vision" illuminates only a homogeneous "infinitesimal" region wherein space, separated from time by consciousness, can be mathematically and so conceptually represented as a tangent vector space. Furthermore, in light of the "new problem of space" posed by the general theory of relativity, even this posit of a homogeneous infinitesimal region is regarded as requiring a rational justification in mathematical/conceptual terms.

5.4 Phenomenological Method

In this section I review the salient methodological aspects of transcendental-phenomenological idealism that may be seen in play in the foundations of Weyl's "pure infinitesimal" geometry and in its use as a "world geometry" for gravitation and electromagnetism in this chapter. In §5.4.1 the phenomenological conception of *Evidenz* is introduced as a "fulfilled intention", "the 'experience' of truth" on which all cognition must ultimately rest. In §5.4.2 the central notions of *essence* (*Wesen*, *Eidos*) are summarized together with the method of identification and exploration of these ideal objectivities through "essential insight" and the corresponding "essential analysis". Finally, §5.4.3 briefly introduces the conception of a "regional ontology" that belongs to Husserl's theory of science. For a given domain of empirical investigation, a "regional ontology" comprises the space of *a priori* possibility, the "pure world" of imaginative objectivities that bound the "constitution" of the actual objects of that domain.

5.4.1 Evidenz

Well-known translational difficulties beset the attempt to find a suitable English equivalent for the phenomenological term *Evidenz*. The obvious candidate *evidence*, while not quite a false cognate, wrongly suggests intersubjectively manifested proof or grounds for belief. But certainly this kind of intersubjectivity, obviously taken for granted within the natural attitude, cannot be fundamental in a phenomenology concerned to show the origins of objectivity in transcendental subjectivity. Rather, it must be established later on, somewhat as Carnap, in §§148–149 of the *Aufbau*, constitutes an intersubjective world from the quasi-phenomenological standpoint of "methodological solipsism".[108] Neither is the English term *self-evidence* completely accurate, for it lacks the connotation of intentional achievement stemming from the coincidence of the object as presented in a "fulfilling act" with the object as intended.

Fundamental to Husserl's account of intentionality, the notion of *Evidenz* is already prominent in the *Logical Investigations*, where it is introduced in distinguishing between the real, and the ideal "theory of *Evidenz*". This distinction is required by the existence of intentional acts revealing the ideal meaning or sense of expressions or judgments (e.g., that "S is P"); such meanings are justified to the extent that they are present as "fulfilling acts" to that immanent vision that Husserl there termed "Ideation". In particular, empiricism and philosophical naturalism are accused of grossly misunderstanding the ideal character of justification of true judgments of logic or mathematics. Considering an arithmetical propositions such as "$2 + 1 = 1 + 2$", Husserl argued that the *Evidenz* regarding such judgments is not at all a feeling causally appended to the judgment or, generally, to any judgment in the class of true judgments. Rather, the particular phenomenological content of such a "fulfilling act" is entirely distinctive for it contains "the 'experience' of truth":

> *Evidenz* is rather nothing but the "experience" of truth [*das "Erlebnis" der Wahrheit*].[109]

It is noteworthy that Weyl reverted to this phrase verbatim in *Das Kontinuum*, in an oft-cited passage where he takes issue with Dedekind's famous injunction that belief, in science, should only be accorded to what is actually proven:

> As if such an indirect collocation of reasons as we designate as "proof" is capable of arousing any belief without our securing the justification of each single step in immediate insight [*unmittelbarer Einsicht*]. This (and not the proof) generally remains the ultimate source of justification of knowledge; it is the "experience of truth" [*das "Erlebnis der Wahrheit"*].[110]

For Husserl, as for Weyl, truth itself is "an idea" (*eine Idee*) encountered in the actual experience of particular judgments that have an originary givenness (*originärer Gegenbenheit*) in "immediate insight"; in fact, the evident judgment itself is just the consciousness of this primal or originary givenness. However murky this appears to contemporary philosophical fashion, the ideal theory of *Evidenz* in the *Logical Investigations* was a crucial weapon in Husserl's attack on psychologism in logic. In the subsequent transcendental orientation of phenomenology, it remained the epistemological foundation of all phenomenological description. Phenomenologists are enjoined to follow the "norm"

> to avail ourselves of nothing but what in consciousness itself we can make essentially evident in its pure immanence [*in reiner Immanenz uns wesensmässig einsichtig machen können*].[111]

In the years between 1917 and 1923, when working on and arguing for "pure infinitesimal geometry" as a basis for field physics, Weyl held to this overtly phenomenological conception of a foundation for mathematical and physical cognition. Thus, in a section of *Raum-Zeit-Materie* entitled "Report on non-Euclidean geometry", noting the warning of Proclus against the "misuse" made of appeals to self-evidence in the justification of Euclid's fifth (parallel) postulate, he remarked:

> But one also must not grow weary of emphasizing that, despite its many misuses, self-evidence [*die Evidenz*] is the ultimate anchoring ground of all knowledge, even of empirical knowledge.[112]

Similarly, it was seen in §5.2.1 that it was Schlick's dismissal of any intuitive basis for cognition that particularly provoked Weyl's ire in his 1923 critical review of Schlick's *Allgemeine Erkenntnistheorie*.

> According to Schlick, the essence (*Wesen*) of the process of cognition is exhausted by [a purely semiotic conception of knowledge]. To the reviewer, it is incomprehensible how anyone, who has ever striven for insight [*Einsicht*], can be satisfied with this. To be sure, Schlick also speaks of "acquaintance" ["*Kennen*"] (in opposition to cognizing [*Erkennen*]) as the mere intuitive grasping of the given; but he says nothing of its structure, also nothing of the grounding connections between the given and the meanings giving it expression. To the extent that he thus ignores intuition, insofar as it ranges beyond the mere modalities of sensory experience, then he outrightly rejects self-evidence [*die Evidenz*] which is still the sole source of all insight [*Einsicht*].[113]

By 1926, Weyl showed many signs of relinquishing an exclusive basis for cognition in intuition, in what is "given to consciousness", a foundation to some extent common to both phenomenology and to Brouwer's intuitionism in mathematics, a doctrine Weyl deemed "idealism in mathematics thought through to the end".[114] In part, the change appears to have been the result of a pragmatic decision in the light of intuitionism's general failure to recapture significant parts of classical analysis. But above all, Weyl seems to have been impressed by the emphasis Hilbert laid, in arguing for the use of ideal assumptions in his "metamathematics", on the analogous holist character of knowledge in theoretical physics.

> [the] individual assumptions and laws [of theoretical physics] have no separate fulfilling sense immediately realized in intuition [*in der Anscuauung unmittelbar zu erfüllender Sinn eigen*]; in principle, it is not the propositions of physics taken in isolation, but only the theoretical system as a whole, that can be confronted with experience. What is achieved here is not intuitive insight [*anschauende Einsicht*] into particular or general states of affairs and a faithfully reproduced *description* of the given [*das Gegebene*], but rather theoretical, ultimately a purely symbolic, *construction* of the world.[115]

While consideration of the matter goes beyond the concerns of the present book, some caution may be urged once again in assessing the apparent assimilation of phenomenology and intuitionism in such passages for the following reasons. First, Weyl returns several times in later writings to the language and, apparently, the standpoint of transcendental phenomenology. Second, it is by no means the case that the intuitionistic and phenomenological conceptions of intuition exactly coincide, nor that phenomenology cannot accommodate the epistemological *cum* semantic holism of space-time geometry coupled with mechanics and physics, already explicitly recognized in the first edition of *Raum-Zeit-Materie*.[116]

5.4.2 *Essence* (Wesen/Eidos); *Intuition of Essence* (Wesensanschauung); *Essential Seeing* (Wesenserschauung); *Essential Insight* (Wesenseinsicht); *Essential Analysis* (Wesensanalyse)

A particularly comprehensive but compressed statement of the nature and method of transcendental phenomenology, one that at once serves also to illustrate the difficulties of entering into the semantical web of Husserlian neologisms, occurs in §66 of the *Ideen I*:

> In phenomenology, which is to be nothing other than a doctrine of the essences [*Wesenslehre*] within pure intuition [*reiner Intuition*], we therefore carry out [acts of] immediate essential seeing [*unmittelbare Wesenserschauungen*] on exemplary givenesses [*exemplarischen Gegebenheiten*] of transcendentally pure consciousness, and fix them *conceptually*, or else terminologically.

The crucial methodological pieces of the Husserlian context requisite to understanding Weyl's conception of a "world geometry" revolve around the notions of *essential seeing, eidetic intuition,* and *eidetic analysis.* Each of these pertains to the basic form of phenomenological method for exploration and investigation of the "pure possibilities of consciousness" uncovered by the phenomenological reduction. The fundamental notion is that of *essence,* an admittedly "new kind of object" that, as the datum of eidetic intuition or essential seeing is analogous to the individual object, the datum of empirical intuition.

Essential seeing [Wesenerschauung] is also precisely intuition [Anschauung].[117]

On the other hand, the method of eidetic intuition is admittedly closely modeled on, indeed considered an extension of "the mathematical style of thinking" directed to, the relationship between individuality and universality, actuality and possibility, the contingent and the necessary, experience and "pure fantasy". In the phenomenological account of intentionality, an individual object of attention or, more particularly, of perception is a never-repeatable "this-here" (*Dies-da*), immediately apprehended or "seized upon" as a bare particular. Yet this mode of apprehension does not give the object "in itself", as a "thus and so". As so qualified, the object has a "specific character"[118] that can be determined by attending to its given accompanying background, its "horizon" of more or less unthematized possibilities of further meanings and semantical or logical connections. Each of these in turn may become an object of pure regard or reflection; considered collectively, they comprise a field of "eidetic possibilities".[119] Within this field lies the essence or *eidos* of each individual object, "its stock of essential predicables which must belong to it making it *necessarily* the thing that it is", and so that determines what "other, secondary, contingent determinations can belong to it".

Experience pairs an essence (*Wesen*) to each particular object, a "this-here" (*Dies-da*), an essential necessity with each factual contingency. Thus, all objects (not only perceived objects) are experienced as instantiations of certain categories or genus forms, *material thing, color, shape, tone,* and so on, comprising a hierarchical framework within which experience unfolds. In view of necessary (formal, or analytic) relations of conceptual entailment, of genus and species, part and whole, semantic compatibility and incompatibility, and so on, between these forms, even this simple perceptual experience has a formal *a priori* structure. This purely formal structure gives rise to such "eidetic truths" as, for example, "All material bodies have spatial extent", while its framework of nonformal connections given rise to such synthetic eidetic truths as "All actual bodies are heavy".[120]

According, therefore, to the *sense* of what a contingent being *is*, every factual contingency has an essence that, expressed in terms of its stock of essential predicables, gives rise to various formal or synthetic *a priori* or "eidetic truths" about it:

> it belongs to the sense [Sinn] of anything contingent to have an essence [Wesen] and consequently to have an eidos to be apprehended purely; and this eidos *comes under eidetic truths of various levels of generality*.[121]

In short, its essence or *eidos* makes the given individual object rational. Essence or *eidos* is the necessary but expandable sense structure against which the contingency of the given can be contrasted, determining the sense in which any particular matter of fact "could have been otherwise". In transcendental phenomenology, knowledge of actuality (*Wirklichkeit*) presupposes knowledge of the space of possibilities that is bounded by the eidetic structures of experience.

> The old ontological doctrine that knowledge of "possibilities" must precede knowledge of actualities [*Wirklichkeiten*] is, in my opinion, so far as it is correctly understood and made use of in the right ways, a great truth.[122]

"Correctly", that is, transcendentally, "understood" means that the "constitution" of an actual object of knowledge, in particular, those of the empirical sciences, must presuppose eidetic structures of experience, the space of possibilities within which alone the essence of the actual object lies. The aggregate of eidetic possibilities (*a priori* objects of possible experience) and of objects of actual experience comprises "the world".

> The world is the "sum-total" of objects of possible experience together with objects of experiential cognition, objects of actual experiences cognized in correct theoretical thinking.[123]

A "world geometry", in Weyl's sense, will then represent, through the constructions of a "pure infinitesimal geometry", a space of eidetic possibilities constituting the unity of sense of the world that contains the actual particulars of field physics.

The simplest cases of eidetic intuition, "seeing" an essence, may arise from within perceptual experience, as may be seen by returning to, and filling out, Weyl's discussion of the perception of a chair. This example, from the introduction to *Raum-Zeit-Materie*,[124] briefly recapitulates the argument for transcendental idealism in *Ideen I* surveyed above in §5.2. The chair is given in perception as an individual object together with all of its contingencies and particular perceptual qualities: location, shade of color, having this physical shape, texture, and perhaps odor, and so on. Perceptual experience of this object does not, however, consist simply in the contingent bundle of properties that it now possesses. Although positivism may maintain such an account of experience, it is deemed incoherent by Husserl and Weyl as ignoring the intentional character of perceptual experience:

> It certainly is in no way the case, as positivists often state, that these experiences consist in the mere stuff of sensations. Rather in a perception, for example, there is indeed an object standing there incarnate [*leibhaft*] before me to which that experience relates in a wholly characteristic, but not further describable, manner....

By directing attention to this perception in a certain way, namely, in a free act of reflection, the perception itself, or any of its aspects, for example, its "style", the shape of a stain on the seat cushion, the recollection of a similar chair, and so forth, can become the object of a new *inner* perception. This new intentional object, together with its act of inner perception, are *immanent* to consciousness, given just as they are, a completely inherent (*reelles*, not "real" as in the Brose translation of *Raum-Zeit-Materie*) component of the flow of consciousness. In its immanence, it contrasts to first object, the chair given in the primary act of perception and so not an object inherent to consciousness but a contingent object (perhaps it was involuntarily noticed, standing in a corner) *transcendent* (not "transcendental" as in the Brose translation) to consciousness. The point is this: An object *immanent* to consciousness, an object of an inner perception or act of reflection, is given *absolutely*; it *is* exactly as I have it *now* in the stream of my consciousness; in this regard " 'pure consciousness' is the seat of the philosophical *a priori*". On the other hand, an object *transcendent* to consciousness, *given* in perception, is given inexactly, one-sidedly or otherwise inadequately, and never "as it is in itself"; such objects have only a *phenomenal* being. Whereas an immanent object has "an *essence* [*Wesen*] that I can eventually bring to immediate givenness [*Gegebenheit*] through acts of reflection," "philosophical clarification of the thesis of reality" reveals that it is the "essence" of nonimmanent objects to have an "inexhaustible content", that the *sense* of their being is that of a "limit idea", and in this lies the empirical character of all knowledge of reality.

> [P]hilosophical clarification of the thesis of actuality [*Wirklichkeitsthesis*] reveals that not one of the experiencing acts of perception, memory, and so on, ascribing existence [*Existenz*] to the perceived object [*Gegenstande*] possesses an ultimate justification...; this justification can always be overturned on the grounds of other perceptions. It lies in the essence of a real thing [*Wesen eines wirklichen Dinges*] to be inexhaustible in its content, to which we can approach only through always new, in part conflicting, experiences whose harmonization [*Abgleich*] is unlimited. In this sense, the real thing is a limiting idea [*Grenzidee*]. On this rests the empirical character of all knowledge of reality.[125]

In general, each act of reflection, or eidetic intuition, is a kind of "seeing"; the datum of eidetic intuition is an essence that itself may be categorially complex, having its place in a hierarchy according to generality and specificity, and ranging from *highest genus* to *eidetic singularities*. Although introduced in analogy with empirical intuition, eidetic intuition has a considerably richer structure and meaning than does the perceptual or merely imaginative representation of an object. Beginning with the simple perceiving, or intuition, of an individual object, an "eidetic insight" (*Weseneinsicht*) that objective reflection on the presenting (*gebende*) consciousness is always possible transforms this experience into "eidetic seeing (ideation)" ["Wesenschauung"]. The data of "eidetic seeing", the phenomena of reflection, "are in fact a sphere of pure, and perhaps, perfectly clear givennesses [Gegebenheiten]".[126] Within this sphere, different categories of essences are distinguished, it seems, by the clarity and distinctness of their "givenness" in intuition, as to whether the "seeing" (Erschauung) is "inadequate" and "one-sided" or "adequate". But whereas an object given in empirical intuition is always given more or less "inadequately", an essence in eidetic intuition can be given "originarily and perhaps even adequately" yet its givenness implies "not the slightest posit of any individual existence [Daseins]".[127]

The guiding idea is of an intuition that "gives originarily", offering the thing itself, an essence or eidos, and not a "representation", a complete harmony between what is intended and what is actually present to consciousness. It is a "seeing" of essence, an "essential seeing" (Wesenserschauung), also termed "essential" (or "eidetic") insight" (Wesenseinsicht), in the experienced necessity and unconditional validity of its immediate transparency to consciousness. The datum of eidetic intuition is a pure essence, a structure of transcendental subjectivity that is immediately present to the mind. What is actually present to consciousness is itself a phenomenon for further analysis and definition. It is the ultimate court of appeal of all knowledge, and indeed the only source of what can be absolutely known, which is what is immediately and directly present to consciousness. Essential intuition is thus the mode of investigation demanded by Husserl's ideal of the ultimate "origin" of cognition in what is given to transcendentally pure consciousness. The task of phenomenology is then to analyze and "conceptually" or "terminologically" fix the essential structures of transcendental subjectivity given through successive acts of "essential seeing".

Essential seeing itself displays the characteristic *noetic-noematic* structure of intentional acts, a subjective structure as an *act*, but an objective structure as an act *intending an object*. As new intentional objects appear with each act of reflection or eidetic insight, an "essential analysis" (Wesensanalyse) is called for to conceptually fix or otherwise determine the levels of structure of each successive act (noesis) and of each act's object, its noema. In view of many misunderstandings on this point, it may be helpful to note that in taking essences as its objects of investigation, transcendental phenomenology proclaims itself a "descriptive science of essence" (deskriptive Wesenswissenschaft) and proclaims metaphysical neutrality, neither asserting nor denying the nonimmanent reality of essences. In this regard, one of several salient analogies, transcendental phenomenology bears comparison with the metaphilosophy of the *Aufbau*, as Carnap indeed recognized.[128]

5.4.3 Eidetic Science; Regional Ontology

Husserl's conception of phenomenology as a "doctrine of essence" or, the more frequently encountered, "pure eidetic science" derives, clearly, from his attempt to model phenomenological inquiry on the modes of thinking characteristic of the "pure eidetic sciences" of pure mathematics: geometry, arithmetic, analysis, and so. "Pure phenomenology" is then a science that is "purified" of any assertions about empirical actuality. This means that the "pure phenomenologist", just like the mathematician, abstains in principle from judgments concerning the actual, dealing only with ideal possibilities and their related laws. Through a method, later termed "eidetic variation", the pure ideal possibilities of an object, the *eidos* or essence alone necessary of the object, is "grasped" or "seized upon" (*Wesenserfassung*) by determining what is universal, *invariant* and pervasively identical, through every imaginable particularization constituted as a *possible* actuality. It is this "mathematical style of thinking", Husserl wrote, that "orient[ed] our concept of the *a priori*".[129] Thus, the pure geometer (to take a favorite example of Husserl), while employing particular figures or constructions as examples, abstracts from the particulars of a given case to turn it into an arbitrary example. From a few basic concepts, such as point, line, plane, angle, and so on, the geometer can "derive purely deductively *all* the spatial shapes 'existing'", that is, all ideally possible shapes in space and their relationships. This is to "imagine them in a world of fantasy", thematizing "pure spatial structures" according to their "ideal possibilities" — "predicatively formed eidetic affair-complexes". Accordingly the grounding act of geometrical cognition lies not in observation and experiment—"predicatively formed actuality-complexes"—but in *"the seeing of essence"*.[130] Indeed, at one point Husserl considered whether phenomenology might be constituted as a "'geometry' of mental processes",[131] although it is admittedly quite distinct from the "formal eidetic sciences" of pure mathematics that treat of "exact essences", essences completely determinable in virtue of precise concepts and pure deduction.

All eidetic sciences are founded upon "the seeing of essence", in acts of "making present" or "presentation" (*Vergegenwärtigungen*) of ideal possibilities created in imaginative free fantasy. Such acts of "essential insight" accordingly have" *a position of primacy over perceptions*"; in fact, presentation "can be so perfectly clear that it makes possible a perfect seizing upon essences and a perfect eidetic insight".[132] But while "pure eidetic sciences" are "pure sciences of essence", the mode of eidetic investigation of giving pride of place to imaginative variation within "free phantasy" and to "eidetic insight" can be extended to the "non-immanent" (to consciousness) essences of the empirical sciences. This is because "every species of being has, owing to its essence, *its* modes of givenness and with that its own mode of cognition".[133]

The eidetic or *a priori* disciplines thus fall within two broad categories: so-called formal ontology and regional ontology. "Ontology" signifies, in each case, the field of pure *a priori* possibility and necessity within which all objects are constituted in "transcendental subjectivity". Formal ontology is inspired by the Leibnizian idea of a *mathesis universalis*. It is the eidetic science of "objectivity as such", comprising the disciplines of formal logic (apophantics), mathematical logic and set theory, the theory of semantic categories, part–whole relations, intensional entailments, and the like. Cognitions pertaining to formal ontology are analytic and hold of all objects whatsoever. A "regional ontology", in contrast, is an eidetic science of a particular material discipline, such as (parts of) physics or chemistry. So, to the various sciences of matter of fact, of actuality, correspond different eidetic sciences or "regional ontologies". Each eidetic science of this kind is oriented toward, and explores, the interdependent connection between individual object and nonimmanent essence in its domain, a necessary composition in both directions.[134] Such a regional essence determines "synthetic" (as not belonging to "formal ontology") eidetic truths; the set of them makes up the regional ontology. Specified through regional axioms and concepts, a regional ontology is the locus of synthetic *a priori* cognitions concerning the region in question. Regional axioms, described by Husserl as functionally equivalent "despite considerable differences in basic views", to Kant's "synthetic cognitions *a priori*",[135] determine the set of regional categories. These concepts express that which is particular to the regional essence or express with eidetic universality that which must belong, *a priori* and "synthetically", to an individual object within the extension of the region.

In the *Ideen I*, the empirical sciences are accordingly regarded as having their ideal foundations in a "regional ontology" conceived as an eidetic science of the basic concepts of the various "regions" of scientific inquiry and investigation. In virtue of the essential connection between factual existence and eidetic possibility, eidetic investigation of these concepts clarifies and terminologically fixes them, providing insight into their "essential structures" of relationships to other concepts and meanings. In this way, a regional ontology becomes the ideal essential structure for the "rationalization" of factual existence, corresponding to the "idea of a completely rationalized science of experience". Indeed, the idea of a "regional ontology" is an extrapolation from the birth of modern science in the 17th century, where the idea of a mathematization of nature was first posed.[136] The epistemological significance of such a "rationalization of nature" lies in giving explicit recognition that the cognition of possibility proceeds cognition of actuality, although this can only be established retrospectively by epistemological critique, through phenomenological investigation of the empirical researches of the particular sciences themselves. Hence, this use of eidetic analysis takes empirical science as given within the scope of the "natural attitude". It comprises a manner of "applied phenomenology", distinct from "pure phenomenology's" task of eidetic analysis of the transcendentally purified consciousness surviving the phenomenological reduction". We now turn, in chapter 6, to Weyl's ideal of a "purely infinitesimal world geometry" for gravitation and electromagnetism, intended as a "regional ontology" for field physics.

6

WEYL'S "PURELY INFINITESIMAL" CONSTITUTION OF FIELD PHYSICS

As far as I see, all a priori statements in physics have their origin in symmetry.

Weyl (1952, 126)

6.1 Introduction

With the background provided in chapter 5, we are now in a position to recognize the particular stages of Weyl's elaboration of the "task of philosophy", to "comprehend the sense and the justification of the posit of actuality" to the space-time world portrayed by the general theory of relativity. In accordance with the thesis of transcendental phenomenology, where "constitution" has the meaning of "sense bestowal", the guiding aim is to exhibit the *sense* of that objective world as that of a "being for a consciousness". For Weyl, this is to say, that "each of its parts, and all determinations in them, are, and can only be given as intentional objects of acts of consciousness". An epistemological reflection of this kind must assume the fundamental correctness of Einstein's theory of gravitation, its "posit of reality", and so must *begin within* the "natural attitude". The default assumption that the Einstein theory portrays an objective mind-independent space-time reality is unobjectionable for the purposes of physics, in its exclusive task of the investigation of nature. But it appears "dogmatic" in the light of critical reflection that finds, already in measurement's necessary reliance upon a coordinate system, an ineradicable trace of transcendental subjectivity within the "geometrico-physical world that reason sifts from the given under the norm of 'objectivity'". Moreover, according to the general covariance of Einstein's theory, such a system is essentially arbitrary, yet even lurking here is a further bit of dogmatism. Choice of a coordinate system entails a choice of unit of scale, but the Riemannian "infinitesimal" geometry underlying Einstein's theory contains the tacit supposition of a global unit of scale, enshrining an epistemological inconsistency that prohibits direct comparisons of direction "at a distance" yet condoning such immediate comparison for magnitudes. Thus, while Husserlian phenomenology indeed sets the epistemological framework of Weyl's reconstruction, its particular steps, mathematical constructions, and especially its "style" manifest very much Weyl's own creative endeavor to interpret and implement the method of phenomenological investigation in the context of a fundamental physical theory.

The centerpiece is Weyl's reconstruction of the metrical concept of congruence, a concept that, on epistemological grounds, must be grasped "purely infinitesimally". The successive steps of Weyl's procedure, outlined in §6.2, couple mathematical construction with an "essential analysis" of geometrical concepts having validity only within "the sharply illuminated circle of perfect givenness" of an infinitesimal space of intuition. As shown in §6.3, the ultimate justification for this reconstruction of the infinitesimal concept of "congruence", the essence of metric, lies in a group-theoretic analysis of the notion of "space" itself, yielding a solution to "the new problem of space" required in the wake of general relativity. Weyl's group-theoretic proof of the uniqueness of the "infinitesimal Pythagorean metric" posited by Riemann grounds the fundamental distinction between the essential and the contingent, the *a priori* "nature" and the *a posteriori* "orientation", of the metric of space. This result is regarded as providing a compelling justification for his infinitesimal geometry but, as well, a "good example" of the "analysis of essence [*Wesensanalyse*] sought in Husserlian phenomenology". Weyl's ensuing theory of "gravitation and electromagnetism", based as it is on a geometry implementing "the epistemological principle of relativity of magnitude", was put forward in two "versions", distinguished mainly by differing strategies of justification, surveyed below in §6.4. As discussed in chapter 4, §§4.2.1 and 4.3, the "official" (Einstein–Pauli) response to his theory was that it was empirically disconfirmed by the constancy of the spectral lines of the chemical elements. In reply, Weyl initially offered a variety of novel and essentially "aesthetic" arguments in its favor, while downplaying the issue of empirical disconfirmation. In a second "version" of his theory, Weyl confronted the Einstein–Pauli objection head-on, proposing a schematic but dynamically plausible origin of "the natural gauge of the world" that is an independent presupposition in Einstein's use of "practically rigid rods" and "perfect clocks" as metrical indicators. In §6.5, I reveal the philosophical reasons behind Weyl's otherwise puzzling opposition, in the 1920s, to Élie Cartan's "moving frame" formalism of differential geometry, from which the modern fiber bundle geometrical formalism of the gauge principle derives. In claiming that Cartan's geometry could not serve as a suitable "world geometry" for physics, Weyl was in effect defending his transcendental-phenomenological conception of a purely infinitesimal "world geometry" as the constitutive framework of the objects of field physics. This chapter concludes with a brief glance at the subsequent development of the gauge principle, now "sedimented" into the "everyday discourse" of quantum field theory. As may be seen from a synopsis of the modern gauge principle, it bears only a formal analogy to Weyl's 1918 "principle of gauge invariance", bereft of all trace of its transcendental-phenomenological origin. One consequence, at least from the perspective of transcendental idealism, is a re-emergence of epistemological naïveté regarding mathematical representation in fundamental physical theory. In losing sight of the basic constitutive role of mathematical concepts in fundamental physics, the new dogmatism has arisen that "the secret of nature is symmetry".

6.2 Constituting the World from the "Purely Infinitesimal"

The general procedure for a transcendental-phenomenological constitution of nature is, according to Weyl, the "construction of objective reality out of the material of immediate experience",[1] an epistemological hurdle also faced by 20th-century sense-datum theorists of the empiricist tradition. However, as "objective reality" to be reconstructed is a physical world that is neither space nor time, but field functions of the four space-time variables, this task appears altogether insurmountable. How can epistemological reconstruction get off the ground at all when general relativity has shown that "nothing of the essence of space and time, given to us in intuition, enters into the mathematically constructed physical world"?[2] Furthermore, how could the generically inhomogeneous space-times of general relativity arise from what is "given to consciousness" in space and time, *forms of intuition*, that are necessarily homogeneous?[3] Weyl's approach to a solution to these difficulties was two-fold. First, he upheld the phenomenological thesis that "the general form of consciousness" is the compenetration of "This" and "Thus", of the data of perception and intuition, an overlap of "matter" and "form", of continuous quality and continuous extension. "Phenomenologically", he noted, it is impossible to "get beyond" this conception.[4] For this reason, "profound understanding" of this "mutual penetration of being and essence" is "the key to all philosophy":

> The world comes into our consciousness only in the general form of consciousness, a compenetration of being [*Seins*] and of essence [*Wesens*], of the "this" ["*Dies*"] and "thus" ["*So*"]. (The thorough understanding of this compenetration is, incidentally, in my opinion the key to all philosophy.) In acts of reflection we are capable of bringing the essence [*Wesen*], the being-thus [*So-sein*] of phenomena into prominence, to be noticed for itself, without *de facto* being able to loosen it from the individual existence [*Sein*] of that intuitive given in which it appears. Here is the origin of concepts![5]

Now the "origin of concepts" lies in locating, through "essential insight", the hierarchy of essential genera, categories and relations, the "essence" or set of "essential predicables", of each contingent phenomenologically given individual "this". The interpenetration of contingency and necessity that is the general form of consciousness also contains the "origin" of the "mathematical treatment of actuality" of theoretical physics, in which "the attempt is made to represent in the absoluteness of pure being the world given to consciousness". Mathematical representation of "actuality" in terms of the "absoluteness of pure being" refashions the description of the contingently given "this" of the physical world in terms of the "thus", the purely immanent objects of ideal mathematical possibility.[6] Such a representation itself retains nothing of the intuitive essence of space and time, nor possesses any sensory or perceptual qualities at all. Yet according to the thesis of transcendental-phenomenological idealism, the sense of this mathematical representation of actuality as a "being for consciousness" follows from the overlap of contingent fact and necessary essence, that form of consciousness that "phenomenologically, one cannot get beyond". Weyl will accordingly refashion existing fundamental physical theory (the theories of gravitation and electromagnetism) from the a priori subsisting eidetic possibilities afforded by a space-time manifold with the field structure of a "purely infinitesimal world geometry". In this way, the description of the actual world of space-time coincidences is cast upon the canvas of ideal geometric possibility. Ultimately, furnishing an ideal space for the design of the theoretical description of the physical world epitomizes the role of geometry in physics.[7]

Second, Weyl restricted the homogeneous space of phenomenological intuition, the locus of phenomenological *Evidenz*, to what is given at, or neighboring, the cognizing ego:

> Only the spatio-temporally coinciding and the immediate spatial-temporal neighborhood have a directly clear meaning exhibited in intuition.... The philosophers may have been correct that our space of intuition bears a Euclidean structure, regardless of what physical experience says. I only insist, though, that to this space of intuition belongs the ego-center [*Ich Zentrum*] and that the coincidences, the relations of the space of intuition to that of physics, becomes vaguer the further one distances oneself from the ego-center.[8]

Weyl alludes here to the limitations in the correspondence between an open neighborhood of a point P of a differentiable manifold M, and a neighborhood of the origin in the tangent space at P, a vector space associated with P, and not a part of the manifold itself. It is a fundamental fact of Riemannian geometry that if T_PM denotes the tangent space at $P \in M$, there is always a neighborhood of the vector $\vec{o} \in T_PM$ which is mapped diffeomorphically (by the so-called exponential mapping) onto an open neighborhood of P in M, a correspondence that breaks down as one proceeds from the origin. But in any case, by delimiting what Husserl termed "the sharply illuminated circle of perfect givenness", the domain of "eidetic vision", to the infinitely small homogeneous space of intuition surrounding the "ego-center", Weyl could restrict consideration to linear relations, since only these arise in passing to the tangent space of a point in a manifold. Linearity, in turn, gave the expectation of "uniform elementary laws".[9] Thus, Weyl initially restricted the concept of a coordinate system to the tangent space covering each manifold point P, essentially assuming a four-dimensional manifold that is Hausdorff, simply connected, and differentiable. Imposition of a local coordinate system is regarded as the original constitutive act of "a pure, sense-giving ego". A coordinate system always bears an indelible mark of transcendental subjectivity; it is "the unavoidable residue of the ego's annihilation in that geometrico-physical world which reason sifts from the given under the norm of 'objectivity' ". In this necessary presupposition of any differential structure, Weyl recognized, as seen shown in chapter 5, §5.3.4, an intimation of the phenomenological postulate that "existence is only given and *can* only be given as the intentional content of the conscious experience of a pure, sense-giving ego".[10]

Guided by the phenomenological methods of "eidetic insight" and "eidetic analysis", the epistemologically privileged purely infinitesimal comparison relations of *parallel transport* of a vector, and the *congruent displacement* of vector magnitude, will be the foundation stones of Weyl's reconstruction. The task of comprehending "the sense and justification" of the mathematical structures of classical field theory is accordingly to be addressed through a construction or *constitution* of the latter within a world geometry entirely built up from these basic geometrical relations immediately evident within a purely infinitesimal space of intuition. A wholly *epistemological* project, it nonetheless coincides with the explicitly *metaphysical* aspirations of Leibniz and Riemann to "understand the world from its behavior in the infinitesimally small".[11] With this, we can turn to consider the fundamental infinitesimal geometrical relations revealed by Weyl's "essential analysis" of the mathematical basis of "infinitesimal geometry".

6.3 Pure Infinitesimal Geometry

6.3.1 "Essential Analysis"

While freely admitting, in his initial reply to Einstein, that he arrived at his theory of gravitation and electricity through "pure speculation",[12] Weyl gave many indications that his mathematical construction was guided by the phenomenological method of attaining "essential insight" into the underlying concepts of "infinitesimal geometry".

This is to be thought of as a manner of "seeing" into essential structures, intentional objects that are completely self-given, without any reliance on sense perception, according to which the fundamental concepts of Riemannian geometry are resolved into successive intentional complexes, nested layers of *noema* or meaning, exhibiting how the data and strata of consciousness are founded on each other.[13] An act of reflection directed upon a given stratum of a geometrical concept, reveals in "essential insight" its meaning or *noema* and its connection to the next stratum; these are then conceptually fixed in mathematical terms. The method's central components are then the meaning-conferring acts of phenomenological *Wesenschau*, "a systematic method for clarification of meaning" by focusing more sharply on the concepts considered by directing attention in a certain way[14] (see chapter 5, §5.4.2).

The goal of the analysis was to place infinitesimal geometry, above all, its central concept of congruence, upon a consistent and rationally justified basis, the immediately evident "purely infinitesimal" relations of comparison of direction and magnitude that depend on a specific choice of coordinates and unit of scale.

Construction of purely infinitesimal geometry is laid out as taking place in three distinct stages of "connection": manifold or "continuous connection" (*stetiger Zusammenhang*), affine connection, and "metric (or, length) connection".[15] The construction itself, "in which each step is executed in full naturalness, visualizability and necessity" (*in voller Natürlichkeit, Anschaulichkeit, und Notwendigkeit*), is "in all essential parts the final result" of the renewed investigation of the mathematical foundations of Riemannian geometry opened up by Levi-Civita's discovery of the concept of infinitesimal parallel displacement.[16] The physical world is then to be distinguished within this "world geometry" through the univocal choice of a gauge invariant action function $S(g_{\mu\nu}, \varphi_\mu)$, where $g_{\mu\nu}$ is the (only conformally invariant) metric tensor and φ_μ is the electromagnetic four potential.[17] However, to Weyl's dismay, it soon became apparent that a number of such functions could be constructed, choice among them being essentially arbitrary.[18]

First stage: continuous connection (topology). Weyl's several discussions of topology in the context of his geometry add little to topology *per se* but take over the modern topological concepts of "point" and "neighborhood", first clarified in his own 1913 book on the concept of a Riemann surface. There is a clearly identified reason for his reticence in extending phenomenological constitution to the manifold, and so to the concept of "continuous connection" itself. With reference to his discussion in *Das Kontinuum* of the "deep chasm" separating the intuitive and the mathematical continuum, Weyl observed that a "fully satisfactory analysis of the concept of the *n-dimensional manifold* is not possible today" in view of the "difficulty of grasping the intuitive essence [*anschauliche Wesen*] of continuous connection through a purely logical construction".[19] Setting that task aside, Weyl simply assumed that in the tangent space covering each manifold point P, there is an affine linear space of vectors centered on P in that line elements dx radiating from P are infinitely small vectors. In this way, functions at P and in its neighborhood (in particular, the displacement functions—see below) transform linearly and homogeneously. Weyl's attention then concentrated on the manifold's *Strukturfeld*, its metric, affine (or conformal, and projective) structures, originating the now familiar machinery of connections in a specifically epistemological context.

Second stage: affine connected manifold. The concept of parallel transference of a tangent vector in a Riemannian manifold M was first developed in 1917 by Levi-Civita (and independently by Hessenberg and Schouten).[20] The hitherto purely analytical Christoffel symbols (of the second kind) of covariant differentiation are equated with the components (relative to a given coordinate system) of the unique affine (henceforth, "Levi-Civita") connection associated with the metric. This furnishes the Christoffel symbols with a geometric interpretation, by relating them to the parallel displacement of a vector along a path connecting a point P to another point P' in the infinitesimal region (tangent space T_P) around P ($T_{P'}M = T_P M$). Covariant differentiation then becomes a means of comparing infinitesimal changes in vector or tensor fields in any given direction from the point in question. Parallel transport is "purely infinitesimal" in the sense that directional comparison of vectors at finitely distant points P and Q can be made only by specifying a path of transference from P to Q and "transporting" to Q a comparison vector defined as "parallel to" the original vector at P.[21] In general, parallel displacement is not integrable; that is, the comparison vector arising at Q will depend upon the path taken between P and Q.

To Weyl, Levi-Civita's concept marked a significant advance of "simplicity and visualizability [*Anschaulichkeit*] in the construction of Riemannian infinitesimal geometry".[22] But whereas Levi-Civita had employed an auxiliary construction, embedding M in a Euclidean space where parallel transport was defined, and then projecting it into the tangent space of P in M, Weyl gave the first intrinsic characterization in terms of bilinear functions $\Gamma(A^\mu, dx)$ since known as the components of a (symmetrical) affine connection.[23] In general, the change δA^μ in a given vector A^μ displaced from P to $P'_{(x^\mu + dx^\mu)}$ is defined

$$\delta A^\mu = -\Gamma^\mu_{\alpha\beta} A^\alpha dx^\beta, \tag{1}$$

whereas the *covariant derivative* of the vector A^μ (a tensor, and so of objective significance) is defined

$$A^\mu_{;\alpha} = \frac{\partial A^\mu}{\partial x^\alpha} + \Gamma^\mu_{\beta\alpha} A^\beta. \tag{2}$$

Parallel transport requires that the components of the affine connection vanish. Next followed the concept of a manifold with an affine connection. A point P is affinely connected to all points P' in its immediate neighborhood just in case it is determined, for every vector at P, the vector at P' to which it gives rise under parallel transport from P to P'. If it is possible to single out a unique affine connection, among all the possible ones at each point P, then M is called a *manifold with an affine connection*. This is essentially a conception of space as stitched together in linear fashion from infinitely small homogeneous patches. To Weyl, parallel transport was the paradigm comparison relation of infinitesimal geometry, for it satisfied the epistemological demand that all integral (and so, not immediately surveyable) relations between finitely separated points must not be posited but must be constructed from a specified infinitesimal displacement along a given curve connecting them. He also introduced the idea of the curvature of a connection $R(\Gamma)$, a $(1, 3)$ tensor analogous to the Riemann-Christoffel tensor of Riemannian geometry, and showed that the calculus of tensors could be developed on the basis of the concept of infinitesimal parallel transport, without any reliance on a metric.[24] However, it was Eddington, not Weyl, who first fully exploited this idea in physics.[25]

According to Weyl, "the essence of parallel transport" (*das Wesen der Parallelverschiebung*) is expressed in that, in a given coordinate system covering P and its neighborhood, the components of an arbitrary vector A^μ do not change as A^μ is parallel-transported from P to a neighboring point P'.[26] Unaltered displacement accordingly depends on a locally Cartesian or "geodesic" (at P) coordinate system, proleptic reference to the fact that at P the $g_{\mu\nu}$ have stationary values, $(\partial g_{\mu\nu}/\partial x^\sigma) = 0$, and so the components of the affine connection vanish. According to the principle of equivalence, such geodesic coordinates at a space-time point always exist. In this dependence on a particular coordinate system, parallel transference of a vector or tensor without "absolute change" is not an invariant or "objective" relation. But a specifically epistemological and nonconventional meaning is intended for the statement that some vector at P' is "the same" as a given vector at P. Namely, from the original vector at P, a new vector arises at P' that, in the purely local comparison made, as it were, by a particularly situated consciousness, is affirmed to be "without change". Despite the subjectivity of the "experienced" condition $(\partial g_{\mu\nu}/\partial x^\sigma) = 0$ required by this construction, such comparison is nonetheless the basis for the invariant relation of covariant differentiation. Obviously, the idea is an analogy formed from Einstein's theory in which the nontensorial gravitational field strengths $\Gamma^\sigma_{\mu\nu}$ (in Weyl's suggestive terminology, the "guiding field" [*Führungsfeld*]) can be locally, but not generally, "transformed away", an observer-dependent "disappearance" of a gravitational field. At the same time, the space-time curvature tensors derived from the $\Gamma^\sigma_{\mu\nu}$ have an objective significance for all observers.

Third stage: metrically connected manifold. In a largely mathematical 1918 paper entitled "Pure Infinitesimal Geometry", Weyl declared:

> a truly infinitesimal geometry [*wahrhafte Nahegeometrie*] should know only a principle of displacement [*Übertragung*] of a length from one point to another infinitely close by.[27]

As the "essence of space" is metric, the fundamental metrical concept, congruence, also must be conceived "purely infinitesimally".[28] Enshrined as "the epistemological principle of relativity of magnitude", a postulate is laid down that direct comparison of vector magnitudes can be immediately made only at a given point P or at infinitesimally nearby points P' ($P' - P = \overrightarrow{P'P} \in T(M_P)$). Just as an affine connection governs direct infinitesimal comparisons of orientation, or parallelism, so a *length* or *metric connection* is required to determine infinitesimal comparisons of congruence. This also requires a vector to be displaced from P to P' and, in general, the "length" l of the vector is altered. Thus, if l is the length of a vector A^μ at $P_{(x)}$, $l_{P_{(x)}}(A^\mu) = ds^2 = g_{\mu\nu}A^\mu A^\nu$, then on being displaced to P', the change of length is defined to be a definite fraction of l,

$$\frac{dl}{dx^\mu} := -l\frac{d\varphi}{dx^\mu}, \tag{3}$$

where $d\varphi = \sum_\mu \varphi_\mu dx^\mu$ is a homogeneous function of the coordinate differentials. The new vector at P', corresponding to A^μ at P, accordingly has the length

$$l_{P'_{(x+dx)}} = (1 - d\varphi)(g_{\mu\nu} + dg_{\mu\nu})A^\mu A^\nu, \tag{4}$$

where $(1 - d\varphi)$ is a proportionality factor, arbitrarily close to 1. In analogy to (1), the change in length of A^μ is defined as

$$\delta l := \frac{\partial l_P}{\partial x^\mu}dx^\mu + l d\varphi. \tag{5}$$

Then, just as the vanishing of its covariant derivative means that a vector has been parallel transported from P to P' without "absolute change" in direction, so here the vanishing of δl indicates that A^μ has been *congruently displaced* from P to P':

$$\delta l = 0 \Leftrightarrow dl = \frac{\partial l}{\partial x^\mu}dx^\mu = -l d\varphi. \tag{6}$$

Up to this point, an arbitrary "gauge" (unit of scale) has been assumed. Recalibrating the unit of length at P through multiplication by λ, an always positive function of the coordinates, multiplies the length $l_{P_{(x)}}$ by λ, $l' = \lambda l$. Then the change in length at P', dl', corresponds to a transformation of the "length connection" $d\varphi$,

$$dl' = d(\lambda l) = l d\lambda + \lambda dl = l d\lambda - \lambda l d\varphi$$
$$= -\lambda l\left(d\varphi - \tfrac{d\lambda}{\lambda}\right) = -l' d\varphi', \tag{7}$$

($d\lambda/\lambda = d\log\lambda$). A *metrically connected manifold* is then one in which each point P is metrically connected to every point P' is its immediate neighborhood through a *metric connection*. In general, length is not integrable for on integration (8) follows from (5),

$$\log l\big|_P^Q = -\int_P^Q \varphi_\mu dx^\mu, \text{ and so}$$
$$l_Q = l_P^{-\int_P^Q \varphi_\mu dx^\mu}. \tag{8}$$

As Pauli demonstrated, displacement of a vector along different paths between finitely separated points P and Q will lead to arbitrarily different results at Q.[29] But when the linear form φ_μ vanishes, the magnitude of a vector is independent of the path along which it is displaced, which is just the case of Riemannian geometry. The necessary and sufficient condition for this is the disappearance of the "tract curvature" (*Streckenkrümmung*) of Weyl's geometry

$$F_{\mu\nu} = \frac{\partial\varphi_\nu}{\partial x_\mu} - \frac{\partial\varphi_\mu}{\partial x_\nu}, \tag{9}$$

just as the vanishing of the Riemann tensor is the necessary and sufficient condition for flat space.

Implementation of the local comparison condition means that the fundamental tensor $g_{\mu\nu}$ of Riemannian geometry induces only a local conformal structure on the manifold. There is then an immediate meaning given to the angle between two vectors at a point, or to the ratio of their lengths there, but not to their absolute lengths. These transform at a point as $g'_{\mu\nu}(x) = \lambda g_{\mu\nu}(x)$. This weakening of the metrical structure has two important consequences. Such a metric no longer determines a unique linear (affine) connection, but only an equivalence class of connections. Yet Weyl required, as the "fundamental fact" of infinitesimal geometry, that there be unique affine compatibility in the sense that the transport of tangent vectors along curves associated with the connection, that is, affine geodesics, leave the vectors congruent with themselves with respect to the metric. Weyl showed that a unique connection, coupled to given choice of a metric tensor, is found by incorporating into its definition the linear differential form φ of his length connection. Then, when the components of φ vanish identically at a point, the connection becomes identical to the "Levi-Civita" connection of Riemannian geometry, as can be seen from comparison of the definitions of the two connections in components:

$$\text{Levi-Civita:} \quad \Gamma^{\sigma}_{\mu\nu} = \frac{1}{2}g^{\sigma\tau}\left[\frac{\partial g_{\tau\mu}}{\partial x^{\nu}} + \frac{\partial g_{\nu\tau}}{\partial x^{\mu}} - \frac{\partial g_{\mu\nu}}{\partial x^{\tau}}\right]; \tag{10}$$

$$\text{Weyl:} \quad \Gamma^{\sigma}_{\mu\nu} = \frac{1}{2}g^{\sigma\tau}\left[\frac{\partial g_{\tau\mu}}{\partial x^{\nu}} + \frac{\partial g_{\nu\tau}}{\partial x^{\mu}} - \frac{\partial g_{\mu\nu}}{\partial x^{\tau}}\right] + \frac{1}{2}(g_{\mu\sigma}\varphi^{\nu} + g_{\nu\sigma}\varphi^{\mu} + g_{\mu\nu}\varphi^{\sigma}). \tag{11}$$

Given a "Weyl connection", it is possible to speak of a manifold with an affine connection where, as in the Riemannian case, there is a unique meaning given to parallel displacement of a vector at every point. Only in the case of "congruent displacement" (*kongruente Verpflanzung*), or displacement without alteration of length, is parallel displacement possible, and so it is that infinitesimal length or "tract" displacement (*Streckenübertragung*), the "foundational principle of metric geometry", brings along also directional displacement (*Richtungsübertragung*). This is to say, according to Weyl, that *"according to its nature, a metric space bears an affine connection"*.[30]

6.3.2 Group-Theoretical Justification of an Infinitesimal Euclidean Metric

The final stage of Weyl's "essential analysis" of infinitesimal geometry was his group-theoretical proof of Riemann's posit of an "infinitesimal Pythagorean (Euclidean) metric". The culmination of his efforts to show that the supposition of the purely infinitesimal character of the geometry underlying field physics was not arbitrary, Weyl regarded the proof as basic confirmation of his entire approach to the problem of space in the context of general relativity, as decisive a result as the calculation of the advance of the perihelion of Mercury had been for Einstein's theory.[31] His group-theoretic result was taken as providing a rational justification for locating the *"a priori* essence of space" in the infinitesimal Euclidean metric at each point of a Riemannian manifold.

From Weyl's "purely infinitesimal" standpoint, the Einstein theory of gravitation reopened the "problem of space", previously thought resolved by the treatment of Helmholtz and Sophus Lie before the turn of the century.[32] That problem concerned the ground of Riemann's assumption that the metric of a manifold is given by a quadratic differential form, as opposed to the other possibilities afforded by the wider class of Finsler metrics.[33] Riemann's own hypothesis, that the distance element can be expressed as the square root of a homogeneous quadratic differential form $dl^2 = g_{ik}dx^i dx^k$ ("infinitesimal Pythagorean metric") was not fully demonstrated; as a result, it had come to be seen as "an article of faith, essentially".[34] As discussed in chapter 3, §3.4, Helmholtz and then Lie, who made Helmholtz's results rigorous through the use of continuous groups, showed that the Riemannian assumption was justifiable in virtue of the "fact" of the existence of freely moveable rigid bodies, together with the assumption that these rigid motions form a (Lie) group. Thus, the Helmholtz–Lie solution is valid in the restricted domain of spaces of constant (including zero) curvature. But this solution is rendered obsolete with recognition of the nonhomogeneous metrical fields in Einstein's theory where, in general, no group of rigid motions is possible. As Weyl first recognized, a new solution to the "problem of space" must be found in this context. In the Helmholtz–Lie tradition, he sought to account for the "uniqueness of the Pythagorean (i.e., quadratic) metric determination" in an n-dimensional differentiable manifold M by treating congruence through a continuous group of motions. At the same time, the group analysis must be "purely infinitesimal", providing "new support" (*neue Stütze*) for the conviction that his geometry, not Einstein's, was the "world geometry" underlying the phenomena of field physics.[35] The problem occupied him for several years in the early 1920s. The result was proven for the case of two and three-dimensional spaces in §18 of the fourth edition of *Raum-Zeit-Materie*, and Weyl there speculated that the proof probably could be extended to n dimensions. This was first outlined in detail in a paper of 1922;[36] the motivations for the proof were made clear in lectures given in Barcelona and Madrid early in 1923, while the proof itself only appears in an appendix to the German text of these lectures later that year.[37] The final lengthy case-by-case proof, which Weyl himself likened to "mathematical tightrope dancing"[38] is cast in very general terms. It is to be valid for manifolds of arbitrary dimension n and for every value of metric signature (or "inertial index"; relativistic space-times have signature $+1$ or -1 depending on sign conventions) and not simply for positive-definite metrics.

Weyl's guiding thought, inspired by the aforementioned "fundamental fact" of infinitesimal geometry, was that the class of Riemannian metrics, as well as the class of Weyl metrics, could be distinguished from the larger Finsler class by seeking those metrics that admit a unique affine connection. Under these conditions, he sought to distinguish "the one, unalterable Pythagorean *nature* of this metric, in which the *a priori* essence [*Wesen*] of space is expressed",[39] from the mutual "orientation" that the metrics in the different points adopt with respect to one another. The former expresses the form of the infinitesimal homogeneous space of intuition:

It lies in the nature of space as a form of appearances that it is *homogeneous*; in a locus as such are founded no inner differences of spatial things.[40]

This *nature* of the metric is *a priori* and is the same at each point of the manifold, since at every point P the metric is represented by a group of linear transformations of the same kind. The "orientation" of the metrics is *a posteriori* and depends, as required by the field equations of Einstein's theory, on the fortuitous distribution of matter and energy. Hence, Weyl saw his task as showing how it is that space has what appears to be conflicting properties but in fact are different and complementary aspects corresponding to, and mathematically characterizing both the purely infinitesimal space of intuition and the variably curved physical space-times of general relativity and the relation between them. As the form of appearances, space is necessarily homogeneous and fully rationally comprehensible; this is the *a priori* "nature of space". On the other hand, space is also variable and, as requiring reference to empirical data, only approximately knowable; this is the *a posteriori* "orientation" of the metrics at various points. Kant, therefore, was not altogether wrong in claiming an *a priori* character for space; only with infinitesimal geometry the boundary between *a priori* and *a posteriori* has shifted.[41] These ideas are implemented in three steps that shall be only informally described here.[42]

Unlike in Kant, Helmholtz, and Lie, for Weyl spatial homogeneity is a condition that obtains only infinitesimally. Homogeneity is mathematically characterized by congruence mappings, but "the concept of congruence, on which metric depends, must be conceived purely infinitesimally".[43] So Weyl began with the notion of a "metric at a point P", defining a "congruent mapping" at a single point P by means of a group of motions (infinitesimal rotations) acting on a vector space, preserving the volume of any parallelpiped spanned by n independent vectors. In the tangent vector space at P, this is a determinate group of homogeneous linear transformations. Furthermore, the group is continuous in that each rotation is an element of a one-dimensional manifold. Then, employing a purely algebraic analysis carried out in the complex domain, Weyl showed that this infinitesimal rotation (orthogonal) group is Euclidean, a distinguished subgroup of the general linear group in that it characterizes the (nondegenerate) quadratic form up to a factor of proportionality. That is, the infinitesimal Euclidean rotation group is distinguished from among all linear groups in that it affords a vector body "free mobility" around a fixed point. This showed that the requirement of homogeneity, necessarily belonging to space as a "form of appearances", lies in the metrical essence (*Wesen*) of space.[44] Next, to show that the metric essence of space is everywhere the same, Weyl demonstrated that the infinitesimal rotation group at P is related to that at some definite neighboring P' by a single linear congruence transformation. This is a similarity transformation in Lie's sense, so the groups share the same abstract Lie algebra, differing only as regards their "orientation". While the nature of the metric is characteristic of space itself, the differing orientations of the orthogonal groups everywhere characterizing the Pythagorean nature of the metric are due to the presence of matter, disturbing, as Weyl would put it, the "state of rest" of the metric field.

Third and finally, Weyl returns to the concept of a "congruent displacement", a "congruent mapping" of a vector from P to an infinitely close by point P', in order to extend the established group property to a "metrical connection" between P and *any* neighboring point P'. Using Scheibe's term,[45] Weyl has here introduced the notion of a *group field*, a field of groups of linear transformations in every tangent vector space by which the congruence of tangent vectors at *each* point may be defined. Such a displacement is infinitesimal if the alteration of the arbitrary vector's components is of the same order of magnitude as are the coordinate difference dx between P and P'. Here he made use of the fact that to every coordinate system at P there corresponds a possible system of parallel displacement of a vector from P to any neighboring point P': transport of the vector without alteration of its components. As shown in §6.3, in a given coordinate system, this parallel displacement is expressed in terms of a symmetric affine connection. Now the variable alteration of the metric field established in Einstein's theory mandates what Weyl termed the "Postulate of Freedom": that the nature of space imposes no restriction on the metrical relationship. So, just as there are many possible concepts of parallel displacement of a vector between neighboring points, the metrical connection between P and any of the points P' in its immediate region must allow for a wide system of infinitesimal congruent displacements of a vector. Then, for a given rotation group at P, the metrical connection to any P' is expressed through a linear equation (all directions are equivalent). Nonetheless, according to "the fundamental fact [*Grundtatsache*] of infinitesimal geometry", that the metric uniquely determines the (symmetrical, i.e., torsion-free) affine connection, there is always one and only one parallel displacement from P to *any* neighboring P' which is at the same time a system of infinitesimal congruent displacements. Assuming the existence of such a unique connection,[46] Weyl then has the result that a "congruent mapping" or "length connection" between P and any nearby P' is uniquely defined. In general, the transition to a purely infinitesimal solution to the problem of space" meant that "the Helmholtz requirement of homogeneity, with which the old conception of the essence [*Wesen*] of the metric of space stands and falls",[47] must be replaced by two completely different postulates of pure infinitesimal geometry: "the Postulate of Freedom" and the requirement that the metric univocally determine the affine connection. Weyl accordingly understood his group-theoretic solution to the "new problem of space" as furnishing *conceptual* justification of the results attained in the conjoint phenomenological *Wesensanalyse* of space, demonstrating

> how high a measure of harmony and inner necessity dwells within the construction of the pure infinitesimal geometry that forms the foundation of the widened relativity theory [*erweiteren Relativitätstheorie*].[48]

Writing to Husserl on 26 March 1921 (see chapter 5, §5.2), Weyl could report that he had compellingly separated "the a priori essence of space [*das apriorische Wesen des Raumes*], . . . from the a posteriori".[49] The philosophical lineage of his achievement was proclaimed in a newly appended passage to the fourth (1921) edition of *Raum-Zeit-Materie*.

> The investigations made concerning space in chapter two [of *Raum-Zeit-Materie*, 4th ed.], culminating in a sketch of his group-theoretic results] appear to me to be a good example of the essential analysis [*Wesensanalyse*] striven for by phenomenological philosophy (Husserl), an example that is typical for such cases where a non-immanent essence is dealt with. We see in the historical development of the problem of space, how difficult it is for us reality-prejudiced humans to hit upon what is decisive. A long mathematical development, the great unfolding of geometrical studies from Euclid to Riemann, the physical exploration of nature and its laws since Galileo, together with all its incessant boosts from empirical data, finally, the genius of singularly great minds—Newton, Gauss, Riemann, Einstein—all were required to tear us loose from the accidental, nonessential characteristics to which we at first remain captive. Certainly, once the true standpoint has been attained, Reason [*Vernunft*] is flooded with light, recognizing and accepting what is understandable out-of-itself [*das ihr aus-sich-selbst Verständliche*].[50]

At the same time, Weyl reproached the impatience of philosophers who believed that "the essence of space could be adequately described on the basis of a single act of exemplary making-present [*eine einzigen Aktes exemplarischer Vergegenwärtigung*]".[51] While "in principle" they may be right, "in practice" it is only after much hard labor that the whole result may be seen at once. "The example of space", Weyl further noted,

> is most instructive for that question of phenomenology that seems to me particularly decisive: to what extent the delimitation of the essentialities

[*Wesenheiten*] rising up to consciousness express a characteristic structure of the domain of the given itself, and to what extent mere convention participates in it.

Returning to this passage some thirty years later, Weyl observed that he still essentially held to its implicit characterization of the relation between cognition and reflection underlying his method of investigation, which combined experimentally supported experience, analysis of essence (*Wesensanalyse*) and mathematical construction.[52]

6.3.3 Transition to Physics

Just as Einstein required the invariance of physical laws under arbitrary continuous transformation of the coordinates (general covariance), Weyl[53] additionally demanded their invariance under the "gauge transformations"

$$g \Rightarrow g'_{\mu\nu} = \lambda g_{\mu\nu}, \text{ and}$$
$$\varphi \Rightarrow \varphi'_\mu = \varphi_\mu - \frac{1}{\lambda}\frac{\partial \lambda}{\partial x^\mu}. \tag{12}$$

And since the first system of Maxwell's equations

$$\frac{\partial F_{\mu\nu}}{\partial x^\sigma} + \frac{\partial F_{\nu\sigma}}{\partial x^\mu} + \frac{\partial F_{\sigma\mu}}{\partial x^\nu} = 0 \tag{13}$$

follows immediately from (9) on purely formal grounds, Weyl made the obvious identifications of his length curvature $F_{\mu\nu}$ with the already gauge-invariant electromagnetic field tensor (of "gauge weight 0"), and his metric connection φ_μ with the space-time four potential. As a mathematical consequence of his geometry, equations (13) are held to express "the essence of electricity"; they are an "essential law" (*Wesengesetz*) whose validity is completely independent of the actual laws of nature.[54] Furthermore, Weyl could show that a vector density and contravariant second rank tensor density follow from the *general* form of a hypothetical action function invariant under local changes of gauge $\lambda = 1 + \pi$, where π is an arbitrarily specified infinitesimal scalar field. These are respectively identified with the four current density $\mathbf{j}^\mu$ and the electromagnetic field density $\mathbf{h}^{\mu\nu}$, through the relation

$$\frac{\partial \mathbf{h}^{\mu\nu}}{\partial x^\nu} = \mathbf{j}^\mu, \tag{14}$$

that is, the second (inhomogeneous) system of Maxwell equations. Thus, Weyl claimed that, even without having to specify a particular action function, "the entire structure of the Maxwell theory could be read off of gauge invariance".[55] Again, using only the general form of such a function, he demonstrated that conservation of energy-momentum and of charge follow from the field laws in two *distinct* ways.[56] Accordingly he asserted that, just as the Einstein theory had shown that the agreement of inertial and gravitational mass was "essentially necessary" (*wesensnotwendig*), his theory showed this in regard to the facts finding expression in the structure of the Maxwell equations, and in the conservation laws, which appeared to him to be "an extraordinarily strong support" for the "hypothesis of the essence of electricity" (*Wesen der Elektrizität*).[57] The domain of validity of Einstein's theory of gravitation, with its assumption of a global unit of scale, was originally held to correspond to $F_{\mu\nu} = 0$, the vanishing of the electromagnetic field tensor. By 1919, Weyl substituted his own "dynamical" account of the origin of "the natural gauge of the world", a possibility afforded by his theory but not by Einstein's.

6.4 Strategies of Justification

Various approaches are recognizable in Weyl's attempt to justify his theory of "gravitation and electricity", motivated, as it is, by implementation of the "epistemological principle of the relativity of magnitude". Since epistemological reconstruction of Einstein's gravitational theory in accordance with this principle begins within the natural attitude, it appeared in the guise of a physical hypothesis. However, as Weyl ultimately admitted, no physical consequences can follow from an epistemological principle whose rationale lies outside the natural attitude on the grounds of transcendental-phenomenological idealism. But neither within the natural attitude could Weyl's theory be confirmed, for there appeared not one shred of direct physical evidence in favor of it. The "two versions" of Weyl's theory show Weyl's shifting strategy in negotiating between these competing and partly conflicting demands of justification, and reflect the theory's ambiguous character as lying in the intersection of physics and philosophy.

Initially, while unsuccessfully attempting to find physical evidence supporting his theory over Einstein's, Weyl tried to fortify his case by adducing formal and aesthetic criteria in its favor. Then, changing tack somewhat, he gave what has been regarded as a "second version" of his theory that met the Einstein difficulty head-on. In this "version", Weyl argued for the explanatory superiority of his theory over Einstein's in two respects. First, it made way for a dynamical account of the "natural gauge" of the world, tacitly assumed in Einstein's theory. Moreover, in so doing, it provided a principled explanation of the existence of the cosmological constant that, appended to Einstein's field equations, appeared *ad hoc*, relying on a "pre-established harmony". But even here difficulties arise in the ambiguous physical status of a conformally invariant *Äthergeometrie*, the geometry of the "rest state of the aether", that is opposed to a *Körpergeometrie*, the "natural geometry" of rods and clocks. In any case, Weyl considered that this explanatory defense of his theory fell sacrifice to the discovery of an "absolute length" in the Dirac theory of the electron. Henceforth, the explanation of the "natural gauge" must lie in the theory of the atom, not in the "detour" into cosmology he had taken. In the end, Weyl retained only a third and final justification for his purely infinitesimal geometry, his group-theoretic proof of the uniqueness of the "infinitesimal Pythagorean metric" (§6.3.2 above). With this result, Weyl sought to give a conceptual, and so, "objective", justification for retaining the infinitesimally homogeneous space of intuition, the form of the "immediate life of intuition", in the context of the finitely inhomogeneous space-time manifolds generically permitted by the general theory of relativity. There is considerable evidence to suggest that Weyl retained faith in this argument long after surrendering all claims in favor of his theory.

6.4.1 The "First Version" of Weyl's Theory

In the first papers on his theory, Weyl quite clearly emphasized its standing as a physical hypothesis, for he argued that Einstein's theory could be valid only in the absence of an electromagnetic field.[58] That his theory could be considered as at least an empirical competitor to Einstein's followed from investigations by Weyl and Pauli. Pauli (at the age of 19, in one of his first published papers) established that for the special case of a static, spherically symmetric field surrounding a material body, a solution of Einstein's field equations is at the same time a solution of Weyl's.[59] As this case is decisive for the two definite confirmations of general relativity known at the time, the precession of the perihelion of Mercury and the bending of light rays in the solar gravitational field, Weyl's theory was at no disadvantage here, despite its generally fourth order field equations. Interestingly enough, Eddington initially considered Weyl's theory as perhaps to be empirically preferred to Einstein's in the third of the classic tests of general relativity, the gravitational red shift.[60] But while lacking any definite physical evidence in favor of his theory, Weyl marshaled a number of aesthetic and "philosophical" arguments that he believed recommended his theory over that of Einstein. Let us consider these arguments in a bit of detail since, in 1918, they were (and are still) novel kinds of argument in physics.

6.4.1.1 WHY THE WORLD IS FOUR DIMENSIONAL

Although briefly treated in "*Gravitation und Elektrizität*", the more extensive argument first appears in "*Reine Infinitesimalgeometrie*" and is based on the new notion of the weight of a tensor which must be included, along with that of rank, and symmetry properties, in the classification of tensors. As the notion of gauge invariance is more restrictive than that of general covariance, not all tensors are gauge invariant. But the degree, as it were, of any tensor's departure from gauge invariance, can be indicated through the notion of the weight of the tensor that classifies how many integral powers of the gauge factor λ are required to make it gauge invariant. Thus, if a tensor becomes gauge invariant on multiplication by λ^e is said to be of weight e, then the conformally invariant metric tensor of Weyl's geometry, as is seen by the gauge transformation, $g_{\mu\nu} \Rightarrow g'_{\mu\nu} = \lambda g_{\mu\nu}$, is of gauge weight $+1$, while $g^{\mu\nu}$ is of weight -1, and $\sqrt{-g}$ in a four-dimensional world is of weight 2 (g is the determinant of $g_{\mu\nu}$ and, in a $(3 + 1)$-dimensional world, negative). On the other hand, due to the arbitrariness of the electromagnetic potential φ_μ, Maxwell's equations remain invariant when $g_{\mu\nu}$ is replaced by $\lambda g_{\mu\nu}$ (as followed from work of Cunningham and Bateman in 1910, although Weyl does not cite this reference until 1921[61]). In this regard, the electromagnetic field strengths $F_{\mu\nu}$ are the prime example of an "absolute tensor" (tensor of gauge weight 0; in the terminology of Eddington, an "in-invariant"—see chapter 8). Stipulating that only tensors of gauge weight 0 have a factual significance,[62] Weyl correspondingly required that only scalar densities W of weight 0 appear as action invariants since only such densities lead to gauge invariant integrals $\int W d\omega$; in Weyl's theory, action is an "absolute invariant", a dimensionless pure number. This arises in the following way. The "volume element" $d\omega = dx^1 dx^2 \cdots dx^n$ in an n-dimensional world is of weight $n/2$, and so in a four-dimensional world of weight 2. As the scalar density **W** is defined as $\sqrt{-g}\,W$, the corresponding scalar W must be of weight -2. This requirement is met, as expected, by the usual variational principle of Maxwell's theory, whose action integral is

$$\int \frac{1}{4} F_{\mu\nu} \mathbf{F}^{\mu\nu} \quad (\mathbf{F}_{\mu\nu} = \sqrt{-g}\, F_{\mu\nu}) \tag{15}$$

and the scalar $F_{\mu\nu} F^{\mu\nu}$ has gauge weight -2, but only in a four-dimensional world. This means "the possibility of the Maxwell theory in our interpretation is bound to the dimension number 4";[63] that is, only in a four-dimensional world does the Maxwell action obtain as a pure number and so the Maxwell equations can be regarded as strictly valid. On the other hand, the action integral whose variation produces Einstein's field equations,

$$\int \mathbf{R} d\omega \tag{16}$$

where the density **R** is formed from the Riemannian curvature scalar R, is not gauge invariant but has gauge weight 1.

6.4.1.2 REMARK ON THE ROLE OF ACTION
        IN WORLD GEOMETRY

While Weyl did not settle on a definitive action invariant for the gravitational equations in his theory, the various possibilities within his geometry were severely restricted by the above requirement. Weyl limited his considerations to scalars that are rational combinations of the components of the curvature tensor, thought still other scalars of weight -2 can be formed. Through Hamilton's principle, the field equations of gravitation and electromagnetism are to be derived in the usual way by the variation of such integrals with respect to the fourteen fundamental field variables, here $g_{\mu\nu}$ and φ_μ. The corresponding action function for Weyl's theory of "gravitation and electricity" is then constructed as an additive combination of the chosen gravitational action invariant **R** and the action invariant $L = 1/4 F_{\mu\nu} \mathbf{F}^{\mu\nu}$ of Maxwell's theory (see below).

It should be pointed out that an action principle plays a crucial role in Weyl's method of "world geometry" for it is the means by which "the actual world is selected from among of all possible four dimensional metric spaces" that are available according to "pure infinitesimal geometry":

> the actual world [wirkliche Welt] is singled out from among all possible four dimen-
> sional metric spaces in that for it, in any containing world region, the magnitude of
> action [Wirkung] takes on an extremal value with respect to variations of the $g_{\mu\nu}$
> and the φ_μ that vanish on the boundary of that region.[64]

and

> the actual world [wirkliche Welt] is one whose aether finds itself in the state of
> extremal action [Wirkung].[65] [Original emphasis in both quotations.]

This new role, of distinguishing the actual world from other geometric possibilities, is characteristic of the program of world geometry. As I will show in chapter 8, action, although not in the form of Hamilton's principle, also plays this key role in Eddington's world geometry, "affine field theory".

Moreover, as the result of the new requirement of gauge invariance, action is now a pure number. Weyl initially argued that the significance of a dimensionless action might be found in providing enlightenment regarding the unit of action in the theory of the atom, thus perhaps enabling matter to be brought within the unified geometric theory. The action of general relativity, not being a pure number, does not allow this possibility. It is worth noting that this argument has receded to the background by 1920, on account of Weyl's skepticism, acquired in the interim, of the ability of a geometric unification of classical fields to encompass matter (see §6.5).

Finally, in Weyl's theory there are fourteen independent fundamental field quantities $g_{\mu\nu}$ and φ_μ, whereas in Einstein's gravitational theory, there are just the ten independent $g_{\mu\nu}$. Because of the general covariance of Einstein's theory, there are four identities between these ten; in Weyl's, due to the additional gauge invariance, there are now five identities between the fourteen quantities. Just as, according to Noether's second theorem (pertaining to "local symmetries"), the freedom to make arbitrary continuous transformations of the coordinates corresponds to conservation of energy-momentum in Einstein's theory, so gauge invariance corresponds to conservation of electric charge.[66] However, Weyl's theory did not account for the difference between "positive" and "negative" electricity (the difference between the "positive electron" (proton) and the electron).[67] Later, in 1929, he showed that conservation of charge is associated with a reinterpretation of the principle of gauge invariance, connecting the electromagnetic potentials φ_μ not with the gravitational potentials $g_{\mu\nu}$ but with the complex phase of the ψ function of the free electron.[68]

6.4.2 The "Second Version" of Weyl's Theory

Weyl's considered response to Einstein appeared only in a third paper. Initially written as an enclosure to a letter to Einstein in November 1918, Weyl also intended it for publication in the *Sitzungsberichte* of the *Akademie*.[69] On account of a wartime paper shortage, the paper could not appear there due to its excessive length.[70] Weyl expanded it further in the latter part of 1918 and published it early in 1919 in the *Annalen der Physik*. In this paper, Weyl elaborates upon the somewhat brief and opaque comment made in his initial response to Einstein that the ideal process of congruent transference of lengths and the behavior of measuring instruments (including atomic "clocks") are conceptually two different processes.[71] Now he will emphasize that one is a physical process while the other is the purely ideal basis for "the mathematical construction of world geometry" so that they have "as such nothing to do" with one another.[72] Such an explicit disavowal of direct physical significance to the basic relation of his geometry had the effect of removing his theory from the immediate empirical disconfirmation pointed out by Einstein. At the same time, with this shift in emphasis, the *physical* standing of his theory became somewhat problematic. In fact, to Pauli and to

Eddington, although apparently not, interestingly enough, to Einstein, Weyl's admission removed empirical status from his theory altogether.[73] Although history has recorded the fate of Weyl's theory otherwise, as falling sacrifice to the empirical disconfirmation pointed out by Einstein, it was actually its lack (witness Weyl's own denial of a direct linkage) of a comprehensible connection to experience that sealed its doom in the eyes of many physicists. Yet despite such a seemingly glaring flaw, Weyl did not surrender his belief that the theory furnished a preferable outlook on all the physical phenomena in the domain of Einstein's theory.

6.4.2.1 EXPLAINING THE "NATURAL GAUGE
OF THE WORLD"

The centerpiece of this new phase of attempted justification is Weyl's account of "the natural gauge of the world". This was a remarkable innovation that, although not ultimately successful in salvaging his theory, even in his own judgment, nevertheless opened up a realm of cosmological speculation picked up by Eddington and others. From the beginning, Weyl had issued a disclaimer of direct physical significance to the mathematical process of congruent transference of vector magnitudes. But this alone hardly sufficed to rule out Weyl's theory as a physical hypothesis altogether, for certainly the connection of theory to experience could be made in some more indirect way, in accordance with the holism of Weyl's "geometrical method".

> The geometrical statements are . . . merely ideal determinations that individually lack a meaning exhibited in the given. Only the entire network of ideal determinations touch experienced reality [*erlebte Wirklichkeit*] here and there, and at these places of contact it must "agree" ["*stimmen*"].[74]

Nonetheless, it was incumbent upon Weyl to show *how* his fundamental hypothesis of the general nonintegrability of transference of space-time tracts ("*Strecken*") *is* compatible with the admitted fact of the constancy of spectral lines. Two issues are paramount. First, the Einstein/Pauli "prehistory" objection is put aside as irrelevant, for Weyl insists that "it lies absolutely not in the sense of our theory that it *a priori* presupposes such a behavior of clocks",[75] because

> [f]rom the epistemological principle of relativity of magnitude [*erkenntnistheoretischen Prinzip von der Relativität der Größe*] it naturally doesn't follow that tract transference [*Streckenübertragung*] through congruent transport [*kongruente Verpflanzung*] is not integrable; from it, in any case, no fact follows.[76]

In this regard, Weyl complained that his theory had simply been misunderstood as attempting to prove by *a priori* speculation what can only be decided on the grounds of experience.[77] Second, the *brute* assumption of integrable tract comparisons must be independently posited in order for Einstein to physically *define* the *ds* as a "line element" by "norming" it to "infinitesimally small" rigid rods or regular periods of clocks. As shown in chapter 3, §3.3, the proffered justification for this assumption, as Einstein freely admitted, was provisional, while the assumption itself tended to fade from view as Einstein turned his own attention from his theory of gravitation to its possible generalization to the unification of gravity and electromagnetism. Weyl attempted to account for this brute fact of a "natural gauge of the world" in an ingenious manner. He argued that the curvature scalar F of his purely infinitesimal geometry (recall that Weyl's affine connection contains a term corresponding to the electromagnetic four potential) leads directly to a "natural unit of length". When incorporated as a action density into a "rational action principle", this scalar provided a justification for Einstein's cosmological constant. This had the surprising effect that if any charges are at all present in the world, this scalar cannot vanish, and for the case of a static world that it is a constant. Furthermore, on the assumption that this curvature scalar is positive, then it automatically follows that the curvature of space is positive, hence a finite and static universe— desirable results Einstein had only purchased through appending an *ad hoc* cosmological constant to his field equations in 1917.[78] Pauli regarded this argument as "a particular merit of Weyl's theory".[79]

A brief consideration of the details will suffice to show how Weyl constructed a rational action function that contains a cosmological term essentially and leads to the agreement of his theory with experience. Various formulations of his action function are given, the differing treatments being one of the primary distinctions between the third through fifth editions of *Raum-Zeit-Materie*. We shall consider only the general approach.

As noted above, the Einstein field equations of gravitation containing the cosmological term provide an opening for this justification of Weyl's theory. In one form these equations run

$$\left(R_{\mu\nu} - \frac{1}{2}g_{\mu\nu}R\right) + \lambda g_{\mu\nu} = kT_{\mu\nu} \tag{17}$$

and, in "empty space",

$$R_{\mu\nu} = \lambda g_{\mu\nu}. \tag{18}$$

Following Einstein,[80] Weyl noted that by contracting (18), one arrives at $R = -4\lambda$ where R is the Riemann curvature scalar. Weyl considered the thus amended field equations as correct, although not for Einstein's reasons. For, attempting to implement a Machian "relativization of inertia", Einstein appended the cosmological term to his equations in order, as Eddington put it, to "abolish infinity" (while retaining staticity); that is, "to make space at great distances bend round until it closed up", thereby avoiding boundary conditions at infinity.[81] However, Weyl could not agree with this rationale for the cosmological term, since, consistent with his view that physical laws are to be found within a world geometry, he held that

the differential equations of the field contain the completed laws of nature and not after imposing instead a further restriction through boundary conditions at infinity or the like".[82]

So Weyl had another view of the necessity of the cosmological term. He considered that the universe had to be in something like statistical equilibrium in order to account for the fact that the stars have not dispersed into infinite space. This meant that it must be possible to reconcile the law of gravitation with a "uniform distribution of stars at rest in a static gravitational field, as an ideal state of equilibrium".

The hypothesis of statistical equilibrium was consistent with both the nonstatic empty universe of De Sitter, as well as the matter-filled static universe of Einstein. Now a reconciliation follows from the recognition that such an ideal state of equilibrium is possible if mass in the universe is distributed with a density λ in which case the resulting space is metrically homogeneous, as corresponds to the "empty space" equations $R_{\mu\nu} = \lambda g_{\mu\nu}$. Weyl could then show that in fact this is the case: these equations are satisfied for a spherical space of a given radius $a = \sqrt{2/\lambda}$, so that space is closed and hence finite. Accordingly, Weyl's interpretation required that the total mass present in the universe stands in a definite relation to the universal constant $\lambda = 2/a^2$ that already appears in his action principle.[83]

At this point it is useful to recall that in his own derivation of the Einstein field equations from an action principle, Hilbert (1915) chose as the gravitational Lagrangian the density $\mathbf{R} = \sqrt{-g}\, R$ constructed from the Riemann curvature scalar. This scalar contains, in addition to the $g_{\mu\nu}$ and its first derivatives, also the second derivatives linearly, and may be obtained from the once-contracted curvature (Ricci) tensor $R = g^{\mu\nu} R_{\mu\nu}$. It plays no distinct role in Riemannian geometry.[84] But in the context of general relativity, this choice was motivated by the fact that Riemannian geometry provides no other invariant (scalar) that contains the derivatives of the $g_{\mu\nu}$ only to the second order, and those only linearly.[85] It is for this reason, according to Einstein, that "the general postulate of relativity" (i.e., principle of general covariance) leads to a very definite theory of gravitation."[86]

Now in sections of *Raum-Zeit-Materie* devoted to a purely mathematical development of his general metrical geometry,[87] Weyl had derived the corresponding curvature scalar of his geometry, $F = F_\nu^\nu = F_{\mu\nu}^{\mu\nu}$, by contracting again the once-contracted curvature tensor. This is a scalar of gauge weight -1 and so has the dimensions of $1/\text{length}^2$. He then observes that if one sets his scalar F equal to a constant ($\neq 0$), a unit length is stipulated whereby vector "tracts" (*Strecken*) at a point P are measured in terms of the "radius of curvature" available at all manifold points (an average of all sectional curvatures passing through a point P). This means, in effect, that the ds is to be measured in units of cosmic dimension. The fact that a globally fixed unit of length may be fixed in his geometry is, he notes, rather remarkable,

> since it stands in a certain opposition to the original conception of length transference in general metrical geometry, according to which a direct comparison of lengths at a distance should not be possible; however, one observes that the length measure mentioned here depends upon the curvature relations of the manifold.[88]

Since this affords the possibility of defining a "natural gauge" of the world, it also provides a "naturally measured" volume (i.e., in terms of the curvature scalar) of the world for constructing an action quantity. This "natural volume" is represented by the integral invariant

$$\int F^2 \sqrt{g}\, dx \qquad (19)$$

where again it is assumed that $F = $ constant. Finally, still in this purely mathematical section of his book, Weyl also provides a general definition (for manifolds of n dimensions) of his curvature scalar, showing its dependence on the electromagnetic potential and of the corresponding Riemannian scalar. When the number of dimensions of the space is 4, the Weyl scalar is

$$F = R - \frac{3}{\sqrt{g}} \frac{\partial(\sqrt{g}\varphi^{\mu})}{\partial x_{\mu}} - \frac{3}{2}\varphi_{\mu}\varphi^{\mu}. \tag{20}$$

Later on, in §40 of *Raum-Zeit-Materie* (5th ed.), Weyl returns to the task of formulating an action integrand for his theory containing the density of this scalar invariant of his non-Riemannian metrical geometry. Weyl proceeds from a schematic *Ansatz* in which the to-be-constructed action is a linear combination of the action invariants for the gravitational and electromagnetic fields,

$$\int (F^2 \sqrt{g} - \alpha L)dx. \tag{21}$$

Here L is the known Maxwell action density, α is a dimensionless positive number, and the quantity *action* is to be composed of the volume that is measured in terms of the radius of curvature, as above. Adopting the "normalization" $F = \text{constant} = 1/4$, corresponding to measurements made with "cosmic measuring rods", the "simplest principle of action" that can be set up is then

$$\delta \int V dx = 0, \qquad V = (G + \alpha L) + \frac{1}{4}\sqrt{g}\{1 - 3(\varphi_{\mu}\varphi^{\mu})\} \tag{22}$$

where G is the Einstein gravitational action R seen above, supplemented by the cosmological term, $G = R + 1/2\lambda\sqrt{g}$. For measurements with "human scale measuring rods", Weyl gives

$$V = (R + \alpha L) + \varepsilon^2 \sqrt{g}\{1 - 3(\varphi_{\mu}\varphi^{\mu})\}, \tag{23}$$

ε reflecting the Einstein cosmological term $\varepsilon^2\sqrt{g}$.[89] Then this gives, on ignoring the "cosmologically small" term, exactly the Einstein-Maxwell equations. In this way, Weyl showed that his theory exactly reproduces Einstein's with only the slight difference of a factor in the (unknown) cosmological term. Weyl can then claim that he not only has provided an explanation of a brute supposition in Einstein's theory, the existence of "the natural gauge of the world", he has also done so by making necessary use in his action principle of a "cosmological term" that appears in Einstein's theory only in an *ad hoc* fashion and, moreover, requires the implausible assumption of "pre-established harmony" between the mass of the universe and its curvature. He has only to add that material objects—rods, clocks, electrons—are to be regarded as always in equilibrium with this "natural gauge" in order to account for the constancy of spectral lines and the observed congruent behavior of rods and clocks.

6.4.2.2 A SKELETAL DYNAMICAL EXPLANATION: "ADJUSTMENT" AND "PERSISTENCE"

Already then, at the end of 1918, Weyl considered that by locating the source of the natural gauge of the world in the curvature scalar of his metrical geometry, he had decisively rebutted Einstein's "prehistory" objection,[90] a position progressively solidified in the successive editions of *Raum-Zeit-Materie*. The first indication of this new strategy, directly confronting Einstein's alleged disconfirmation, occurs in the third (1919) edition of *Raum-Zeit-Materie*, the first to provide a presentation of his theory:

> With the same degree of plausibility as in the Einstein theory, we may conclude from our results that a *clock* in quasi-stationary motion indicates that proper time ∫ *ds* which corresponds to the normalization *F* = constant. If, in the motion of a clock (an atom) with infinitely small period, the world tract [*Weltstrecke*] covered by it during one period is congruently transferred from period to period in the sense of our world geometry [*Weltgeometrie*], then two clocks proceeding from the same world point A with the same period, (that is to say, during their first period they cover congruent world tracts at A) would in general possess different periods on coming together again at a later world point B. In any case, accordingly, the circulation [*Umlauf*] of electrons in an atom cannot be carried out in such a manner since atoms emit spectral lines of definite frequency, independent of their prehistory. However, it also absolutely doesn't lie in the sense of our theory that it *a priori* presupposes such a behavior of clocks; rather this is to be decided only on the basis of the laws of nature. And in the action principle now presupposed as valid it is shown that this process consists not in a congruent transference but instead: the world tract covered during a period maintains a constant relation to the curvature radius of the universe . . . present at each place.[91]

The refinements of this response bring an additional notable conceptual distinction, between the two different modes of behavior in which a physical quantity is, as it were, fixed to have the value it is found to have on measurement. The first is the purely differential conception of a "persistence" (*Beharrung*)—a physical quantity persists in having the same magnitude it had at the immediately preceding instant. The other is that of the "adjustment" (*Einstellung*) of a physical magnitude to a field quantity in the space-time region where it currently is; in particular, lengths and durations (rods and clocks) will be said to be in "adjustment" to the curvature scalar of Weyl's metrical geometry, taking up the "natural gauge of the world" where they are. Persistence is the "tendency of guidance" imparted from instant to instant to a body by the *Führungsfeld* of inertia and gravitation once its initial direction or state of motion is fixed. Thus, the direction of a Foucault pendulum, once released from initial constraint, is carried over, from moment to moment within its circle, by a continually acting tendency of persistence. On the other hand, the fact that all electrons and protons (*Wasserstoffkerne*) have the same charge and mass testifies to another mode of behavior in that the initial state of a body cannot be set arbitrarily but reflects a constant equilibrium state. So also for (atomic) clock periods and measuring rods.[92]

6.4.2.3 *AETHERGEOMETRIE* AND *KÖRPERGEOMETRIE*

Now these two modes of dynamical behavior are relevant to the basic relations of pure infinitesimal geometry and to the fact that infinitesimal parallel displacement of a vector (manifesting the tendency of persistence) brings with it a congruent transference of length (manifesting the tendency of adjustment). As shown in §6.3, this coupling of transport of direction and length, that the metric uniquely determines the affine connection, is termed the *Grundtatsache* of infinitesimal geometry and on this basis Weyl upholds the *a priori* standing of metrical geometry. Interpreted physically, the *a priori* metric means the reality of an *Aethergeometrie*, an actual geometry of the "rest state of the aether".[93] The *a priori* metric field does not cease to exist in an "empty world" (where $\varphi_\mu = 0$). But this "rest state" of the metric field is metrically homogeneous, the mutual orientations of the infinitesimal orthogonal groups characterizing the "infinitesimally Pythagorean" *nature* of the metric (see §6.3.2) do not differ, for they are not "disturbed" by the "agency" of matter, the "spirit of unrest". While such a homogeneous structure field could be seen as the state of space far from all electrical fields, or, as in the De Sitter hyperbolic universe, without matter at all, it is nonetheless an "absolute" and nondynamical structure, an apparent vestige of space-time substantialism—if interpreted literally.[94] If so, then the physical reality of two distinct but complementary geometries of space-time, pertaining to matter-free and matter-filled regions respectively, is asserted. There is the *Körpergeometrie* of measuring instruments in which congruent trans-ference of *material* lengths must take into account an instantaneous corresponding compensation in the "adjustment" of these material objects at any location to the governing curvature radius of the world where they are.[95] On the other hand, there is underlying metrical structure of *Äthergeometrie*, empirically ascertainable in principle only in empty regions and so effectively "hidden" by the use of material measuring appliances. In the last analysis, Weyl can defend the epistemologically superior conceptual resources of his "purely infinitesimal geometry" only by pos-iting the *a priori* existence of an effectively empirically undetectable metrical structure.

The primary purpose of all of the above arguments of Weyl should be clearly understood: it is not that of developing physical consequences of his theory, which, by his own admission, have turned out to be paltry.[96] Nor was Weyl challenging the basic facts about measuring bodies that Einstein has cited against his theory. Rather the sole justification he could offer is that his theory recasts Einstein's gravitational theory and the theory of electromagnetism together in a common world geometric representation that, in satisfying the principle of gauge invariance, obeys the desired epistemological principle of relativity of length (or magnitude). The theory featured a rationally constructed action principle giving results that stand in agreement with all known experimental facts, while rendering superfluous the treatment of electromagnetism as a separate field in addition to the metrical field.[97] Weyl's emphasis has distinctly shifted: at the end of the fifth edition (the last in his lifetime) of *Raum-Zeit-Materie*, he wrote that his theory is not so much a physical hypothesis as "a theoretically very satisfying epitome [*Zusammenfassung*] and interpretation of our whole knowledge of field physics".[98]

Still, at the end of his authoritative *Physikalische Zeitschrift* (1921) assessment of the aesthetic and explanatory advantages his theory offered over Einstein's, Weyl displayed some understandable sensitivity about his "epistemological principle of relativity of magnitude". Such a postulate, he allowed, in itself indeed seems ar-bitrary; however, his group-theoretical investigation of the nature of metrical space (taken over from the *Raumproblem* of Helmholtz-Lie) "shows that this is not so". Despite all the arguments (none of them truly empirical) Weyl could advance on behalf of his theory, in the end he apparently felt compelled to acknowledge that his attempted extension of relativity theory was still perceived by physicists as abstractly formal and speculative. Even so, he made a final plea for consideration of his theory on the grounds that in it, the necessary *a priori* component involved in all physical theory is given a prominent position:

> Physicists who distrust speculations will probably find that the entire question of an extended theory of relativity, which encompasses the electromagnetic phenomena in an organic fashion, is not yet ripe for discussion, since no experiences could be brought into play to decide it, as long as influences of cosmological smallness elude observation. But one must not forget that in all knowledge of actuality [*Wirklichkeitserkenntnis*], in addition to the collection of typical facts of experience, the *a priori* element, the formation of suitable intuitions and concepts, with whose help the facts are to be interpreted, plays an unavoidable role.[99]

Ultimately, by the end of the decade, Weyl did abandon his theory, although not because of Einstein's objections, which by 1921 he considered to have definitely and definitively rebutted. However, the "principle of relativity of length" was abandoned after the discovery, in the Dirac theory of the electron, of an "absolute length", the so-called Compton wavelength of an elementary particle $l_C = \hbar/mc$, a universal length that, where m is the mass of the particle, is definable in terms of the universal constants $\hbar$ and c. In fact, this only meant that Weyl's account of "the natural gauge" in terms of the adjustment of material bodies must be somewhat modified; the explanation of a "natural gauge" lies solely within the province of the field physics of particles and not through a "detour" through cosmology.[100] As late as 1949, Weyl reiterated his critique of the unsatisfactory reliance of Einstein's gravitational theory on a presupposition that the lengths of rods and the periods of clocks simply "persist". This unexplained tendency stood in stark contrast to the account of "the systematic theory", where the observable behavior of rods and clocks is in principle regarded as arising through the definite ration of "adjustment" of the Compton wavelength of their constituent particles to a field quantity.[101]

6.5 A Philosophical Critique of Cartan

It is Élie Cartan, rather than Weyl, from whom derives the modern differential geometric formulation of the gauge configurations underlying quantum field theories in terms of connections on a principal bundle (or an associated vector bundle), and indeed the notion of a fiber bundle itself, a central concept in modern differential geometry, topology, and algebraic geometry. Cartan's basic idea of a "moving frame" was inspired by Einstein's Principle of Equivalence.

> Basically, I have borrowed from Einstein the idea that an observer freely falling in a gravitational field and carrying a frame of reference undergoing a translation, should find, in his neighborhood, the same laws of physics as if the frame were motionless and the gravitational field were absent: the motion of such an observer satisfies the principle of inertia. In other words, the observer regards certain frames as *parallel* even though they are not in the classical sense of the term; one views space-time as a manifold with an affine connection.[102]

In the first appearance of his new geometric ideas, in 1922–1923, Cartan significantly extended Weyl's notion of an affine connection to a broader class of linear connections in a "generalized affine space". The core of Cartan's idea is a new law of parallel transport based on the idea of a *"repére mobile"* ("moving frame") connecting two infinitely adjacent points P and P' along a curve γ in M on which both lie. The "moving frame" involves parallel displacement of the tangent space $T_P(M)$ from P to P' along γ. Each of these tangent spaces is an affine space with an orthonormal vector basis, and the parallel displacement is a linear isomorphism between them. Cartan's treatment requires a generalized connection such that in parallel displacement the components of the connection do not normally vanish but induce a rotation of the parallel-displaced vector. This is the "torsion" of the connection. It thus introduces a new element of structure, but it is an element that does not connect organically with the metrical structure and related invariants of a Riemannian manifold.[103]

Later differential geometers (in particular, Ehresmann) saw in Cartan's notion of a "moving frame" the first implicit example of a fiber bundle. Recall that a "bundle" is a triple (E, M, π) consisting of two manifolds E and M, and a surjective mapping $\pi: E \to M$. In the cases of interest here, E itself is a "bundle" of fibers (anticipating Weyl, below, these are identical tiny "Klein spaces" Σ_P), one associated with each point P of the "base manifold" M via the "projection map" π, which sends the fiber Σ_P into each $P \in M$. This bundle is characterized by a "structure group" G determining those structures of the fiber that are invariant under transformations of the group. If the base manifold is n-dimensional, and each fiber is m-dimensional, then the "fiber bundle" E is $(n + m)$-dimensional, and the manifold is decomposable into its various fibers. A fiber bundle is then defined as a bundle with "typical fiber" Σ, the structure group G, and a "section", a map from M into E, $\sigma_\alpha: U_\alpha \to \pi^{-1}(U_\alpha)$ where U_α is a coordinate patch on each of the family of open sets $\{U_i\}$ covering M. A section introduces a local coordinate system for the portion of E above the coordinate patch U_α on M. As M (a differentiable manifold) does not admit a single global coordinate system, so also there is no universal coordinate system covering all the fibers in E, also a differentiable manifold. Significant for our discussion is the fact that a fiber bundle is "locally trivial"; that is, the fibers are all "local product spaces" $U_\alpha \times \Sigma_P$, that is, a point φ in $U_\alpha \times \Sigma_x$, is the ordered pair $\langle x, u \rangle, x = x^\alpha \in U_\alpha, u \in \Sigma_x$. Expressed in terms of Cartan's "moving frame", the "bundle" is the family of orthonormal frames (linearly independent vectors expressed in a particular basis in the tangent space) associated with the points of a "base" manifold M, here, four dimensional space-time. From this beginning came the modern insistence on "doing differential geometry on the bundle of orthonormal frames over M".[104] However, Cartan's conception readily generalizes to bundles of other kinds.

After this digression, it is of some interest then to note Weyl's objection in the 1920s to Cartan's conception, and to speculate on its possible grounds. The earliest expression of Weyl's opposition appears in a work written in 1925 but not published until 1988. According to Weyl, the Cartan generalization undermines the notion of a "world geometry" since the associated Klein space (fiber) is "not a pure product of the manifold M". As a result, the "Cartan schema" is unsuitable as a "world geometric foundation for physics":

The space (Σ_P) is not a pure product of the manifold M (as is the tangent space); it requires grounds of determination lying still outside M. As long as that is the case, the Cartan schema is quite out of the question as a theory of a single manifold and accordingly as a world-geometric foundation for physics.[105]

Weyl's criticism was made more specific in a 1929 article, one of the first he wrote in English. There Weyl noted that Cartan attempted to apply Klein's group-theoretic treatment of geometry (as outlined in the latter's famous Erlangen program) to the tangent plane and not to the n-dimensional manifold M itself. But, Weyl objected, the whole scheme lacked what was essential in order even to justify the designation of T_P as "the tangent plane to the point P", namely, *how it is determined from within M*. So it appeared necessary to amend Cartan's geometry by outlining how Cartan's T_P might be embedded in M. The initial step is to choose a definite point O as the center of T_P that, by definition, covers P. But this leads to a restriction in the choice of a coordinate system on T_P and so to a restriction of T_P's (structure) group G, since G must then be chosen so that it has a subgroup of representations leaving O invariant. It is further required that line elements of T_P must "coincide" with those of M, issuing from P, that is, that there must be a one-to-one affine representation of P and T_P. Now these conditions are satisfied only if G is the affine group, but not a more extensive group, a considerable hindrance to the supposed generality of the Cartan approach. Summarizing his criticism of Cartan, Weyl noted that the "tangent plane" T_P is not as yet uniquely determined by the nature of M, and

> so long as this is not accomplished we can not say that Cartan's theory deals only with the manifold M. Conversely, the tangent plane in P in the ordinary sense, that is, a linear manifold of line elements in P, is a centered affine space; its group G is not a matter of convention. This has always appeared to me to be a deficiency of the theory....[106]

It should be pointed out that Weyl in the same year followed his own advice regarding the unique suitability of the affine group G, when giving gauge transformations their modern meaning as pertaining to a factor of "phase", rather than of scale. In this 1929 paper, Weyl set the Dirac theory of the electron into the frame of general relativity by systematically employing the *vierbein* or tetrad formalism for the representation of two component spinors in four-dimensional space-time. In Weyl's conception, a tetrad at P' arises from one at P through parallel transport of a vector according to the affine connection uniquely associated with the metric $g_{\mu\nu}$ of M. Then the tetrad at P' is related to the one associated with P by an infinitesimal rotation that is linearly dependent on the displacement PP' or its components. This meant that the class of local orthonormal tetrads, connected together by the infinitesimal rotation group, is determined by the metric; only a single member is selected arbitrarily.[107] In fact, one can look at the situation rather differently. Arguably, the principal interest of the tetrad formalism is that tetrads are determined by the metrics only up to the local transformations of a six-parameter internal Lorentz group, where the physics of the Dirac theory chiefly lie. Thus, Weyl's attempt to bind the new physics to his "prejudices" regarding infinitesimal geometry may appear somewhat forced.

In any case, Weyl's criticism of Cartan's failure to clearly establish a closer connection between the coordinates of the spaces Σ_P and the manifold coordinates x^α

of P was somewhat tempered in a review article of 1938. Later on, the mathematical fecundity of Cartan's approach became undeniable. By 1949, Weyl had wholeheartedly embraced it, even to the point of deeming it "imperative" for bringing the Dirac theory of the electron into general relativity. At that time, he made what could be said to be a full penitence.

> It is not advisable to bind the frame of reference in Σ_P to the coordinates x^i covering the neighborhood of P in M. In this respect the old treatment of affinely connected manifolds is misleading.[108]

It remains to try to reconstruct the reasoning behind Weyl's eventually discarded objection to Cartan. From the discussions earlier in this chapter, it would appear that a philosophical contention, indeed, a phenomenological one, underlies the stated mathematical reasons that kept him for a number of years from concurring with Cartan's "moving frame" approach to differential geometry. In view of the fact that the Cartan approach underlies the modern geometric formalism of gauge fields, it is of some interest to consider whether Weyl's objections appear to apply here as well.

A rather straightforward speculation is that behind Weyl's mathematical objection are concerns pertaining to the reconstructive program of transcendental-phenomenological "constitution" of physics through a "world geometry". In an obvious way, the geometries of Cartan's generalized spaces cannot serve as "world geometries", that is, "regional ontologies", for classical field physics since nothing in classical physics extends beyond M (i.e., possibilities of all space-time coincidences) alone. But a further "phenomenological" reason underlies the unsuitability of Cartan geometries to play the role of "world geometries". Constitution in virtue of "sense-bestowal" must originate in what is given, in essential insight, within consciousness, in the homogenous, linear infinitesimal spaces that are the form of intuition. But recall from §6.2 above that "the general form of consciousness" is a "compenetration of "This" and "Thus", of sensation and intuition, an overlap of continuous quality and continuous extension" that "phenomenologically, one cannot get beyond". But then we see that the *new conception of an event*, however "local", cannot be grounded entirely on such evident meanings exhibited through the forms of intuition, space, and time. In the Cartan conception, a property or an aspect of *quality* of a spatiotemporal event does not overlap but lies "above" the manifold of continuous extension, the continuum of spatiotemporal coincidences. Then the event itself cannot be constituted from within consciousness because its "general form" requires the "insoluble unity of intuition and sensation".

The lack of overlap of extension and quality appears all the more clearly in the fiber bundle representation of gauge field theory where a gauge group is a particularly important bundle, a "principal fiber bundle" whose fibers are just a given (Lie) gauge group G. Given a principal bundle, the interaction dynamics of the matter field along a timelike curve γ in M can be represented by first associating the various associated bundles of fibers Σ_P with the principal bundle; for the matter fields of quantum field theory, these are all vector bundles D whose fibers are vector spaces. Now there are two fibers Σ_P associated with each $P \in M$, one belonging to the principal bundle B, and one belonging to the associated vector bundle D. Each is interpreted as an entity or event at the space-time point P_x, whereas the points along each fiber Σ_P are possible characteristics of this event. Here, let us say, the points $\psi(x)$ along the curve $\bar{\gamma}$ in the principal bundle B represent the interaction potential A_μ while the points $\phi(x)$ in the associated vector bundle D represent the phase factor of the wave function of the matter field. Then a "connection" in the principal fiber bundle, determines the phase of matter field by picking out a particular point $\phi(x)$ in D. The gauge field itself is the "curvature" of this connection, determined by going round a closed loop from fiber to fiber in the principal bundle B. As one can readily see, the interaction dynamics are no longer locally characterized uniquely *within the manifold M*. In the fiber bundle characterization, events have two "dimensions", a "qualitative" character (of A_μ, of phase) and a dimension of extension (represented by the space-time coordinates x^α). Thus in the new conception, the aspects of *extension* and *quality* clearly are *no longer an "insoluble unity"* but relate rather simply (in the local trivialization of fibers) as an *ordered pair*. Lacking the richer "compenetration" of the "This" and "Thus" that is the "general form of consciousness", there appears to be *an insurmountable barrier to transcendental-phenomenological constitution of gauge field theories.* Countenancing possibilities beyond the reach of *Evidenz* of phenomenological intuition would *prima facie* appear to be completely unsupportable in anything like Weyl's program of a world geometric constitution of field physics.

However, it is not at all clear that "phenomenological intuition", together with the whole phenomenological method of "essential analysis", need be so closely yoked to the geometry of physical space and time. While not surrendering the core of transcendental idealism that space and time are necessary forms of intuition, there seems to be no reason, except the weight of Kantian tradition, to tie transcendental constitution of geometrical objects to constitution in three-dimensional physical space or indeed to the four-dimensional manifold of space-time. There is little in Husserl's phenomenology that would endorse such a limitation; to the contrary, there is much (going back as far as the "categorial intuition" of the *Logical Investigations*) to speak against it. In the event, such a limitation rules out the possibility of continuing the program of transcendental idealism to the physics of the present. What is needed is a general phenomenological account of "extension" and its relation to "quality" that encompasses as well the "internal spaces" of quantum gauge field theory. One analogy is found in the tetrad formalism of general relativity, since in that case there is a "solder form" that permits an identification of the tetrad frame bundle in terms of the manifold M, largely in the way suggested already by Weyl above.[109] While such a form is lacking for other bundles, a possible construction is already indicated by the notion of a "principal connection", pointedly depending on the affine connection of the metric of space-time,[110] built up from the transcendental-phenomenological core of Weyl's "pure infinitesimal geometry".

6.6 The Gauge Principle—A Sedimented History

6.6.1 Modern Gauge Invariance: The Basic Idea

The term "gauge theory" is a collective name for a variety of quantum field theories of fundamental interactions having the common feature that their physical predictions remain invariant under a group of transformations ("local gauge transformations") of the basic variables of the field theory. The fundamental idea can be clearly seen already in classical electromagnetism, "the archetype of a gauge theory". Taking as field variables the electromagnetic four potential $A\mu$, the physical predictions of the Maxwell theory remain invariant under the "gauge transformation" of the four potential that involves the addition of a gradient of a scalar function

$$A_\mu \Rightarrow A'_\mu = A_\mu + \partial_\mu \alpha \qquad (24)$$

where $\partial_\mu \alpha = (\partial \alpha(x^\mu)/\partial x^\mu)$. The transformation *introduces* an *arbitrary* (suitably differentiable) scalar function $\alpha(x)$ of space-time. Not only the Maxwell equations but also the electromagnetic field tensor

$$F_{\mu\nu} = \frac{\partial A_\nu}{\partial x^\mu} - \frac{\partial A_\mu}{\partial x^\nu} \qquad (25)$$

is invariant under substitutions of this kind, a fact already exploited by Weyl. In other words, the (physically measurable) field strengths are given by the derivative of the vector potential A_μ, the field tensor $F_{\mu\nu}$, and not by the (absolute) values of the vector potential. The introduction of an arbitrary position-dependent function is characteristic of "local gauge transformations" ("gauge transformations of the second kind") and the reason why the "gauge principle" is sometimes called the *principle of local symmetry.*

From this simple case, an illustration of how the "gauge principle" or "principle of local symmetry" is implemented in a quantum gauge field theory is readily understandable.[111] In its usual form, a "gauge argument" is invoked, requiring the invariance of the field Lagrangian density $\mathscr{L}$ in passing from a global to a local internal symmetry. This mandates that an "internal parameter" characterizing the global symmetry is required to vary as a function of space-time. As the simplest example, in quantum electrodynamics, one begins with a free electron field $\Psi(x)$ that is determined up to a phase factor θ, its free electron (Dirac) Lagrange density is invariant under the *global* phase transformation,

$$\Psi(x) \Rightarrow \Psi'(x) = e^{i\theta}\Psi(x). \qquad (26)$$

The requirement of local symmetry demands that the phase parameter θ vary as a function of position x, so that the phase invariance is *local*:

$$\Psi(x) \Rightarrow \Psi'(x) = e^{-i\theta(x)}\Psi(x) \qquad (27)$$

Any "realistic" Lagrangian $\mathscr{L}$ depends not only on the field magnitudes or potentials but also on their derivatives. However, in requiring local symmetry, the derivative $\partial_\mu \Psi(x)$ picks up an extraneous term $\partial_\mu \theta(x)$ in its transformation, and as $\theta(x)$ is a function of space-time, it is not a covariant object. In order to cancel this term, a "gauge covariant derivative" is introduced,

$$\partial_\mu \Rightarrow D_\mu := \partial_\mu + ieA_\mu, \qquad (28)$$

where the new derivative transforms as

$$D_\mu \Psi \Rightarrow D_\mu' \Psi = e^{ie\theta(x)} D_\mu \Psi \qquad (29)$$

and $A_\mu(x)$ is an "invented" vector field required to transform as

$$A_\mu(x) \Rightarrow A'_\mu(x) = A_\mu(x) - \partial_\mu \theta(x)/e \qquad (30)$$

The resulting Lagrangian for the field is then invariant under the joint local transformation of $\Psi(x)$, given in (27), *and* of $A_\mu(x)$ in (30), the added partial derivative exactly compensating the extraneous term of position-dependent variation of the phase factor. Yet in imposing the requirement of local symmetry, the free electron field is coupled to the electromagnetic field through the introduction of the four potential $A_\mu(x)$. This is a canonical illustration of the statement that *local symmetries dictate the form of the interaction.*[112]

6.6.2 *Sedimentation*

The renaissance of interest in the origins of the gauge principle since the 1970s has brought considerable historical attention from physicists and philosophers of physics to Weyl's 1918 theory of "gravitation and electricity", the locus of origin of the gauge principle.[113] Befitting their emphasis on present developments, most of these investigations invariably assess Weyl's idea of gauge invariance as "a classical case of a good idea that was discovered before its time",[114] the precursor of a conception that has occupied such a prominent role (because gauge-invariant quantum field theories are renormalizable) in recent fundamental physical theory. Naturally, the transcendental-phenomenological impetus behind Weyl's "pure infinitesimal geometry" has been effaced in these studies, and its detectable signs in Weyl's 1918 theory have been understandably ignored or downplayed. Instead, the perceived continuities between Weyl's theory (1918–1923) and later gauge theories rest on a mere formal analogy: Invariance of field laws under a "gauge transformation" containing an arbitrary function $\Lambda(x)$ of space-time:

$$A_\mu(x) \Rightarrow A'_\mu(x) = A_\mu(x) - \partial_\mu \Lambda(x) \qquad (31)$$

and on Weyl's *deliberate* retention of the term "gauge" in 1929 for a theory involving an arbitrary function of position.[115] Entering the world as "the epistemological principle of relativity of magnitude", the centerpiece of an attempt to constitute the world of space-time physics as having the sense of an "intentional being constituted for consciousness", the gauge principle today "is generally regarded as the most fundamental cornerstone of modern theoretical physics".[116]

Undoubtedly, the striking empirical success of the gauge field theories comprising the so-called Standard Model of fundamental interactions is one of the glories of 20th century physics, and on this basis, leading theoretical physicists have proclaimed their belief that "the secret of nature is symmetry".[117] But how is such a claim to be understood? In view of the hardly empirical status of the gauge principle in contemporary field theory, it would appear that there are less extreme options than the mathematical Platonism suggested or implied by some theoreticians and (even) philosophers. In this regard, Weyl's original example of a transcendental constitutive account of the *a priori* constraints of "reasonableness" placed upon fundamental physical theory serves at least as an illustrative example, although the gauge principle has since acquired what Husserl termed a "sedimented history". Through assimilation to the "everyday discourse" within the "natural attitude" of contemporary theoretical physics, the original epistemological intent of the principle has been subverted, all taint of its transcendental constitutive origin completely sanitized. Practicing "the Galilean style" of the mathematization of nature, contemporary gauge theorists, from the viewpoint of Husserl (and the gauge principle's creator), have carried out

> the surreptitious substitution of the mathematically substructed world of idealities for the only real world, the one that is actually given through perception, that is ever experienced and experienceable....[118]

Undoubtedly this has been certainly a fruitful maneuver for fundamental physics. But in view of the mélange of philosophical perplexity currently surrounding the significance of "local symmetries",[119] that it is similarly beneficial in philosophy of physics would appear to be altogether another matter.

7

"WORLD BUILDING"

Structuralism and Transcendental
Idealism in Eddington

> Thus the order and regularity in appearances that we call Nature, we
> introduce ourselves, and indeed we could not find them there had
> not we, or the nature of our mind, put them there originally.
>
> Kant, *Critique of Pure Reason* (A125)

> In the end what we comprehend about the universe is precisely that
> which we put into the universe to make it comprehensible.
>
> Eddington (1936, 328)

7.1 Introduction

A. S. Eddington's enormous stature in early 1920s Britain as *the* prominent representative of the new relativity theory considerably postponed critical response to his emerging philosophy of physical science. But from the publication of his 1927 Gifford lectures onward,[1] what was vicariously taken to be his philosophy consistently met with severe reproach, if not outright hostility. Wittgenstein, to cite an extreme example, is reported as contemptuously scowling that "he would rather be in Hell by himself than in Heaven with Eddington".[2] Others have been more explicit regarding the grounds of indictment. A. O. Lovejoy took issue with Eddington's claim that "the stuff of the world is mind-stuff", judging it the expression of a subjective idealism akin to Berkeleyan phenomenalism.[3] Bertrand Russell and

Philipp Frank invoked ulterior motives, declaring Eddington's "idealist" account of scientific knowledge compromised through subjugation to the demands of religious belief.[4] Norman Campbell, a renowned experimental physicist, accused Eddington of "distorting science" by encouraging confusion between laws and theories and of sliding into overstatement about the epistemological significance of "pointer-readings".[5] To Herbert Dingle, Eddington was a "modern Aristotelian" (together with E. A. Milne and P.A.M. Dirac), worshiping at the idol of a rationalistic cosmophysics, the arch-practitioner of a dangerous rite posing the "radical danger" to the foundations of science of prostrating observation before the golden calf of mathematical invention.[6] Nearly all took exception to Eddington's infamous assertions to the effect that "there is nothing in the whole system of laws of physics that cannot be deduced unambiguously from epistemological considerations".[7] His repeated declarations that the fundamental laws and constants of nature are ultimately epistemological in nature became the target of widespread ridicule, even occasioning a satirical note in a learned scientific journal.[8] Nevertheless, a small current of informed opinion has long resisted the Bœotian outcry, seeking to reconstruct or restate the argument Eddington's difficult last works, in the hope of clearly pinpointing where it goes off the rails. It must be admitted that, despite the efforts of physicists and mathematicians running from Schrödinger,[9] E. T. Whittaker,[10] H. Jeffreys,[11] through E. Bastin and C. Kilmister,[12] these efforts have met with only partial success. While none of his critics could question Eddington's brilliance nor his enormous scientific contributions, the mathematical arguments of *The Relativity Theory of Protons and Electrons* (1936) and the posthumously published *Fundamental Theory* (1946) have not inspired overmuch confidence that his epistemological conclusions actually follow from stated premises. For their part, the philosophical community assessed his overtly philosophical writings as at least confusing and, many have concluded, confused. Stebbing's book-length critique of 1937, widely read at the time, complained that Eddington had "nowhere expounded his philosophical ideas in non-popular language", even as she recognized that these ideas were intimately bound up with technical publications and works, she freely admitted, she was not competent to evaluate.[13] The observation that "essentially Kantian notions"[14] are to be found at the basis of Eddington's philosophy of physical science more or less underscore what Eddington himself averred. Yet the nature of his connection to Kantian philosophy, or more generally, to transcendental idealism, has remained largely unexplored, despite the revealing clue that "the idealistic tinge in my conception of the physical world arose out of mathematical researches on the relativity theory".[15] To be sure, philosophers and physicists alike have largely accepted on faith that scant little, if anything, of transcendental idealism might survive in the aftermath of the theory of relativity. For this reason alone, the statement above appears all the more puzzling.

It is now many decades since Eddington was a topic of controversy within philosophy of science (he died of cancer in 1944, at the relatively young age of 62); indeed, the last serious attempts to survey his philosophy as a whole stem from the period 1958–1965. In the absence of countervailing exegesis or analysis, the damning indictment of the prewar period has remained in place, even though now only known at second- or third-hand. To be sure, a recent reawakening of interest as emerged among contemporary "structural realists" who relish Eddington's explicit championing of the "structural thesis", that physics is competent only to provide knowledge of the structure of an external world but nothing of its content, or essential nature.[16] Herbert Feigl, some time ago, linked Eddington with alleged proto-structural realists (such as Schlick and Russell), with the ultimate aim of finding an adequate resolution of the mind-body problem.[17] While the agenda of contemporary structural realists is not so ambitious, a still relevant, although subsidiary, matter concerns whether Eddington's "structuralism" similarly falls prey to the fatal objection to Russell's structuralism raised by the Cambridge mathematician M.H.A. Newman in 1928. I think not, for reasons discussed at the end of this chapter. But contrary to structural realism, my primary objective in this chapter, and in the next, is to show that Eddington's "structural thesis" is a fundamental plank of that "idealistic tinge in [his] conception of the physical world" arising expressly in the context of his "mathematical researches on relativity theory". In perhaps an unexpected manner, Weyl's 1918 theory of gravitation and electromagnetism, introducing the constitutive character of a "world geometry", will be a highly significant component of that context.

The epistemological conclusions that Eddington drew from the theory of relativity—of which the "structural thesis" is but one—were set in train by Weyl's own treatment of general relativity in the frame of a "world geometry". I have already shown that, as early as 1920, Weyl became skeptical of the maximal goal of the unified field theory program, the derivation of the particulate structure of matter within the frame of continuous field functions of space-time geometry. From then on, elementary particles appeared in Weyl's theory to be true singularities of the field (as literally, "beyond the field"), marking the limits of classical field theory. Nonetheless, Weyl never surrendered his epistemologically motivated demand for explanation of the behavior of rods and clocks by a "systematic theory" of matter-field interactions. In chapter 8, we shall see that Eddington's 1921 generalization of Weyl's theory of gravitation and electromagnetism had a similar epistemological motivation, aiming not at the "unknown laws" of matter (viz., the "non-Maxwellian forces" holding electrons together), but at throwing "new light . . . on the origin of the fundamental laws of physics". My aim in the present chapter is to set in place the several steps of Eddington's complex epistemological argument forming the background of his 1921 paper and, indeed, laying the foundations of his controversial philosophy of physical science. In particular, I will examine his argument that relativity theory has transformed the concept of physical knowledge as pertaining only to the external world's *structure*, showing how this conclusion is necessary to his epistemological account of the *origin* of the fundamental field laws. Eddington's explanation of this origin—citing the mind's activity in selectively cloaking the bare skeleton of geometric "world structure" with measurable physical quantities—gives expression to a transcendental idealism *updated* by relativity theory.

Relying on the "structural thesis" Eddington found demonstrated in relativity theory, his epistemological argument crucially employs a deductive reconstruction of that theory termed "world building". Extensively developed in his aptly named text of 1923, *The Mathematical Theory of Relativity* (hereafter, *MTR*), "world building" can be, and usually has been, viewed simply as an informal axiomatic presentation of space-time theory from primitive posits of differential geometry. Clues to its specifically epistemological character, however, can be found in several of Eddington's semipopular and philosophical writings in the 1920s, works that, with hindsight, prove invaluable in identifying the epistemological intent of this method. Accordingly, after first introducing Eddington as "The Apostle of Relativity" in Britain, in §7.3 I consider the argument for the "structural thesis" as informally presented in his semipopular book on relativity, *Space, Time and Gravitation* (hereinafter, *STG*), as well as other nontechnical writings of the early 1920s. I then outline, in §7.4, how the structural thesis is requisite to "world building" through an informal discussion of that method largely based on two publications in *Mind* in 1920. In turn, this provides a discursive roadmap for portrayal, in §7.5, of the explicit reconstruction of relativity theory that issues in the notorious claim that the Einstein field equations are merely "definitions". This apparently outrageous conclusion is to be interpreted as Eddington's idiosyncratic way of expressing the essential content of the Kantian dictum that "reason has insight only into that which it produces after a plan of its own". That Eddington, as Kant, thought there are definite limits to scientific knowledge serving to demarcate it from other kinds of experience ("nonmetrical" to use Eddington's term), will only be briefly touched upon in §7.6.2. Nor is this the place to track the subsequent development of Eddington's epistemology (alternatively termed "structuralism" and "selective subjectivism") from its origins in his engagement with relativity theory in the early 1920s to his later attempts to build an epistemological "bridge" between general relativity and quantum theory.[18] But I distinguish, in §7.6.1, Eddington's "structural thesis" from apparently similar views articulated by Russell, and so from recent discussions assimilating both to an epistemic variety of "structural realism".

7.2 The "Apostle of Relativity"

First published in 1920, *STG* is still widely regarded as one of the most successful lay expositions of the theory of relativity, its considerable popularity assisted by a delightful prose style. Reading it in English and not his native tongue, the Munich theoretical physicist Arnold Sommerfeld declared to Einstein that he was "captivated" (*entzückt*) by it, confessing "I know of no book that is as well-written".[19] But in fact, *STG* and, as I will show, *MTR* present the theory of relativity from a distinctive perspective that was entirely Eddington's own, incorporating Weyl's epistemological principle of "relativity of length" as an already essential component of "the relativity standpoint". Many if not most readers of these works, each a classic in its own way, have not been aware of just how heterodox this perspective was, and still remains. In no small measure, the failure is due to the fact that in Britain, and hence in the English-speaking world, Eddington's name was indelibly linked to the theory from the very beginnings of professional and lay awareness of its existence. The reasons for this are worth considering more closely.

During the World War, Eddington had been the first in Britain to learn of the completed general theory, receiving copies of Einstein's 1915 and 1916 *Proceedings of the Berlin Academy* publications through the astronomer Willem de Sitter in neutral Holland. Eddington provided a first report of the theory to British scientists in an issue of *Nature* in December 1916, notably declaring that, in the new theory, "the space and time of physics are merely a mental scaffolding in which for our own convenience we locate the observable phenomena of Nature".[20] He then arranged for De Sitter to write a series of papers on the theory and its cosmological implications that appeared in 1917 and 1918 in the *Monthly Notices of the Royal Astronomical Society*. Together with the expository papers of H. A. Lorentz, buried in the *Proceedings* of the Dutch Academy in 1916 and 1917, these were the first detailed publications on the theory to appear in English. In 1918, at the behest of the Physical Society of London, Eddington, the first second-year student to have won (in 1904) the distinction of "Senior Wrangler" at Cambridge, wrote the first comprehensive account of general relativity to be generally available in English. The *Report on the Relativity Theory of Gravitation* quickly sold out; a second edition appeared eighteen months later.[21] In the assessment of the distinguished astrophysicist S. Chandrasekhar in 1982, this is a work "written so clearly and yet so concisely that it can be read, even today, as a good introductory text by a beginning student".[22]

Of course, in the storied annals of relativity theory, Eddington is best known for his pivotal role in providing the first empirical confirmation of Einstein's theory. Under wartime conditions rendered even more difficult by Eddington's utter indifference to the likelihood of prosecution for his conscientious objection to the war, the Royal Astronomer Sir F. W. Dyson managed to organize and obtain funding for a joint Royal Society and Royal Astronomical Society expedition to test Einstein's prediction of the deflection of starlight passing near the sun. Within a few months of the end of the war, teams were sent out on 8 March 1919 to the equatorial regions—Sobral in Brazil and Principe in the Gulf of Guinea off West Africa—to make the famous solar eclipse observations of 29 May 1919. Eddington himself led the group going to Principe. On the fateful day, the observations at Principe were almost completely obscured by cloud cover. But several days later Eddington experienced "the greatest moment of my life" when measurements on one 10×8 inch photographic plate "gave a result agreeing with Einstein".[23] After months of data analysis, the expedition's results were announced in London at a joint session of the Royal Society and the Royal Astronomical Societies on 6 November 1919, a meeting whose "whole atmosphere of tense interest", according to Whitehead, "was exactly that of the Greek drama".[24] In the clamor that followed, without precedent and as yet unmatched by a purely scientific development lacking technological or military implications, Eddington quickly acquired in Britain a reputation as "the apostle of relativity theory". By December 1919, as he wrote to Einstein, Eddington had lectured to a "huge audience" at the Cambridge Philosophical Society, with hundreds more turned away at the door.[25] But Eddington was not merely the leading evangelist of the new theory in Britain; he increasingly was identified with it, both as technical expositor and as popularizer, to an extent that is somewhat difficult to comprehend today. P.A.M. Dirac's reminiscence of these early years of general relativity provides decisive testimony. Emphasizing the extent of Eddington's fame in association with the theory, Dirac recalled near the end of his life that

Einstein was rather a remote figure in a foreign country. A person who was much more present was Eddington. He was the leader of relativity in England at that time. He was the great authority whom everyone listened to with the greatest respect, and he was rather regarded as the chief exponent of relativity. Einstein was in the remote background.... [Eddington] was the fountainhead of relativity so far as England was concerned. Einstein was just too remote to count.[26]

In these early years of general relativity, Dirac elsewhere noted, the only "proper understanding" of Einstein's gravitational theory available to one who didn't read German had to be acquired from Eddington's *MTR*. This is also a book widely praised for the clarity and elegance of its exposition. Yet it customary to find sincere appreciation of *MTR* as a masterful treatment of general relativity coupled with a more critical or cautious attitude toward the "philosophy" the book contained, as Dirac also revealed:

> We really had no chance to understand relativity properly until 1923, when Eddington published his book, *The Mathematical Theory of Relativity*, which contained all the information needed for a proper understanding of the basis of the theory. This mathematical information was interspersed with a lot of philosophy. Eddington had his own philosophical views, which, I believe, were somewhat different from Einstein's, but developed from them.[27]

One could wish that Dirac had revealed the grounds for his intriguing assessment that Eddington's philosophical views "developed from (Einstein's)". Nonetheless, the comment is instructive, and probably representative, for its juxtaposition of praise for Eddington's mastery of (not to mention original contributions to) the theory with an implied incomprehension of, or distancing from, his "philosophical views", particularly as these became to be recognized as heretical, beginning in the mid-1920s.

In short, in the early years of relativity theory, the very heterodox character of Eddington's conception of the physical knowledge, transformed by what he termed "the relativistic outlook", would not have been readily recognizable by an English-only reader in the absence of other baseline presentations of general relativity. As a result, in the first decade of the general theory in Britain, Eddington's pronouncements on relativity were, to cite only the example of Russell,[28] widely taken to be authoritative *regarding both the theory and its philosophy*. Even so, there was no general agreement as to just what this philosophy was nor how, exactly, it followed from the physical theory of relativity. One cannot but agree with those critics who contend that at least some of the responsibility for this failure must be laid on his shoulders. Lay readers, limited to *STG*, or other of Eddington's best-selling popularizations of science, were expected to grasp the philosophical implications of modern physics through elegantly crafted parables, illustrative analogies and clever tropes. It has been uniformly acknowledged that this strategy could hardly succeed, and it famously elicited the attack of L. Susan Stebbing, who railed that "in his desire to be entertaining, [Eddington] befools the reader into a state of serious mental confusion".[29] On the other hand, the explicitly philosophical passages in his technical treatises, such as *MTR*, could be, and so usually were, ignored and in any case were so pithily expressed as to seem merely quixotic. The result is that the popular or semipopular writings and the technical treatises appeared to be complementary representations, impossible to hold simultaneously before the mind.

To a considerable extent, Eddington's failure to clearly communicate his philosophical views stemmed from the fact that he was neither trained as a philosopher, nor even widely read in philosophy. Not until his Tarner Lectures in 1938, when he ventured a certain loose parallel with Kant, did he provide much of orientation for situating an epistemology originating apparently in reflections on the new physics.[30] This independent development is yet a further fundamental source of difficulty in comprehending Eddington's philosophy of physical science which, in the 1930s, was increasingly linked to heretical researches on a "bridge" between general relativity and quantum theory.[31] But by his own admission, his philosophical views emerged in the course of his mathematical contributions to relativity theory:

> [T]he idealistic tinge in my conception of the physical world arose out of mathematical researches on the relativity theory. In so far as I had any earlier philosophical views, they were of an entirely different complexion.[32]

As noted above, my account of how Eddington's epistemology of physical knowledge developed out of the theory of relativity will proceed by considering these "mathematical researches" against the backdrop of the more expansive voicing of this "idealistic tinge" in his less technically forbidding works. With this assistance, his conception of the physical world can be identified as a form of transcendental idealism, specifically tailored by what Eddington referred to generally as "the relativity standpoint", that is,

> a discarding of certain hypotheses, which are uncalled for by any known facts, and stand in the way of an understanding of the simplicity of nature.[33]

It is worth pointing out that the deductive presentation of "world building" and the accompanying transcendental idealism is entirely absent from Eddington's first work on relativity theory, the above-mentioned *Report on the Relativity Theory of Gravitation*. Rather the stimulus provided by the mentioned "mathematical researches" concerns his reconstruction of gravitational and electromagnetic theory in the context of a "world geometry", a conception that Eddington discovered in Weyl only after the *Report* was written, and the first edition published. In the considerable, now mostly dated, literature on Eddington, the determinative stimulus of Weyl's notion of a "world geometry" and its guiding "principle of relativity of length" on the formation of Eddington's epistemological views has remained largely unrecognized.[34]

7.3 The Point of View of No One in Particular

With hindsight, the initial step of Eddington's argument is already recognizable early on in *STG* in the guise of a question setting out the epistemological theme of the book, and indeed of what thereafter will be variously termed "the relativity standpoint" or "the relativistic outlook". Is it possible to form a conception of the physical world from "the point of view of no one in particular", a completely "impersonal picture of the world"?[35] Doubtless, the question appears familiar enough, the customary rhetorical device epitomizing just what general relativity theory, as ordinarily understood, has achieved—a characterization of the physical world independent of the state of motion of any particular observer. Such an impersonal picture is glossed as one comprehending the viewpoints of "all conceivable observers". Notably, but nearly imperceptibly, the net of "conceivability" is cast wider than that of the Einstein theory. Besides including observers having any relative positions or motions, it also comprises observers who may be "shrinking" or "expanding", such as are allowed by Weyl's "principle of relativity of magnitude". Such observers, however nonexistent, are conceivable, while failure to take into consideration "all conceivable observers" means taking "the side of the Inquisition against Galileo".[36] Then the relative circumstances of any particular observer (a notion comprehending "all his measuring appliances") must include as well his "gauge of magnitude" and the sought-for point of view must not give any preference here either.

> The circumstances of an observer which affect his observations are his position, motion, and gauge of magnitude. . . . Position, motion, magnitude-scale—these factors have a profound influence on the aspect of the world to us. Can we form a picture of the world which shall be a synthesis of what is seen by observers in all sorts of positions, having all sorts of velocities, and all sorts of sizes.[37]

Now the "synthesis" in question is to be a conception independent of the particular circumstances of any given observer and so might well be considered what physicists believe to be "inherent in the external world". On the other hand, it can plausibly be regarded, Cassirer similarly observed in 1920, as a reasonable construal of what is actually *meant* in speaking this way; namely,

> just this independence from the arbitrary standpoint of the observer . . . is meant in speaking of a determinate object of "nature" and of determinate "laws of nature".[38]

Such a conception of the world may only be "grasped" and not really "pictured" since our senses do not fully equip us for forming such a picture.[39] Indeed, this will become a fundamental postulate of Eddington's epistemology of physical science. A standpoint-independent world can be represented only conceptually and structurally, although it arises as a "synthesis" of the spatiotemporal determinations of all conceivable observers, the aggregate of all sensible representations, as it were.

7.3.1 The Necessary Role of Synthesis

Interspersed by several chapters of sparkling semipopular exposition of the usual topics (the four-dimensional world of events, non-Euclidean geometries, space-time curvature, the Einstein law of gravitation and its empirical tests, the Einstein and De Sitter cosmologies) the quest for an answer to this question is pursued through the book. The argument seeks to show that the relativity standpoint is a requirement that physical knowledge, based as it is on measurable physical quantities, in particular *lengths* and *durations*, must be stated in a form that takes into account the possible measurements of wider and wider classes of observers. Thus the four dimensional geometric representation of the world of Minkowski is a "synthesis" required for taking into account the measurements of observers in all states of uniform motion with respect to one another, and so not privileging any single inertial observer. Then, when Einstein "extended the synthesis" to include all possible relative states of motion, encompassing accelerating and rotating observers, the geometrical representation required the non-Euclidean variable curvature geometry of Riemann. Finally, as the perspectives of observers varying in magnitude in any way are taken into account, the synthetic representation can no longer be contained in Riemannian geometry, but requires the broader geometry of Weyl. (Eddington's still more general geometry was to be introduced in the following year.) It is allowed that this last step has not met with universal acceptance, and in any case, it is not taken up until the penultimate chapter of the book, devoted to Weyl's theory. But there, at least, a "natural halting-place (is) reached", although no claim is made that further steps will not be forthcoming.[40] Weyl's geometry provided the synthetic means for completing the desired "impersonal" conception of the world, one capable of taking into account the physical aspects presented to "all conceivable observers", or at least all observers conceivable from the vantage point of the physics of the early 1920s.

In this way, adoption of the "relativity standpoint" affords a *definition* of "the real world of physics" or "physical reality" or "the external world of physics" (terms all used interchangeably, to the annoyance of Stebbing, and Dingle[41]). Such a world is "capable of precise definition" as "the common element abstracted from the experiences of individuals in all variety of physical circumstances."[42]

Talk of abstraction is not to be taken literally; the vast majority of these observers, and their purported experiences, do not exist except in conception. The more apt metaphor is that of "synthesis": this impersonal external world or "physical reality" is defined as a "synthesis" of all its "possible physical aspects" or "possible points of view": "Physical reality is the synthesis of all possible physical aspects of nature";[43] "The external world is a synthesis of appearances from all possible points of view.... (rightly or wrongly) the result of the synthesis [is] to be the real external world".[44]

Ultimately, if the "synthesis" is performed correctly, physical quantities must be represented by tensors in four-dimensional space-time; tensors (or, as I will show, scale-invariant "in-tensors") alone enable intrinsic properties and objects to be distinguished from those that are frame (and scale) dependent. Such a proscription for the form of a physical knowledge gleaned from the measurements of all possible observers takes all aspects into account equally but corresponds to no one of them, or rather, it corresponds to all of them jointly. The necessity of a tensor formulation of physical knowledge underscores Eddington's point that this mode of *definition* of the "external world" stems not from philosophy or from some metaphysical conception of reality. Rather, the intent is that it be recognized to be a definition *for physics*, in accord with the broad principle of definitions in science, "that a thing must be defined according to the way it is in practice recognised and not according to some ulterior significance that we imagine it to possess".[45]

This "domestic definition of existence for scientific purposes" is accordingly regarded as comprehending all that physics *can* meaningfully convey in speaking of "the real world" or "the external world" or "physical reality". The question of the "reality" of such a world is just the question of whether "the rules for forming the synthesis have been properly followed". Any further concern as to the "reality" of the external world, or as to "a knowledge of the world, which does not particularize the observer, but does not postulate an observer at all",[46] is of dubious intelligibility but in any case not of the slightest interest to physics. The truth sought in science is just the truth about this external world, as the latter has been *defined*; physical statements, if unambiguous, are statements about this defined world as the subject of inquiry, and so are either true or false.[47]

It is Eddington's *definition* of an impersonal external world as a "synthesis" of its relative aspects that attracts our attention, recalling Kant's statement (in the Transcendental Analytic) that "the most general meaning of synthesis" is "the action (of thought) of putting different representations together with each other and grasping their manifoldness [*Mannigfaltigkeit*] in one cognition" (A77/B103). Of course, the Kantian cognitive synthesis is intimately bound up with an 18th century psychology of "active" and "receptive" faculties of cognition, that is, of the understanding, and reason, as opposed to sensibility and intuition.[48] As emphatically affirmed in the opening paragraph of the B Deduction, *any* cognition of an object by the human mind necessarily involves the combination (*Verbindung*) by the *understanding* of a collection of diverse elements of information, including those *given* to *sensibility*, into a "synthetic unity". The object of knowledge only arises through such combination for "we cannot represent to ourselves anything as combined in the object which we have not ourselves previously combined" (B130). Moreover, the very concept of combination must precede any particular act of synthesis and is itself made possible by a representation of synthetic unity that Kant will identify as the transcendental unity of apperception, the "I think" that must accompany all my representations. This prototypical activity of the understanding is *a priori* and spontaneous, largely unconscious or, better, "*preconscious*" in the sense that it is the condition of the possibility of consciousness of an object, and so logically prior to it.[49]

Of course, the "synthesis of all relative aspects" neither takes place unconsciously nor involves active and passive faculties or strict claims of *a priori* necessity and universality attending the Kantian account. But the synthetic *definition* of the external world of physics abides this "most general meaning of synthesis"; taking place in a higher register, it is a *second-order synthesis* that relativity theory has shown to be required according to the *regulative* demand of completely impersonal objectivity. In the strictly Kantian sense, a synthesis *constitutive* of the object of knowledge necessarily involves the combination of a sensible manifold under the rule of a concept or combination of concepts. In such an account, a model of empirical cognition evidently stemming from a view of mathematical knowledge wherein objects are "constituted" by "construction in pure intuition", regulative principles, "prescribing to the understanding a direction of which it has itself no concept" (A326/B383) can play no role. Now the scale and clock readings of any particular observer might well be understood as the resultant of a synthesis of concepts and intuitions in the strictly Kantian sense of constitution. However, relativity theory has shown that, however necessary to cognition from a particular "point of view", from the impersonal standpoint demanded by relativity theory these can comprise at most "relative knowledge", knowledge that, as observer dependent, is not fully objective. Although Kant was cognizant of the purely kinematic Galilean relativity of Newtonian physics, the ideal of completely impersonal objectivity that is the conceptual core of relativity theory mandates conceptual or functional, rather than spatiotemporal, cognitive representation, breaking the confines of the Transcendental Aesthetic.

Nonetheless, as recounted in chapter 2, neo-Kantian currents prior to relativity theory already recognized that the original Kantian treatment of synthesis, and so of the *constitution* of the object of knowledge, required modification in the light of turn-of-the-century physical and mathematical science. With the theory of general relativity, Cassirer in particular emphasized, the "synthesis" of spatiotemporal frameworks according to the demand of general covariance becomes a *regulative ideal* or methodological norm, a further qualitative step toward a general structural account of physical objectivity.

In exact, but unwitting, agreement with Cassirer, Eddington, also in 1920, argued that in the light of relativity theory, the concept of object of "the external world of physics" mandates this further synthesis, *nello grosso modo*, of the manifold of all these relative physical aspects. Indeed, the aspects are "relative" precisely because they presuppose an initial cognitive synthesis (a measurement of length or duration) within some arbitrary partitioning of space-time into space *and* time, a representation in spatiotemporal intuition. Then the further synthetic combination of the collective manifold of "relative aspects" comes through the posited algebraic transformations between these "equally justified frames of reference". The original Kantian account of cognitive synthesis, however, does not allow what relativity theory mandates: that a further and "higher act of thought" is required to combine diverse spatiotemporal represented objects of knowledge (lengths, durations) into the desired "synthetic unity" of fully objective knowledge, that is, knowledge of a nature *defined* as satisfying the completely impersonal viewpoint. Relativity theory has demonstrated the need for such a synthetic extension beyond individual cognition by showing the frame dependence of "all the familiar terms of physics". Furthermore, the resultant of the "synthesis", common to all observers, is the *space-time* interval or other geometrical objects, invariant under the specified group of transformations linking the conceivable frames of reference. In sum, relativity theory has transformed the conception of physical knowledge, showing the necessity of this higher order synthesis by showing that spatiotemporal physical measurements can yield only "relative aspects" of an "absolute" world. On the other hand, such an absolute world is indeed not mind independent for it is *composed* by the "synthetic unity" of all of its relative aspects. By its very construction, it is a unity not representable in space and time, but only as a *structure* in a geometrical *conceptual space*. However, before further elucidation of Eddington's structural theme, we need to consider Eddington's often-misunderstood remarks on "pointer-readings".

7.3.2 Pointer-Readings

Relativity theory has called attention to "what it is that we really observe", demonstrating that the actual observational basis of physical knowledge is much poorer than usually supposed. What is actually seen ("at least in all exact measurements") in any observation, as opposed to what is inferred from what is seen, are "point-events", or as he will later state, "pointer-readings". Paradigmatically, such knowledge arises from observation: The pointer of a dial coincides with a number on the dial face of an instrument, or a mark on a measuring rod is brought into coincidence with the end point of a physical body. In space-time discourse, pointer-readings are intersections of world lines, providing "absolute knowledge independent of the observer", on which all observers can agree.[50] However, with their necessary reference to a *possible* observer, Eddington's talk of pointer-readings appears to make a characteristic mistake of positivist accounts of general relativity. In these (e.g., of P. Frank), the spatiotemporal "point-coincidences" that are the last vestige of the objectivity of space and time, the intersections of world lines invariant under all diffeomorphisms of the space-time manifold, are identified with coincidences of sense experience.[51]

One manner of realist response to this conflation, that of Reichenbach, distinguished the order of coincidences among physical things (light rays, world lines of particles) in "objective reality" from conditions of consciousness or perception.[52] This is not an option for Eddington. Like Cassirer, Eddington could understood talk of "objective reality" only as a reference to the aforementioned regulative ideal governing the mathematical synthesis of all possible measures made by all conceivable observers. At least within physics, insofar as physics is based on measurable quantities, that is all the physicist can mean by such talk. The ensuing geometrical theory of the world is, as I will show, just the world regarded from the point of view of no one in particular, that is, as a *structure*. But this is not a conception of objectivity particularly congenial to a positivism exclusively privileging the role of sense experience in cognition. Furthermore, the required synthesis, although based on observers' direct knowledge of pointer-readings, is also testimony that such immediate knowledge, while "absolute", is insufficient for the purposes of physics. Rather it must be recognized that measurable physical quantities, *lengths* and *durations* and the physical quantities dependent upon them, such as *mass*, *motion*, *velocity*, *force*, and *energy*, all make tacit reference to a specified observer and so by definition comprise "relative knowledge," a "relative aspect" of the impersonal world.[53] The synthetic conception of an impersonal objective reality only results if the "rules" are properly carried out regarding how the synthesis of all these relative aspects is performed. These rules encode the transformations between an ever-expanding set of conceivable observers, mathematical knowledge expressing *purely local constraints* on physical objectivity. In stark contrast to both positivism and to Reichenbach's separation of "subjective" and "objective" coincidence, and despite the terminology of "relative" and "absolute", the object of relativity theory is not to "attempt the hopeless task of apportioning responsibility between observer and the external world". It is rather to show that "the two factors are indissolubly united" through the synthesis.[54] The necessary synthetic unity will come through a "world geometry" axiomatically constructed to ensure that its invariant objects satisfy the objectivity postulate of "the point of view of no one in particular".

7.3.3 A Geometrical Theory of the "Real World of Physics"

In 1920, the synthetic definition of physical reality is framed in a metrical geometry of a four dimensional continuum of possible "point-events" that is neither space nor time. The aggregate of possible "point-events" is the "absolute world"; the geometry of this world posits an intrinsic quantitative relation of extension, the space-time interval ds^2 between any two neighboring events, an "absolute" relation postulating no particular observer. The interval may be thought "something intrinsic in external nature" defined by its geometrical properties, but these are "not to be taken as a guide to the real nature of the relation, which is altogether beyond our conception".[55] Without distances and durations, the operative meaning of space and time for physical measurement, the "absolute world" cannot be pictured or intuitively represented, but it can be "graphically represented" by geometrical structures. Doing so leads "to a geometrical theory of the world of physics". No claim is made that a geometric representation of the "absolute world" is final, or cannot be improved upon. Thus, Weyl's geometry appears to be a further improvement upon the Riemannian geometry of Einstein, as it allows, as the latter does not, "shrinking" or "expanding" observers, encompassing a wider class of possible points of view. But in virtue of the manner in which such a synthetic representation is fashioned, it represents "the real world of physics, arrived at in the recognised way by which physics has always (rightly or wrongly) sought for reality". The search for reality is just the quest for a completely impersonal viewpoint, and its unexpected result is a "geometrical theory of the world": "We did not consciously set out to construct a geometrical theory of the world; we were seeking physical reality by approved methods, and this is what has happened".[56]

The successive extension of the synthesis of viewpoints to include wider and wider classes of observers constitute objects that can only be accommodated within broader and broader geometries. This "genetic" route to a complete geometrical theory of the world roughly corresponds to the progressive development of the function concept in physics that Cassirer traced from Galileo and Newton to Einstein. But having arrived at such a terminus, a deductive approach is naturally suggested, a path Eddington deemed Weyl to have first explored. Under the provocative heading of "world building", it is the framework within which Eddington will develop the transcendental idealist conclusions of his emerging epistemology of physical science.

7.4 World Building: An Informal Dress Rehearsal

In the final chapter of *STG*, beguilingly entitled "The Nature of Things", and in two philosophical publications of the same year in the journal *Mind*, Eddington provided an informal dress rehearsal of the treatment of relativity theory according to the method of "world building" in *MTR*. Necessarily, in these lay venues, most mathematical details are omitted. The result is an extremely compressed cryptic argument drawing a number of surprising, and seemingly subjectively idealist, conclusions from the above conception of physical knowledge as constituted by a "synthesis" of all physically measurable "aspects" of the objects of an "absolute" world. In fact, these claims only emerge from reconstruction of the Einstein theory via the method of "world building", but they are nonetheless striking: that the Einstein law of gravitation is *definitional* in nature and that "matter can scarcely be said to exist apart from mind".[57] The deductive approach of "world building" is viewed as complementing the usual presentation of the theory of relativity, wherein the interval is stipulated to have an immediate physical interpretation in terms of measurements made by rods and clocks. However, its ultimate aim is rather larger, a transcendental idealist conclusion "that the mind's search for permanence has created the world of physics". Fleshing out this informal argument in some detail will prove useful in understanding the formal treatment of world building in §7.5.

7.4.1 The Epistemological Significance of Tensors

Up to this point, the argument threading through STG has been that, in giving rise to an impersonal, locally symmetrical, and completely geometrical, conception of "physical reality", relativity theory has also shown that the frame (and scale) dependent measurable quantities of physics can present only "aspects" of this reality. As a result, a gaping epistemological abyss appears to separate "the real world of physics", abstractly conceived, according to "the relativity theory of nature", as geometrical "world of point-events with their primary interval relations" from the familiar world of physical quantities and measurement.[58] How, then, is knowledge of this "real world of physics" possible and what manner of knowledge must it be? The problem may be compactly located in the ten independent components of the "metric tensor" $g_{\mu\nu}$ appearing in the mathematical definition of the interval ds^2. For the ten independent $g_{\mu\nu}$ "are concerned, not only with intrinsic properties of the world, but with our arbitrary system of identification numbers" (i.e., they are functions of the coordinates) from which numbers may be computed to express the "intensity" of the different components.[59] Hence, any particular values of the $g_{\mu\nu}$ irremediably blend together information about intrinsic properties of the "real world of physics" with the extrinsic characteristics of a single observer (reference frame).

Eddington's question concerning the possibility of knowledge of "the real world of physics" is then to be addressed in part by requiring that physical quantities be expressed *only in tensor form*. A physical magnitude expressed as a tensor has definite components once a basis is given in a chosen coordinate system, but, abstractly considered, it stands for its components in all coordinate systems. In this way, tensors furnish a means of winnowing the wheat of "intrinsic information" about the absolute four-dimensional world from the extrinsic chaff of particular perspectives ("mesh-systems"). Precisely because physical measurement presupposes choice of reference frame, tensor formulations alone cannot express measures of intrinsic conditions of the absolute world:

> A tensor does not express explicitly the measure of an intrinsic quality of the world, for some kind of mesh-system is essential to the idea of measurement of a property. . . .[60]

However, the *vanishing* of a tensor (since by definition, if the components of a tensor vanish in one coordinate system, they do in every system) or an *identity* between two tensors defined in the same region, does express intrinsic information about absolute properties:

But to state that a tensor vanishes, or that it is equal to another tensor in the same region, is a statement of an intrinsic property, quite independent of the mesh system chosen. Thus by keeping entirely to tensors, we contrive that there shall be behind our formulae an undercurrent of information having reference to the intrinsic state of the world.[61]

Accordingly, information regarding "the intrinsic state of the world", that is, the "real world of physics" in the impersonal conception constructed within pure geometry, must be represented in the form of *tensor equations*, asserting either the vanishing of a tensor or an identity of two tensors. Tensor equations are deemed essential to a portrayal of "the nature of things" according to the "relativistic outlook" on physical theory in 1920.

However, the surpassing significance of tensor equations for "world building" lies in a further fact. For, in addition to the Einstein physical definition of the interval through scale and clock readings, tensor equations furnish a means of establishing another, and complementary, bridge between deductive mathematical theory and physical knowledge based on observation and measurement. In "world building", on one side of a tensor equation will stand a purely geometrical object, built up within a "world geometry" from the primitive posit of a relation of extension between neighboring points, a relation defined locally in a four-dimensional continuum not necessarily to be interpreted as space-time. On the other side is another tensor, the "synthetic" expression of a physical quantity occurring in the existing classical field laws of gravitation and electromagnetism. World building *presupposes* Einstein–Maxwell theory, and so the usual physical interpretations, while its intent is to be a reconstruction of existing theory for some yet to be clarified explanatory purpose. Then to the extent that a mathematically identical tensor can be generated within "world geometry" for each tensor quantity of the classical field laws, existing fundamental physics is effectively embedded within the wider geometrical theory of "world structure". In this way, a "world" is constructed from the geometrical relations derivable within that theory that shall have the same laws as those of the known physical world. But for what purpose?

7.4.2. *The Einstein Field Equations as "Definitions"*

The additional mode of connection of (deductive) mathematical theory and known physical theory through tensor identities is the hallmark of world building. Recall that Einstein simply posited that rods and clocks, understood as primitive concepts independent of gravitational theory, provide an immediate physical significance to the space-time interval ds^2. This was *the* bone of contention with Weyl, and so it is with the reconstruction of relativity theory within "world building", concerned as it is with "the origin of the fundamental laws of physics".

> [I]n the usual presentation of the theory . . . the interval is at once identified with something familiar to experience, namely the thing that a scale and a clock measure. However advantageous that may be for the sake of bringing the theory into touch with experiment at the outset, we can scarcely hope to build up a theory of the nature of things if we take a scale and clock as the simplest unanalysable concepts.[62]

A "theory of the nature of things" is just the conception of the physical world from the point of view of no one in particular, graphically represented in purely structural terms through a "world geometry" wherein rods and clocks cannot be primitive concepts. Within its "strict analytical development",

> the introduction of scales and clocks before the introduction of matter is—to say the least of it—an inconvenient proceeding. Thus in our development $R_{\mu\nu}$ is not merely of unknown nature but unmeasurable.[63]

Now $R_{\mu\nu}$ (the Ricci tensor) makes its first appearance within the "strict analytical development" as an object of "world geometry". World building begins from a primitive relation of comparison at neighboring point-events in a four-dimensional continuum. From this primitive material, any number of more complicated structures of relations and qualities, in fact, a fundamental series of tensors, can be derived via the operations of the tensor calculus. One such series follows from stipulating that the primitive relation is the "interval". Another, broader, series of such structures, as Eddington later showed in 1921, arises from choice of a more primitive relation of extension, that of comparison of infinitesimal "displacements" at infinitely close points. But whatever the primitive relational structure, what is to be built from it is a physical world that "functions like the known physical world". Limiting discussion for the moment to the more familiar first case, the analytical development begins with the fundamental "world tensor" $g_{\mu\nu}$ appearing in the formal definition of the invariant extension between neighboring points called "interval". The development of "world tensors" continues through successive differentiation of the Levi-Civita connection $\Gamma^{\sigma}_{\mu\nu}$ uniquely associated with $g_{\mu\nu}$ (since, with the usual semicolon designation for covariant differentiation, $g_{\mu\nu;\sigma} = 0$, abruptly halting further analytical development), yielding in the familiar way the Riemann-Christoffel curvature tensor $R^{\rho}_{\mu\sigma\nu}$, and then as its first contraction $R^{\sigma}_{\mu\sigma\nu} = R_{\mu\nu}$. From the latter set of quantities follows $g^{\mu\nu}R_{\mu\nu} = R$, the Riemann curvature scalar. Finally (via the contracted second Bianchi identity), a new tensor $G_{\mu\nu} \equiv R_{\mu\nu} - (1/2)g_{\mu\nu}R$ can be built up from the already constructed terms with the significant property that its covariant divergence vanishes identically ($G^{\mu}_{\nu;\mu} = 0$).

Merely originating as a term within "world building", $R_{\mu\nu}$ is "unmeasurable", indeed "of unknown nature", since the $g_{\mu\nu}$ and its associated connection, are regarded as purely analytical quantities without physical significance; hence, $R_{\mu\nu}$ is "of defined *form*" (as derivable from the $\Gamma^{\sigma}_{\mu\nu}$) "but of undefined *content*".[64] For it or any other geometrical property of curvature to acquire physical significance, appeal must first be made to some material or optical appliance for measurement, yet nothing so far constructed has been identified as "matter" or "light", and so even tacit appeal to such appliances contravenes the constructional order. However, a physical interpretation can be sought through identification with known tensors of existing physics whose observational basis presupposes such appliances. Since world building is a reconstructive procedure, it may presuppose existing physical theory and so find mathematically identical tensors in the Einstein field equations. In the source-free case of "empty space", these equations (without the cosmological constant) state that the Ricci tensor vanishes identically,

$$R_{\mu\nu} = 0 \qquad\qquad (1)$$

On the other hand, in the presence of "matter" (a classical, and purely phenomenological "matter", that is, a neutral, pressureless "dust"), the equations run, in one form,

$$G_{\mu\nu} \equiv R_{\mu\nu} - \frac{1}{2}g_{\mu\nu}R = \kappa T_{\mu\nu}, \tag{2}$$

where κ is a coupling constant and $T_{\mu\nu}$ is the so-called stress energy tensor, compactly summarizing the gross mechanical properties of this "matter" (momentum, stress, pressure, etc.). But when the obvious identifications are made, associating physical content with certain purely analytical objects of the world geometry, both sets of equations can be regarded *from the perspective of world building* simply as *definitions*. Hence, *within world geometry, $G_{\mu\nu} \equiv R_{\mu\nu} - (1/2)g_{\mu\nu}R = \kappa T_{\mu\nu}$* serves to define "matter" in virtue of the fact that the covariant divergence of $T_{\mu\nu}$ also vanishes, as is required by conservation of energy (see §7.5.1). Correspondingly, $R_{\mu\nu} = 0$, which implies $R_{\mu\nu} - (1/2)g_{\mu\nu}R = 0$, defines in perceptible terms a "vacuum" or "empty space", the absence of "matter".

The point of this heterodox interpretation of the Einstein field equations is purely explanatory, as may be seen by contrasting it to "the usual view":

> I suppose that the usual view of these equations is that the first of them expresses some law inherent in the continuum—that the point-events are forced by some natural necessity to arrange themselves so that their relations accord with this law. And when matter intrudes, it disturbs the linkages and causes a rearrangement to the extent indicated by the second equation.[65]

The "usual view", when made explicit, as it rarely is, holds that physical laws somehow *govern* states of affairs, compelling behavior into the pattern proscribed by law. While philosophers of an antinominalist persuasion may pass the explanatory buck to a favored candidate metaphysics of "natural necessity", the "usual view", according to the standpoint of physical theory, is normally the halting point for explanation, a dead end unless, seemingly oppositely, recourse is made to various anthropomorphic arguments. World building, however, also adds the criticism that the usual view of physical law as *governing* events introduces "a kind of dualism", the incongruity of introducing "matter" (and so, rods and clocks) as an independent postulate.

> [T]here is something incongruous in introducing an object of experience (matter) as a foreign body disturbing the domestic arrangements of the analytical concepts from which we have been building a theory of nature.[66]

Foregoing the comforts of "the usual view", explanation of laws can be pushed yet further. Within the "analytical development" of a "world geometry", both the source and the source-free Einstein equations appear as "definitions of the way in which certain states of the world (described in terms of indefinables) impress themselves on our perceptions".[67]

Within world geometry, the Ricci tensor $R_{\mu\nu}$ is a mere "empty form" denoting "a definite and absolute condition" of curvature. But to state that it vanishes, $R_{\mu\nu} = 0$, i.e., that some region of the world is curved in no higher than the first degree, is interpretable in "the familiar terms of experience" as a matter-free region of "empty space". Similarly through the identity $G_{\mu\nu} = \kappa T_{\mu\nu}$, the world geometric intrinsic curvature $G_{\mu\nu}$ indicates in these familiar terms the presence of matter in some region of the world. In measuring the mass and momentum of matter in some frame of reference we are in fact measuring certain components of world curvature. In this way, each Einstein equation serves as a "bridge" linking geometrical objects of pure mathematics to empirically confirmed field laws of existing physics through the mathematical identification of tensors. This *reconstructive tie* of world geometrical theory to experience ("experience" as encapsulated in Einstein's theory) takes place at a meta-level *presupposing* the physical validity of the Einstein theory (and so its ties, however attained, to physical objects of experience, such as rods and clocks) in accordance with the avowed goal of coming to a deeper explanation of physical law. This must be kept in mind in evaluating Eddington's heterodox claims, for example:

> We need not regard matter as a foreign entity causing a disturbance in the gravitational field; the disturbance is matter.[68]

and,

> Matter does not cause an unevenness in the gravitational field; the unevenness of the field *is* matter.[69]

If taken out of the intended explanatory context of the method of "world building", such statements indeed appear to be "a fairly full anticipation of what Wheeler was later to call geometrodynamics".[70] But in fact these are the second order conclusions of Eddington's epistemological attempt to explain the laws of existing physics, not bold first order hypotheses of new physics. They are not what they seem: a purported theory according to which all physical phenomena are to be interpreted as manifestations of space-time geometry. Instead, the requisite identifications are of tensors that "the mind" recognizes as "empty space" and as "matter".

Once such identifications are made, however, a series of questions can be entertained *in crescendo*: Why *these* tensor identifications? *How* does it happen that the Einstein field equations serve to connect the analytical theory with experience? Why is it the *Einstein tensor* in the series of tensors developed in world geometry that is recognized as "matter"? After all, in a passage illustrating that Eddington's structuralism is not subject to the criticism that devastated Russell's (see §7.6.1): "Out of the primitive events making up the external world, an infinite variety of "patterns" can be formed".[71]

The problem is similar, Eddington suggested, to picking out the major constellations from among the vast numbers of stars visible on a clear night, and the solution is similarly anthropomorphic. Naturally, there are formal constraints. The identity stated in Einstein's law between $G_{\mu\nu}$ and $T_{\mu\nu}$ requires that the tensors be of the same valence and rank and, in addition, that each has the property that its covariant divergence vanishes. Physically, this condition is required for satisfaction of conservation energy and in fact, Einstein chose this tensor for the left-hand side of his field equations, after scouting other unsuitable possibilities, essentially for this reason.[72] But there is an explicitly anthropomorphic explanation for the identity, assumed correct, between $G_{\mu\nu}$ and $T_{\mu\nu}$. A thing characterized by a tensor whose covariant divergence vanishes identically (i.e., there is no net flux across the boundary of the region where the tensor is defined) possesses a property recognized by the mind as the *quality of permanence*. There is an understandable adaptive interest in things possessing permanence; only such objects have measurable aspects and so can become objects of physical science (see §7.5.2 below).

While the measurable aspects are necessarily represented in one or another perceptual space and time, the "synthesis" of all of them that *is* the object is not. What world building now adds is an explanation of *why* matter is identified (via known field equations) with this geometrical structure amidst the manifold other possibilities derivable within a geometrical theory satisfying the conception of an external world from "the point of view of no one in particular". In this regard, the choice of the Einstein tensor as "matter" is the near-inevitable outcome of "mind's search for permanence". Namely, it is the nature of the mind that has singled out "the world of physics" (of "matter" and of "empty space") from other geometrical possibilities within the overall structure afforded by the world geometry: "Is it too much to say that mind's search for permanence has created the world of physics?"[73] Still further,

> Our whole theory has really been a discussion of the most general way in which permanent substance can be built up out of relations; and it is the mind which, by insisting on regarding only the things that are permanent, has actually imposed these on an indifferent world.[74]

7.4.3 The Significance of the Structural Thesis

It must be pointed out that the structural thesis has been essential to this conclusion, one we take to be an expression of transcendental idealism (see §7.5). Analytical reconstruction of existing physical theory (representing "the real world of physics") in world building begins with "a primitive relational structure", of relations defined on the crude relata of point-events, although "practically anything would do for that purpose [i.e., as relata] if the relations were of suitable complexity".[75] This beginning is of course due to general relativity, as is also the idea that the desired conception of a completely impersonal world is only expressible as a geometrical structure. Thus, relativity theory (taken as including something like Weyl's extension of the class of conceivable observers), by implying the structural thesis, has completely overturned the older conception of an external world as substance or material: "The relativity theory of physics reduces everything to relations; that is to say, it is structure, not material, which counts".[76] Accordingly, in as much as physical knowledge is viewed "in regard to the nature of things", it is "only an empty shell—a form of symbols. It is knowledge of structural form, and not knowledge of content".[77]

Much emphasis above was laid on Eddington's claim that knowledge regarding the "absolute world", the physical world from the point of view of "no one in particular", is necessarily expressed by tensors, analytical quantities without any intuitive or "visualizable" content, and so "knowledge of structural form, and not knowledge of content". But the fundamental step comes next. It is only the fact that physical knowledge of an impersonal external world can be put in a form amenable to "graphical representation" as structures within a world geometry that affords any possibility of an explanation of Einstein's gravitational law. The relational structure in question, although pertaining to, in fact *defining*, an impersonal world, can hardly be thought independent of mind. Rather, as "the aggregation of relations and relata that form the building material for the physical world", it is "mind-stuff" (Eddington borrows W. K. Clifford's term; see §7.6.2) from which the mind fashions "a habitation for itself".[78] As in Weyl, the actual world is to be reconstructed as a *selection* from a wider *conceptual space* of possibilities delimiting the conception of physical object, of what can be an object of a completely impersonal world. But, as I will show in §7.5.2, there are important differences in how this selection is to be made, crystallized in differing assessments of the significance of an action principle in field theory. The startling conclusion, reached already in 1920, is that

> [t]he intervention of mind in the laws of nature is, I believe, more far-reaching than is usually supposed by physicists. I am almost inclined to attribute the whole responsibility for the laws of mechanics and gravitation to the mind, and deny the external world any share in them.[79]

This is the meaning of the cryptic parable famously ending *STG*:

> We have found a strange foot-print on the shores of the unknown. We have devised profound theories, one after the other, to account its origin. At last, we have succeeded in reconstructing the creature that made the foot-print. And Lo! It is our own.[80]

7.5. *The Mathematical Theory of Relativity*

Just as Weyl's *Raum-Zeit-Materie* cannot really be considered a textbook of general relativity, so characterization of Eddington's *MTR* as a "text" is misleading. Yet as opposed to Weyl's classic, from which few might actually "learn" general relativity without already knowing quite a bit about it, Eddington's book is a masterful self-contained exposition eminently suitable for individual study. Three decades after its first appearance, Einstein was reported to consider it as still "the finest presentation of the subject in any language".[81] Although rendered somewhat dated by later developments and more modern mathematical techniques, *MTR*'s wide influence was recognized by the astrophysicist Sir William McCrae as late as 1991:

> More people must surely have learned general relativity through that book— either by reading it themselves or learning it from someone who had learned it from the book—than in any other way.[82]

Even so, the book's success is a remarkable fact for, like Weyl's, it is written from an epistemological vantage point that is neither Einstein's nor yet that of Weyl. While regarding "Weyl's theory of the relativity of gauge" as "an essential part of the relativistic conception" until the end of his life,[83] Eddington fundamentally reinterpreted its meaning. The very title obliquely signals Eddington's distinctive reconstructive method of "world building", although this signification is surely lost on readers who believe that it is simply descriptive of the book's thorough presentation of the mathematical framework relativity in 1923. However, as Eddington remarked in the *Preface*, he intended his book to be a more systematic and comprehensive treatment on the mathematical theory of relativity ... (to) meet the needs of those who wish to enter fully into these problems of reconstruction of theoretical physics.

Not exposition but "problems of reconstruction of theoretical physics" is the book's declared purpose. As to what these problems might be, several preliminary hints and indications are given. First it is noted that "the reader is expected to have a general acquaintance with the *less technical* discussion of the theory given in *Space, Time & Gravitation*" (emphasis added). This may be plausibly taken as a reference to the *more philosophical* parts of that book, and in particular to its argument that relativity theory has transformed the concept of physical knowledge. Second, Eddington noted that his "task" has been "to formulate mathematically" the "new conception of the world" brought by the theory of relativity and "to follow out the consequences to the fullest extent". The former remark is a prolepsis to the deductive method of "world building"; the latter, to the second half of the last chapter of *MTR* treating his generalization of Weyl's theory of gravitation and electromagnetism. Finally, as relativity has led "to an understanding of the world of physics clearer and more penetrating than that previously attained", the book's specific aim is "to develop the theory in a form which throws most light on the origin and significance of the great laws of physics". To this end, the most distinctive feature of *MTR*, its methodology of "world building", is devoted.

Obviously with a different readership in mind, Eddington did not begin the book by picking up where the philosophical themes of the last chapter of *STG* left off. Instead, a subtler means to the same end, concerning the mind's predominate role in the origin of the field laws, is adopted. To catch the attention of the physicists, the book begins with an introductory meditation on relativity theory's transformation of the concept of a physical quantity. In the seven formula-free pages of its introduction, two principal epistemological conclusions are laid out. The first of these, running against customary realist prejudice, holds that a physical quantity is a "manufactured article":

> The physical quantity ... is primarily the result of ... operations and calculations; it is, so to speak, *a manufactured article*—manufactured by our operations. But the physicist is not generally content to believe that the quantity he arrives at is something whose nature is inseparable from the kind of operations which led to it; he has an idea that if he could become a god contemplating the external world, he would see his manufactured physical quantity forming a distinct feature of the picture.[84]

The primordial physical quantity is a *length* or *distance*, a quantity that, before relativity theory, was generally regarded as referring to an objective property of extension inhering between points or events in a mind-independent world. By way of planting an initial doubt to this default hypothesis, Eddington contrasted the completely artificial quantity "cubic parallax"; surely a *length* is a natural property of the world whereas a *cubic parallax* is an artificially contrived notion, without significance in the physical world. Not at all. Each is an "indication" of "some existent condition or relation in the world outside us" yet neither can be rightly understood as resembling that condition or relation.

The physicist would say that he *finds* a length, and *manufactures* a cubic parallax; but it is only because he has inherited a pre-conceived theory of the world that he makes the distinction. We shall venture to challenge this distinction.... Indeed, any notion of "resemblance" between physical quantities and the world-conditions underlying them seems to be inappropriate.[85]

The message of this passage seems much like an anticipation of the doctrine of "operational techniques" formulated several years later by Bridgman, affirming that a physical quantity is defined through a prescription of operations and calculations always producing an unambiguous result within the practice of measurement. In this way, "there can be no question as to whether the operations give us the real physical quantity or whether some theoretical correction (not mentioned in the definition) is needed". But if read closely, the gist of Eddington's brief for operational definition is quite different from Bridgman's requirement of an operationalist treatment for *all* physical concepts.

I should be puzzled to say off-hand what is the series of operations and calculations involved in measuring a length of $10^{-15}cm.$; nevertheless I shall refer to such a length when necessary as although it were a quantity of which the definition is obvious. We cannot be forever examining our foundations; we look particularly to those places where it is reported to us that they are insecure.

To the charge that this cavalier treatment licenses terms possessing no definite observational meaning, including among physical quantities "things not the results of any conceivable experimental operation", Eddington simply responded,[86]

By all means explore this criticism if you regard it as a promising field of inquiry. I here assume that you will probably find me a justification for my $10^{-15}cm.$; but you may find that there is an insurmountable ambiguity in defining it. In the latter event you may be on the track of something which will give new insight into the fundamental nature of the world.

Clearly Eddington does not demand an operationalist definition of physical concepts. Moreover, in Bridgman's operational treatment of relativity theory, "absolute significance" attaches to a "definite, unique physical operation" whereas "covariance plays no necessary part".[87] With Eddington to the contrary, tensors are a necessary means of synthesizing the wheat of "intrinsic information" about the absolute four-dimensional world from the extrinsic chaff of particular perspectives and particular physical operations. It is only with the assistance of the tensor calculus that physics can express "simultaneously the whole group of measure-numbers associated with any world condition" thus "enabl[ing] us to deal with the world-condition in the totality of its aspects without attempting to picture it". Still more, even in tensor formulation the fundamental distinction between physical quantities and world conditions, a meaningless distinction for Bridgman, remains untouched. The former, "the results of our own operations (actual or potential)", do yield "some kind of knowledge of the world conditions, since the same operations will give different results in different world conditions". Still these experiments or observations provide only an "indirect knowledge", a representation of a "condition of the world" through its influences on these operations, and "it seems that this indirect knowledge is all that we can ever attain". Any attempt to otherwise directly describe a world-condition "is either mathematical symbolism or meaningless jargon".[88]

The denial of any relation of resemblance between physical quantities and world conditions may not seem particularly exceptional to the reader. It might be, and no doubt has been, concluded that once again Eddington here simply underscored relativity theory's relativization of measures of lengths and durations to an observer's frame of reference. To this effect, he wrote, in a more popular vein, that "the constancy of a measuring scale is the rock on which the structure of physics has been reared" and that the "Fitzgerald contraction...bring[s] the whole structure of classical physics tumbling down".[89] But that is not the whole of it. Eddington is tracking a further relativization not hitherto explicitly recognized, a relativity between "world-condition" or object surveyed and measuring appliance, termed elsewhere "the relativity of field and matter".[90] That is, a complete standardization of measurement apparatus, including specification of relative motion between apparatus (frame of reference) and object measured, brings in its train an additional relativization regarding the notion of a physical quantity: "Physical quantities are not properties of certain external objects but are relations between these objects and something else".[91]

The latter relativity "goes still further" than that involved in the relativistic treatment of length and duration. Since a quantity providing knowledge of any condition of the external world must be based upon the aggregate of possible measurements of that condition or object, "any *intrinsic* property" of that world-condition or object "must appear as a uniformity or law in these measures". The usual view of this uniformity is to regard it as corresponding to a determinate relation in a mind-independent world, even while recognizing relativistic variability of measures of it. Hence, the uniformity among such measures as expressed by covariant laws is regarded as having its seat in a invariant mind-independent reality. It is precisely this ordinary view of physical law that is targeted by the further "relativity of field and matter". For remembering that physical quantities are "relations between certain external objects and something else", namely, the apparatus of measurement and observation, the converse comparison can be made: Uniformity among measures of physical quantities is equally attributable to standardization of measuring appliances. Hence, the second, and principal, epistemological conclusion is that the "great field laws" of gravitation and electromagnetism arise from the fact that the apparatus that measures the world is itself part of the world.

> When one partner in the comparison is fixed and the other partner varied widely, whatever is common to all the measurements may be ascribed exclusively to the first partner and regarded as an intrinsic property of it. Let us apply this to the converse comparison; ... keep the measuring-appliance constant or standardized, and vary as widely as possible the objects measured—or, in simpler terms, make a particular kind of measurement in all parts of the field. Intrinsic properties of the measuring-appliance should appear as uniformities or laws in these measures. We are familiar with several such uniformities; but we have not generally recognised them as properties of the measuring-appliance. We have called them *laws of nature!*[92]

The origin of field laws will thus be precisely traced to the standardization of measuring appliances, in conformity with customary practices of mensuration (see §7.5.4).

Eddington's decision to *begin* the book with these controversial conclusions concerning the relative nature of physical quantities, and the origin of the field laws, is undoubtedly provocative. Without reference to the "relativity of field and matter", his discussion of physical quantities resembles operationalism or positivism. Neither viewpoint fits. As astronomer, Eddington was forthrightly not inclined to positivism.

> For the reader resolved to eschew theory and admit only definite observational facts, *all* astronomical books are banned. *There are no purely observational facts about heavenly bodies.*[93]

Then again, he delighted in taunting experimentalists by inverting their "good rule" not to "put overmuch confidence in a theory until it has been confirmed by observation":

> I hope I shall not shock the experimental physicists too much if I add that it is also a good rule not to put overmuch confidence in the observational results that are put forward *until they have been confirmed by theory.*

This remark signals no commerce with operationalism, instrumentalism, or any other form of anti-realism.[94] In any case, few readers might be expected to endorse the heretical conclusions of the introduction without first being persuaded by convincing argument. In fact, *MTR* is an intricate book-length argument for these general epistemological conclusions, stemming from a reconstruction of relativity theory concluding that Einstein law of gravitation is a near-inevitable consequence of the construction of a complete cycle of reasoning between deductive theory and the physical world of measurement and observation. While the general strategy for "world building" was informally surveyed above, its use in *MTR* to cast light on "the origin and significance of the great laws of physics" may be briefly summarized as a guide to the remaining sections of this chapter.

The idea guiding Eddington's treatment of the "problems of reconstruction of theoretical physics" is "to construct . . . a world which functions in the same way as the known physical world". Doing so is considered the touchstone of success of deductive theory, for "it is difficult to see how anything more could be required of it".[95] Now as a reconstruction of existing theory, in "world building", the Einstein and Maxwell field equations may be presupposed in their customary physical interpretations. Thus, for example, the metric in general relativity may be physically defined by linking the mathematical interval ds^2 to measurements made with rods and clocks. The reconstruction of existing theory ("the known physical world") proceeds by fashioning "world building material"—tensors or (as I will show) more general invariants—deductively stemming from one of several possible primitive relational structures of events and extensions in a four-dimensional continuum. Depending on the degree of generality of this primitive relation structure, various "things" can be built, perhaps leading to an *embarras du choix*. If so, a selection from these possibilities will have to be made by appealing to the known field equations, but considering them now as *definitions*. Thus, the Einstein field equations for the gravitational field produced by a material system enables the world geometrical ("Einstein") tensor $G_{\mu\nu}$ derived within the deductive theory to be *defined* as "matter", through its mathematical identity with the physically manifested stress-energy tensor. With this identification, the cycle of reasoning that began with the physical interpretation of the mathematical interval ds^2 in terms of the readings of rods and clocks is closed. For these appliances now can be considered composed of "matter" as derived and identified by the Einstein tensor "world geometry". Various world geometries have been proposed—Weyl's, Eddington's own "theory of the affine field, Einstein's—and many others are possible. Each is a purported characterization of "world structure", essentially (in the language of *STG*) a representation of the world from "the point of view of no one in particular", as that (now) locally symmetrical "point of view" is understood. Each is a purely mathematical theory, a "graphical representation" of the world, principally constrained only by "the ingenuity of the mathematician", until it has been connected with measurement, and so with the notion of a physical quantity.[96] But the completion of the cycle of reasoning demonstrates that the apparatus used to measure the world (e.g., Einstein's "practically rigid rods" and "perfect clocks") is indeed part of the world (as having been "built up" within that world geometry) and not some foreign excrescence introduced from without. This at least means that gauge of magnitude, or unit of scale, cannot be a property of the apparatus of measurement alone, fixed independently, but must be instead a relational property obtaining between that apparatus (abstractly represented as "matter") and the rest of the world. In turn, this implies that the origin of the "natural gauge of the world" presupposed by Einstein's standardized rods and clocks must lie wholly within the "world geometry". Accordingly, the Einstein law of gravitation in empty space, amended to include the cosmological constant, may be simply taken as the world-geometric statement of that relation, a "gauging equation". As the expression of a certain symmetry and homogeneity of the world revealed by measurement, the Einstein law is not a "governing" law but arises from the way the uniformities are manufactured, that is, through measurement with standardized rods and clocks. Since the cycle of reasoning is closed, there is no way leading outside it, and so no basis for the usual realist view of physical quantities as "standing for" or "mirroring" some "world condition" in a mind-independent reality.

7.5.1 General Relativity Reconstructed: "World Building"

The discussions in *MTR* fall naturally into two parts; the first, comprising the great bulk of the book, is occupied with the deductive reconstruction of Einstein's theory of gravitation from the fundamental posit of the interval. It culminates, in chapter 6, in the explanation of Einstein's law obtained by closing the cycle of reasoning. The second part, in chapter 7, turns to Weyl's and then Eddington's own generalizations, each reconstructing the Einstein theory from a broader geometric basis. This will be taken up in chapter 8 of this volume. After several preliminary chapters on the special theory of relativity and a primer on the calculus of tensors, the reconstruction of Einstein's theory properly begins in chapter 3 and is continued in chapters 4 and 5. Chapters 3 and 4 provide separate derivations of Einstein's law of gravitation, on the grounds of different connections of the deductive theory to experiment and observation. In so doing, each provides a novel interpretation of that law and establishes an essential link in the chain of reasoning that is not closed until chapter 4, §66. There, assuming two distinct connections of the Einstein theory to experience, Eddington provides a "new interpretation" of the Einstein law of gravitation as the relation between these two modes of connection to experience, thus closing the cycle of reasoning. In this guise, the Einstein law appears as a gauging equation of the world, "the almost inevitable outcome of the use of material measuring-appliances for surveying the world".[97]

Eddington's reconstruction of relativity theory starts with the "fundamental hypothesis":

> Everything connected with location which enters into observational knowledge—everything we know about the configuration of events—is contained in a relation of extension between pairs of events. This relation is called the interval, and its measure is denoted by ds.[98]

Point-events and the interval are the primitive relata and the primitive relation for building up a relational structure (a "world structure") within which the gravitational/mechanical physics of Einstein's theory will be situated. While the "fundamental hypothesis" states nothing about the standard interpretation of rod and clock readings as providing measures of the interval, it guarantees that the same observational knowledge of the relations among events can be obtained from things identically constructed within this primitive extensional structure.

> If two bodies are of identical structure as regards the complex of interval relations, they will be exactly similar as regards observational properties, if our fundamental hypothesis is true. By this we show that experimental measurements of lengths and duration are equivalent to measurements of the interval relation.[99]

Scales and clocks are "rather elaborate appliances" only to be introduced at a later stage in the deductive treatment. Following Weyl, the theory of relativity can be connected to experience through two more primitive postulates valid in all systems of coordinates, that the paths of freely moving particles are geodesics, and that the path of a light ray is a geodesic satisfying the equation $ds^2 = 0$. However, Eddington observed that these two postulates depend on the truth of empirical laws, of the motion of a body, and of light propagation. While satisfying from the point of view of enabling the theory to specify its own ties to observation, "[I]n a deductive theory this appeal to empirical laws is a blemish which we must seek to remove later".[100] For this same reason, the Principle of Equivalence plays a very diminished role in Eddington's reconstruction of relativity theory:

> The Principle of Equivalence has played a great part as a guide in the original building up of the generalised relativity theory; but now that we have reached the new view of the nature of the world it has become less necessary. Our present exposition is in the main deductive. We start with a general theory of world-structure and work down to experimental consequences, so that our progress is from the general to the special laws, instead of vice versa.[101]

Derivation of the Einstein field equations is the task of chapter 3 ("The Law of Gravitation"). Assuming the relation of extension whose measure is ds is given by the "fundamental hypothesis", its centerpiece (in §37, "Einstein's Law of Gravitation") is the deduction of the source-free gravitational field equation for "empty space". This is stated in two forms, first as the vanishing of the once-contracted Riemann-Christoffel tensor (i.e., the Ricci tensor),

$$R_{\mu v} = 0, \tag{3}$$

and second, in the observationally indistinct form incorporating Einstein's 1917 cosmological constant, hereafter Eddington adopts as canonical,

$$R_{\mu v} = \lambda g_{\mu v}, \tag{4}$$

where λ is a very small universal constant (see §7.5.4). The "deduction" here is informal and entirely relies (without citation) on Einstein's heuristic argumentation in motivating the law, namely, a desire for second order equations and the fact that any tensor not containing derivatives higher than second order must be compounded from the metric tensor and the Riemann–Christoffel tensor. Hence, "the choice of a law of gravitation is very limited, and we can scarcely avoid relying on the tensor $R_{\mu v}$".[102]

What *is* novel about this "deduction" is initially hinted by the interpretation of the law Eddington provided: "Einstein's law of gravitation expresses the fact that the geometry of empty region of the world is not of the most general Riemannian type, but is limited".[103] Now in the most general type of Riemannian geometry that is consistent with the fundamental hypothesis, the $g_{\mu v}$ are arbitrary continuous functions of the coordinates of the points of the manifold. However, as just shown, the Einstein law for "empty space" asserts that the possible values of the $g_{\mu v}$ are restricted to those which satisfy the ten differential equations $R_{\mu v} = \lambda g_{\mu v}$. This restriction on the possible values of $g_{\mu v}$ is therefore a limitation on the "intrinsic" natural geometry of a Riemannian space, such a restriction appearing as a "field of force". Just as in Newtonian mechanics the term "force" is given to anything responsible for a body's deviation from relative rest or uniform motion in a straight line, so here "a field of force arises from the discrepancy between the natural geometry of a coordinate system and the abstract Galilean geometry attributed to it".[104]

The natural geometry of the coordinate system is the fully general Riemannian geometry implicitly contained in the fundamental hypothesis, whereas the abstract Galilean geometry is the restricted geometry wherein the $g_{\mu v}$ are constrained to satisfy the equations $R_{\mu v} = \lambda g_{\mu v}$ and the "field of force" is gravitation. In this way, the law of gravitation appears as a restriction on the possible natural geometry of the world specified by the postulates of world building. Why the modality of necessity? In "world building", field laws *must* appear as limiting the geometrical possibilities within the complex relational structure built up upon the adoption of one or another differential geometric primitives. In this way, the laws can be exhibited as a selection from a wider set of tensor equations, a selection attributable to "mind" or some attribute of consciousness. From the perspective of "world building", it is "a fatal flaw" of the Newtonian inverse square law that it does not appear as "a restriction on the intrinsic geometry of space-time", the fully general Riemannian geometry.

In the remainder of chapter 3, the law of gravitation for "empty space" is applied in the familiar astronomical applications (e.g., planetary orbits, motion of the moon), and to rehearse the status of the various confirmations of general relativity (advance of perihelion of Mercury, deflection of light by the gravitational field of the sun, gravitational red shift). The gravitational field equations for space-times filled with continuous matter are introduced only at the very end of the chapter (in §46, "Transition to Continuous Matter"), exploiting the analogy with the Newtonian gravitational potential equations when transformed from empty space to one filled with continuous matter.

The objective of Eddington's deductive reformulation in chapter 4 ("Relativity Mechanics") is the chain of reasoning culminating in the "new derivation of Einstein's law of gravitation" in §54. The central point of this derivation is the *identification* of a "world-geometric" tensor $G_{\mu\nu}$ (belonging to the fundamental series derived from $g_{\mu\nu}$) with the stress-energy tensor. *Hence, the Einstein field equations may be regarded as giving a world-geometric definition of matter.*[105] This completes the first half of the "complete cycle of reasoning". Its significance is not fully explored until the second section (§66, "Interpretation of Einstein's law of gravitation") of chapter 5 ("Curvature of Space and Time") when the second half of the cycle is completed and the cycle closed. The overall objective is "the explanation of gravitation", an explanation presupposing that the correct form of the empty space field equations includes the cosmological term.

To begin, Eddington recapitulates what has been assumed already in Einstein's theory: that the covariant divergence of the energy tensor of matter vanishes, expressing the conservation of energy, stress, and momentum,[106] that is,

$$T^{\nu}_{\mu;\nu} = 0. \tag{5}$$

The task is to connect this second rank physical tensor possessing the mathematically prominent property that its (contracted covariant) divergence vanishes identically with the only second rank tensor possessing this property within the purely mathematical development. Now (according to the methodology of "world building"), if, beginning with the fundamental hypothesis, a world is to be geometrically constructed that "functions in the same way as the known physical world", there must be "certain analytical quantities in the deductive theory" whose "vulgar names in the known physical world" are mass, momentum, stress, and so forth.[107] It is a matter of finding such a second-rank tensor possessing a vanishing covariant divergence (hence is mathematically identical to the stress-energy tensor) in the fundamental series deriving from $g_{\mu\nu}$ and its unique associated affine connection, and calling *it* "matter. Unaware of the Bianchi identities,[108] Eddington first gives "a clumsy analytical version" of a proof of the contracted second Bianchi identity, stated as the vanishing covariant divergence of a second rank tensor that lies in the fundamental series. I will not recapitulate Eddington's proof, but note that this tensor G^{μ}_{ν} (equivalently, $G_{\mu\nu}$ the Einstein tensor, although Eddington does not use this designation) is reached as satisfying the contracted second Bianchi identities,[109]

$$G^{\mu}_{v;\mu} \equiv \left(R^{\mu}_{v} - \frac{1}{2} g^{\mu}_{v} R \right)_{;\mu} = 0 \tag{6}$$

This result, for reasons to become more apparent below, is denominated "the fundamental theorem of mechanics". The key step in the "new derivation" is then an appeal to what is termed a "Principle of Identification": Tensor expressions known from existing physics may be equated with tensors occurring in the deductive series stemming from the metric tensor on the grounds of mathematical identity. Then, the fact that the covariant divergence of these two tensors of the same rank vanishes identically suffices to set them as *mathematically identical*, so the result is

$$R^{\mu}_{v} - \frac{1}{2} g^{\mu}_{v} R = \kappa T^{\mu}_{v}, \tag{7}$$

Of course, these are just the Einstein field equations in the presence of a continuous matter (a pressureless, neutral "dust"[110]). It is of utmost significance that this mathematical identity is an identity of tensors, and so tensor equations alone, through such identities, provide the only means of representing and expressing the knowledge of the "absolute world" that is possible, that is, knowledge of its relational structure.

According to its "world-geometric" reconstruction, "Einstein's law of gravitation does not impose any limitation of the basal structure of the world".[111] Rather, it serves as *either* a *definition* of a (classical) vacuum (in the case of $R_{\mu v} = 0$), *or* as a "gauging equation" (in the form $R_{\mu v} = \lambda g_{\mu v}$; see §7.5.4), *or*, finally, as here, an analytical *definition* of "matter" (rather "the mechanical abstraction of matter which comprises the measurable properties of mass, momentum and stress sufficing for all mechanical phenomena"[112]). As "matter" has been given a world-geometric definition, it has been shown to be something inherent in the world (i.e., in the world built up from the fundamental hypothesis) and not something extraneous to that world. This is just about the pinnacle of achievement for the employment of deductive theory in physics (and so for "world building"):

> If the (deductive) theory provides a tensor which behaves in exactly the same way as the tensor summarizing the mass, momentum, and stress of matter is observed to behave, it is difficult to see how anything more could be required of it.[113]

On the face of things, this declaration may appear, as it did in the judgment of R. B. Braithwaite, a frequent critic of Eddington, merely an idiosyncratic description of the hypothetico-deductive method of confirming a theory by its consequences. If so, it is difficult to see quite what all the excitement is about.[114] But it should be clear by now that this is not at all Eddington's objective in using the method of "world building", a *reconstruction* of known physics seeking to shed light "on the origin and significance of the great field laws". Now, with the analytical definition of matter in place, this purpose may be further elaborated; its full significance requires the introduction of "Hamiltonian derivatives" and emerges in two separate steps.

7.5.2 Hamiltonian Derivatives: Expressing Mind's Regard for Permanence

Having now built up an analytical expression for "matter", through the above mathematical identity with a conserved quantity physically standing for the gross mechanical properties of matter, the question may be posed: why is *this* expression to be recognized as "matter"? Why, amidst the other analytical possibilities (that also may have vanishing covariant divergence) thrown up by a world geometry consistent with "the fundamental hypothesis" should "matter" be identified as $R_v^\mu - (1/2)g_v^\mu R$? The answer is that "it is a feature of our attitude toward nature that we pay great regard to that which is permanent". While this attitude displays "the idiosyncrasy of our practical outlook", it is nonetheless "an outlook adopted by the mind for its own reasons", a "tendency...the mind itself may have developed...through contact with the physical world". In general, intelligence in man and animals may be deemed the result of idiosyncratic traits produced through the operation of natural selection.[115] But whatever the provenance of its selection, $R_v^\mu - (1/2)g_v^\mu R$ has been singled out from "the passive field of space time" by mind on the grounds of its permanence. This claim is to be clarified by regarding $R_v^\mu - (1/2)g_v^\mu R$ as a kind of generalized differential quotient, a "Hamiltonian derivative", exploiting the fact that an invariant density, unlike a tensor, can be legitimately considered to occupy a volume. But to grasp the significance of Eddington's use of "Hamiltonian derivatives", it is first necessary to refer to his unusual view of "Hamilton's principle".

Now the method of application of a variational or Hamiltonian principle that almost always occurs in general relativity, or any field theory for that matter, is as a principle of stationary or least action. In other words, the variations of an invariant density or Lagrangian with respect to arbitrary small variations of the fundamental field variables and their first or second derivatives are stipulated to vanish. Let the Hilbert gravitational action I_G, considered over some arbitrary compact region of space-time Ω, be defined by a volume integral,

$$I_G = \int_\Omega \sqrt{g}R d\omega, \quad (d\omega = dx^1 dx^2 dx^3 dx^4) \tag{8}$$

Here $g = \det g_{\mu\nu}$ and R is the Ricci or curvature scalar, defined by $g^{\mu\nu}R_{\mu\nu} = R$. It is then required that the action be stationary for arbitrary small variations $\delta g_{\mu\nu}$ of the field variables $g_{\mu\nu}$ with certain derivatives of the $g_{\mu\nu}$ fixed on the boundary of Ω; that is,

$$\delta I_G = 0 \tag{9}$$

From these conditions, as Hilbert outlined in 1915 following his "axiomatic method", generalized Lagrangian expressions may be derived from which a further calculation yields the explicit form of the Einstein field equations, $R_{\mu\nu} - (1/2)g_{\mu\nu}R = \kappa T_{\mu\nu}$. But the derivation of the field equations from an action principle is *toto coelo* different from the "deduction" of the Einstein field equations through the method of "world building". Recall that in "world building", Einstein's law of gravitation may already be *assumed* whereas the task at hand is to exhibit it as a restriction or limitation within the wider field of geometrical possibilities provided by the "world geometry". This is the heart of Eddington's account of "the origin and significance of the great laws of physics". While the derivation of field laws from a variational principle appears in what might be called the "context of discovery" of general relativity, that context, like the Principle of Equivalence, may simply be taken for granted by the reconstructive procedure of "world building". Yet Eddington expressed skepticism that the principle of least action might be validly used as a tool for the discovery of field laws, because he doubted that "the vanishing of the variation for *all* small changes of the parameters is a possible form for a law of nature".[116] The reason is that the condition of stationary action strictly obtains *only* when the stress-energy tensor vanishes, that is to say, in "empty space". This leads to a general skepticism about the principle of least action itself:

> In fact action is only stationary when it does not exist—and not always then. It would thus appear that the Principle of Stationary Action is in general untrue.[117]

Accordingly, Eddington deemed it "unfortunate" that Hamilton's method of variation of an integral "is nearly always applied in the form of a principle of least action". On the other hand, variation of the action integral with respect to small variations of the $g_{\mu\nu}$, or other fundamental field variables, gives rise to its "Hamiltonian derivatives, a kind of generalized differential coefficients, that "may be worthy of attention even when they disappoint us by failing to vanish". In this case, also considering arbitrary small variations $\delta g_{\mu\nu}$ that vanish at and near the boundaries of the region, Eddington uses the gravitational scalar density $R\sqrt{g}$ and some calculation to obtain

$$\delta \int_\Omega R\sqrt{g}d\omega = -\int \left(R^{\mu\nu} - \frac{1}{2}g^{\mu\nu}R \right) \delta g_{\mu\nu}\sqrt{g}\, d\omega \qquad (10)$$

The term $-(R^{\mu\nu} - (1/2)g^{\mu\nu}R)$ is called the *Hamiltonian derivative* of R with respect to $g_{\mu\nu}$. Clearly it is a functional, or generalized derivative, and may be written symbolically:

$$\frac{\mathbf{h}R}{\mathbf{h}g_{\mu\nu}} = -\left(R^{\mu\nu} - \frac{1}{2}g^{\mu\nu}R \right) = 8\pi T^{\mu\nu} \qquad (11)$$

While the use of "Hamiltonian derivatives" in general relativity is not standard (although Schrödinger followed suit[118]), their *raison d'être* in the context of "world building" is reasonably clear. Since the $g_{\mu\nu}$ comprise the "primitive relation structure" from which, ultimately, "matter" is built, the nonvanishing of the Hamiltonian derivative indicates that in the region in question (where the $g_{\mu\nu}$ have particular values) there is something having the quality of "permanence". The Hamiltonian derivative accordingly "follows the kind of dissection which we ourselves have made when we analyse the world into *things* existing in *space* and *time*".[119] It has a "creative quality", being "the natural method of deriving physical quantities prominent in our survey of the world, because it is guided by those principles which have determined their prominence".[120] On the other hand, the condition of "empty space" is that the tensor identified as "matter" vanishes, which is just to say that $(\mathbf{h}R/\mathbf{h}g_{\mu\nu})$ vanishes. Accordingly, as "the natural method of deriving physical quantities prominent in our survey of the world", "the Hamiltonian derivative...stands out in our mind as an active agent working in the passive field of space-time".[121] They are therefore a natural means of expressing conserved quantities (via a theorem: *The Hamiltonian derivative of any fundamental invariant is a tensor whose divergence vanishes*[122]), for these are associated with Hamiltonian derivatives that nowhere vanish.

7.5.3 Closing the Cycle of Reasoning

With the analytical definition of matter, and the suggestive representation of matter in terms of the mind's "selective activity" through Hamiltonian derivatives, there are now two connections between the deductive theory, based on the fundamental hypothesis, and the world of known physics. The first connection, the identification of the geometrical term ds "with a quantity which the result of practical measurements with scales and clocks", was that taken by Einstein and deductively rehearsed in chapter 3 of *MTR*. The second connection, just described, is the mathematical identity of the physically manifested energy tensor with a tensor in the deductive theory. It is the latter step that gives "a great lift forward".[123] But with both of these in place, there is an opportunity to close the "cycle of reasoning", the completion of which provides Eddington's "explanation of the law of gravitation". This will require connecting each of these points of contact with the other, that is, "matter as now defined by the energy-tensor" must be connected "to the interval regarded as the result of measurements made with this matter".[124]

The Complete Cycle of Reasoning

Deductive Theory		Known Physical World
		(e.g., Einstein gravitational theory)
$ds^2 = g_{\mu\nu}dx^\mu dx^\nu$	$=$	measured lengths, durations
(fundamental hypothesis)	(physical definition)	(material apparatus of measurement)
(1) $\downarrow$		(2) $\uparrow$
$R_{\mu\nu} - (1/2)g_{\mu\nu}R$	$\equiv$	$T_{\mu\nu}$
("matter")	(mathematical identity)	(physically manifested tensor of matter)

The second half (2) of the cycle, completed in chapter 5 (§66), provides "the explanation of gravitation". Eddington's argument, "closing the cycle", is a consequence he has drawn from Weyl's principle of the *relativity of magnitude* (or "relativity of length, as Eddington preferred). Now Einstein regarded the extant empirical confirmations of general relativity as resting upon physical identifications of the interval with measurements made by "infinitesimally rigid" rods and perfect clocks, in violation of the principle of relativity of length. As I showed in chapter 6, §6.4.2.1, from Weyl's point of view, these identifications merely exploit the serendipitous existence of a "natural gauge of the world", a contingent fact to be explained by a theory competent of treating space, time, *and* matter. However,

Weyl's account of the "natural gauge of the world" introduced a distinction between *Einstellung* and *Beharrung* ("adjustment" and "persistence") designating the different manners in which physical quantities may be determined in a space-time continuum. Weyl's strategem indeed salvaged his principle of "relativity of length" in the face of the existence of a "natural gauge" but at the cost of restricting the unconditional validity of the principle to the ideal matter-free realm of an *Äthergeometrie*. There it quite literally had "nothing to do" with the actual behavior of rods and clocks and so its physical relevance appeared questionable. Eddington judged Weyl's attempt to vindicate the principle of relativity of length seriously flawed; the full account of his dissatisfaction requires discussion of his "affine field theory" and so is shown in more detail in chapter 8. For present purposes, his revised understanding of the principle of "relativity of length" may be considered as resulting from an argument that ties the analytical definition of "matter" provided within "world building" with material rods and clocks (as posited by Einstein) that measure the $g_{\mu\nu}$. As "world building" dictates, the measuring appliances used in surveying the world must themselves be part of the world constructed, they must also be brought within the "world building" framework that is, to be composed of "matter", key to Eddington's "explanation" of the law of gravitation: "The explanation of the law of gravitation thus lies in the fact that we are dealing with a world surveyed from within".[125] I will show that, according to the explanation, the Einstein law of gravitation is then to be interpreted as a "gauging equation".

7.5.4 Einstein's Law of Gravitation: A "Gauging Equation"

The final stage of Eddington's complex epistemological argument begins with several mathematical preliminaries. In the preceding and purely mathematical §65 titled "Curvature of a Four-Dimensional Manifold", Eddington noted (following Levi-Civita and others) that a four-dimensional Riemannian manifold can be represented as a surface in a Euclidean hyperspace of ten dimensions. In such a representation, the four-dimensional surface may possess "curvature" in any or all of the additional six dimensions. To lessen the complexities attending this fully general notion of curvatures, Eddington first considers the manifold embedded in a five-dimensional Euclidean space, noting that this will not in general suffice. But the simpler example provides the invariant (scalar) R a simple geometric interpretation in terms of the principal radii of curvature; thus, it generalizes to four-dimensions the Gaussian curvature of a two-dimensional surface. Consideration all the possible curvatures of the surface requires a full ten-dimensional hyperspace. But again reverting to the simpler illustration of a curved three dimensional manifold embedded in five dimensions, Eddington introduces a more complex invariant quantity formed from the Gaussian curvature, termed "the radius of spherical curvature", that is, "the radius of a hypersphere which has the same Gaussian curvature as the surface considered". Taking the three-dimensional space formed by the section of the surface where $x^1 = 0$, this quantity has the form

$$R_{11} - \frac{1}{2}g_{11}R = \frac{1}{2} \quad \text{Gaussian curvature (less terms with index 1).} \quad (13)$$

This method may be readily extended to the general, ten dimensional case.[126] When this is done, an invariant "quadric of curvature" is obtained,

$$\left(R_{\mu\nu} - \frac{1}{2}g_{\mu\nu}R\right)dx^\mu dx^\nu = 3, \quad (14)$$

whose radius, by analogous construction, "is equal to the radius of spherical curvature of the corresponding three-dimensional section of the world". A universally present "natural standard of length" for every material object occupying an extension can then be derived from this hypercurvature.

Doing so requires an implicit appeal to "the principle of relativity of length", a principle Eddington understood as stating that length is relative to a standard of comparison. On assumption of such a principle, the correct form of the Einstein field equations for "empty space" *must* contain the lambda ("cosmological" or, in the older style of Eddington, "cosmical") term introduced by Einstein in 1917,

$$R_{\mu\nu} = \lambda g_{\mu\nu}, \quad (15)$$

since the ordinary form, $R_{\mu\nu} = 0$, can provide no such standard, one geometrical term of the comparison being null. However, λ must be exceedingly small, so as not to upset the observational confirmation of the original form of the Einstein field equations for empty space (which were used in calculation of the two successful empirical tests of the theory known in 1923, the advance of Mercury's perihelion and the deflection of light passing close to sun's surface).

Now the revised condition for empty space can also be written in terms of the Einstein tensor, recalling the world-geometric "definition of matter",

$$R_{\mu\nu} - \frac{1}{2}g_{\mu\nu}R = -\lambda g_{\mu\nu} \quad (16)$$

As noted in chapter 6, §6.4.2.1, from the field equation for empty space $(R_{\mu\nu} = \lambda g_{\mu\nu})$, it follows that the value of the Riemann scalar curvature is

$$R = 4\lambda \quad (17)$$

Substituting twice into the invariant quadric (14) yields

$$-\lambda g_{\mu\nu}dx^\mu dx^\nu = 3 \quad (18)$$

or

$$-ds^2 = \frac{3}{\lambda} \quad (19)$$

This is just to say that the "quadric of curvature" in empty space is a sphere of radius $\sqrt{3/\lambda}$. Hence, the radius of curvature in every direction (i.e., the radius of curvature of the three-dimensional section of the world at right angles to that direction) and at every point in empty space has the constant length $(3/\lambda)$.[127] This omnipresent constant enables the required relativization for the notion of length, and in particular for such measurements made in the confirmation of Einstein's law of gravitation in empty space.

Length is not absolute, and the result can only mean *constant relative to the material standards of length* used in all our measurements and in particular in those measurements which verify $R_{\mu\nu} = \lambda g_{\mu\nu}$.[128]

Uniformities revealed in measurement can, as Eddington already emphasized in the introduction, be interpreted in different ways. According to the method of world building, however, they are to be understood in terms of the "relativity of field and matter", that is, of "empty space" and such matter as is found in the apparatus of measurement. With this in mind,

the precise statement of our result is that the radius of curvature at any point and in any direction is in constant proportion to the length of a specified material unit placed at the same point and oriented in the same direction.

The statement gains in perspicuity on inverting the comparison:

The length of a specified material structure bears a constant ratio to the radius of curvature of the world at the place and in the direction in which it lies.[129]

In short, the ratio of the meter to the radius of curvature of empty space is determined by λ. If λ is zero, we are left with a space that does not fulfill the first conditions of a medium of measurement. For if empty space is to be a physical concept at all, it must not be conceived as independent of the measurements that can be made of it; such measurements, according to the "principle of relativity of length" in turn presuppose a notion of length that is not *absolute*, but *relative*. This is just what is attained the above argument, concluding that the unit of length everywhere is a constant fraction of the directed radius of curvature of the world at that point. Which, Eddington insists, is just to say that the Einstein law of gravitation (in empty space) is a "gauging equation", simply the statement that the world-radius of curvature furnishes the ubiquitous standard in relation to which all lengths are gauged (in the absence of other matter). On the other hand, when space is not empty, the directed radius of the world at each point is not a constant. But then space has other characteristics besides metric, and "the metre rod can then find other lengths besides curvature to measure itself against".[130]

The argument at first bears a certain analogy to Weyl's derivation of a constant "natural gauge" of the dimensions of a length everywhere present in the manifold. In fact, it is a completely different account of the origin of the "natural gauge" and of the meaning of the principle of relativity of length. For Weyl supposed that the composition of material structures must involve an unknown dynamical mechanism of "adjustment" whereby material bodies are in constant instantaneous equilibrium with the omnipresent "natural gauge". By instantaneously "adjusting" their *lengths* to the ubiquitous standard, the fiduciary behavior of rods and clocks supposed by Einstein is in principle accounted for, at least in schematic dynamical outline. But Eddington's treatment of the origin of "the natural gauge of the world" differs from Weyl's just as the frame dependence of "length" in special relativity differs from the "length contraction" posited by Lorentz and Fitzgerald, namely, it is a *kinematical* not a *dynamical* explanation. In particular, Eddington regarded the notion of *"length"* as having no physical meaning in a world (e.g., that of Weyl's *Äthergeometrie*) where there is no comparison standard, without which there can be no meaning to measurement, and so no physical meaning to the very notion of such a world. The standard must exist, even in "empty space" if we regard the Einstein law of gravitation as obtaining there. And, as "world building" necessarily considers measurement apparatus as constituted of the "matter" that was derived from the $g_{\mu\nu}$, so its standardization as represented by the "gauging law" does not reach beyond a closed deductive procedure.

> Matter is derived from the fundamental tensor $g_{\mu\nu}$ by the expression $R_{\mu\nu} - (1/2)g_{\mu\nu}R$; but it is matter so derived which is initially used to measure the fundamental tensor $g_{\mu\nu}$.[131]

With this double linkage of field and matter, the cycle of reasoning is closed, showing that the source-free Einstein law of gravitation, as supplemented with the cosmological constant, no longer has reference to an empty continuum but is interpreted as

> a law of material structure showing what dimensions a specified collection of molecules must take up in order to adjust itself to equilibrium with the surrounding conditions of the world.[132]

Two general observations about this argument may be made. First, Eddington regarded its conclusion as so important that he included it in an unpublished list of his considerable scientific achievements compiled for some or another purpose, probably between 1931 and 1937, and found among his papers after his death in 1944.[133] Second, the argument makes clear the reasons for Eddington's subsequent "passionate attachment" to the cosmological constant;[134] without it, since the Einstein law in empty space has the form $R_{\mu\nu} = 0$, the demonstrated ratio of geometrical characteristics disappears. It is this ratio alone that, by implementing "the principle of relativity of length", first constitutes space, as it were, as the medium of measurement required by physics; in sum, "whatever embodies this comparison unit is *ipso facto* the space of physics".[135] Indeed, after 1931, when Einstein retracted his support for the cosmological constant, allegedly calling it "the biggest blunder of my life", Eddington would have none of it, particularly as, in the new context of the expanding universe, it could be given the further meaning of the amount of cosmological repulsion at unit distance from the observer.

> It was a defect of Einstein's original theory, first remedied by H. Weyl, that it implied the existence of an absolute standard of length—a conception as foreign to the relativistic point of view as absolute motion, etc. To set $\lambda = 0$ implies a reversion to the imperfectly relativistic theory—a step which is no more to be thought of than a return to the Newtonian theory.[136]

In a more dramatic vein, Eddington declared, "*To drop the cosmical constant would knock the bottom out of space*" (original emphasis).[137] Eddington's insistence on the cosmological constant is accordingly really only understandable from within the epistemological perspective of "world building", with its accompanying "principle of relativity of length" incorporated into the requirement of the "point of view of no one in particular". Without this context, the proffered "explanation" of the law of gravitation is nothing less than enigmatic, and many have judged it more harshly. For example, in a recent overview of the curious history of the cosmological constant, John Earman has responded to Eddington's claim that the constant was essential to the general theory of relativity by objecting that since space is *in fact* not empty, it hardly follows that dropping λ (Earman writes 'Λ') would knock the bottom out of space (or spacetime), even if we accept Eddington's debatable doctrine that length is relative to a standard of comparison found in the features of spacetime.[138]

From the above, however, it will be clear that, as phrased, this objection misfires; for if "Eddington's debatable doctrine" *is* accepted, then an "empty space" *without* the cosmological constant will not be a physically meaningful notion, and so indeed, dropping it "knocks the bottom out of space (or spacetime)". On the other hand, considered within the epistemological context of world building, Eddington's argument has shown that the degree of isotropy and homogeneity of space-time affirmed by the Einstein field equation for empty space need not have a counterpart in the world-geometric relational structure suitable for representing the world from the point of view of no one in particular. Rather, this law simply expresses the presupposition of measurement with standardized material appliances, indeed, that the apparatus measuring the world must itself be considered part of the world. Reconstructed in "world building", the Einstein law appears as a "gauging equation" and not a law "governing" the point-events and their associated intervals. Of course, since the physical validity of the Einstein theory was presupposed at the beginning, Eddington's explanation of the Einstein law provides a complementary, not an exclusive, perspective on that law. Its intent is solely that of accounting for the "origins" of the fundamental field laws rather than simply accepting an arbitrary differential equation (or system of equations) as just "the way the world is".

Ultimately, the fact to be accounted for in world building is that certain mathematical structures generated within a world geometrical conceptual space are given special attention. With regard to the Einstein law of gravitation, in the form of the Einstein field equations for "empty space" (supplemented with the lambda term), the law simply expresses the presuppositions of measurement with standardized rods and clocks. In the case of matter sources, the Einstein field equations express that a certain structure within world geometry is singled out on account of its "permanence", reflecting an idiosyncratic tendency of the human mind due to natural selection. In either case, the operative principle is the human mind and this, in the last analysis, is the reason that physics (or more precisely, the notion of physical quantity) is necessarily based on measurement and observation. Taken together, these arguments suggest a novel *transcendental idealist* understanding of physical theory and physical law as a formulation of conditions presupposed by experimental procedures, and not of theory and law as determined empirically by those procedures.[139] From 1920 on, this inversion became a permanent fixture of Eddington's epistemology of physical science.

7.6 Structuralism

7.6.1 Eddington, Russell, and Newman

As discussed above, at the end of *STG*, Eddington concluded that relativity theory had shown that only the structure of the external world is knowable, that it is only structure "that counts":

The relativity theory of physics reduces everything to relations; that is to say, it is structure, not material that counts. The structure cannot be built up without material; but the nature of the material is of no importance.[140]

In support of this conclusion, Eddington approvingly quoted a passage from Russell's book of the previous year, *Introduction to Mathematical Philosophy* (1919), a passage subsequently also celebrated in several major works of logical empiricism.[141] It occurs on page 61 of Russell's book; the text in brackets is omitted by Eddington:

> There has been a great deal of speculation in traditional philosophy which might have been avoided if the importance of structure, and the difficulty of getting behind it, had been realized. For example, it is often said that space and time are subjective, but that they have objective counterparts; or that phenomena are subjective, but are caused by things in themselves, which must have differences *inter se* corresponding with the differences in the phenomena to which they give rise. Where such hypotheses are made, it is generally supposed that we can know very little about the objective counterparts. In actual fact, however, if the hypotheses as stated were correct, the objective counterparts would form a world having the same structure as the phenomenal world[, and allowing us to infer from phenomena the truth of all propositions that can be stated in abstract terms and are known to be true of phenomena. If the phenomenal world is Euclidean, so must the other be; and so on]. In short, every proposition having a communicable significance must be true of both worlds or of neither: the only difference must lie in just that essence of individuality which always eludes words and baffles description, but which, for that very reason, is irrelevant to science.

Russell had introduced the notion of structure (or "relation number") as an equivalence class of relations (in modern terms, isomorphism of models) in a previous chapter (on "Similarity of Relations"). In this passage the notion finds employment in the context of the causal theory of perception that previously informed his earlier theory of knowledge (1912) but had largely remained out of view throughout the constructivist phase of his "external world program" of 1914–1915. By appeal to the identity of "structure" between the phenomena of perception and their extraphenomenal causes, Russell sought to retain the common sense belief in naive realism in the face of the undoubted epistemic priority of firsthand perceptual "data" ("knowledge by acquaintance").[142] In these general terms, the bridge between perception and physics is envisaged as secured by the purely logical notion of sameness of structure. To be sure, the linkage is not spelled out until later on, in *The Analysis of Matter* (1927). There, suitably impressed by the "abstractness" of the new physics that is encapsulated, above all, in Eddington's "definition of matter" as $R_v^\mu - (1/2)g_v^\mu R$ [see equations (7) and (11) above], Russell returned to the causal theory of perception to produce a revised version of the doctrine he termed "neutral monism".[143] In the new version, Russell adopted an explicit form of what has recently been called "epistemic structuralism": Of the neutral particulars termed "events" that are the fundamental constituents of the world, we can know the *intrinsic nature* or *quality* only of those occurring in regions where there is a brain ("percept events"), whereas our knowledge of nonpercept events is limited to knowledge of their *structure*. In this incarnation, Russell's structural thesis was shown fatally flawed by the Cambridge mathematician,

M.H.A. Newman, in 1928. As has been much discussed in the recent literature on "structural realism", Newman pointed out that the claim there can be only structural knowledge of the unperceived events of the world is entirely trivial. For to say of two aggregates that they share a specified structure W of relations R is merely to assert that they have the same cardinality, since "any collection of things can be organized so as to have the structure W, provided there are the right number of them".[144] On pain of an absurd conception of physical knowledge restricted to questions of cardinality, the thesis that *only* the structure of the nonperceived parts of the world can be known must then be false. Indeed, in reply to Newman, Russell conceded that he had always supposed knowledge of the external world included something *more* than knowledge of its mere structure.[145]

Except for the bracketed text that is always elided, the above Russell passage is quoted *verbatim*, several more times in Eddington's published writings, ample testimony of its deemed significance.[146] However, the omitted lines are crucial to Russell's conception of the structural thesis, but not to Eddington's on account of their express reference to inductive inferences from the data of perception resulting in structural knowledge of the nonperceptual external world, inferences allegedly licensed by the causal theory of perception. As made abundantly clear in this chapter, Eddington's thesis that knowledge of the world of physics is knowledge of its structure is not supported by inferences based on the causal theory of perception. Rather, physical knowledge, the attribution of physical quantities to properties of a body or field, arises in the response of metrical indicators. Such responses provide "aspects" of the reality in question, which itself is *defined* as the (correctly conducted) "synthesis of appearances from all possible points of view", namely, of measurements and observations of all conceivable observers. Such a "real world of physics" corresponding to "point of view of no one in particular" can only be mathematically characterized; in 1920, this is as the absolute four-dimensional geometrical world of events and intervals to which frame-dependent measures do not literally apply. Knowledge of the "real world of physics" is purely structural since an individual observer can only be "acquainted" with "aspects" (particular measures), not the totality of measures that requires representation by the widest possible group of admissible transformations. Any correspondence with the absolute four-dimensional geometrical world can only be a mathematical or structural correspondence, given a precise meaning within "world building" as an identity between tensor expressions. This is because tensors, the inherent geometrical language of the four dimensional world of events and intervals, also provide the appropriate means of representation of the measures of particular observers. Such measures purport to pertain to a completely impersonal world, and they do so as tensor components within a given coordinate system. Eddington's claim that there is only a structural knowledge of physical reality stems from relativistic spacetime's necessary invariant theoretic characterization. Inferences based on the causal theory of perception serve no legitimate epistemological purpose in Eddington's conception, but rather indicate a kind of "category mistake".

Newman's objection to Russell therefore does not touch Eddington's understanding of the structural thesis. However, the full context of Eddington's thesis concerns the attempt to extend physical explanation to the notion of physical law itself, striving for the explanatory "ideal".

> To show, not that the laws of nature come from a special construction of the ultimate basis of everything, but that the same laws of nature would prevail for the widest possible variety of structure of that basis.[147]

A demonstration of this kind would result if, by generalizing as widely as possible the "primitive relation structure" which furnishes the bricks and mortar of world building, mathematically identical counterparts can be built to all the known physical laws. In chapter 8 I consider Eddington's further generalization beyond mere tensors to gauge invariant tensors, and later on in the 1930s Eddington builds on the basis of the theory of groups.[148] But the explanatory quest is precisely the same. The wider the possible variety of structure that can be built upon a given basis, the more pressing becomes the question of how a given selection is to be made from amidst this variety of patterns. As I have shown, Eddington's explanation was that the mind "select[s] and endow[s] with substantiality one particular quality of the external world" in such a way that "practically no other choice was available for a rational mind". For this reason, the actual laws of nature are accounted as "not inherent in the external world, but were automatically imposed by mind when it made the selection".[149] In coming to this view of selection of a given pattern or structure, Eddington invoked, already in *STG*, notions first introduced fifty years previously by W. K. Clifford.

7.6.2. Eddington, Clifford, and the Metaphysics of "Mind-Stuff"

Several pages prior to the above quotation of Russell, Eddington reproduces a long and somewhat mysterious passage from an obscure publication of William Kingdom Clifford, concluding with the speculation that "matter and motion may be described in terms of extension only". Clifford, who first translated Riemann's 1854 *Probevorlesung* into English, was the brilliant British mathematician who prematurely died of tuberculosis in 1879, just two months before his thirty-fourth birthday. While written under the influence of Riemann's essay, his neo-Cartesian conjecture of a purely extensional theory of matter and motion, "On the Space-Theory of Matter" (1870), was fully original, and since the 1920s his name has been linked jointly with that of Riemann as a precursor of Einstein.[150] However, the curious Clifford passage appearing in *STG* stems not from this well-known source but from a lengthy review essay in 1875 of a period work of theistic physics that, in Clifford's assessment, mixes "wide and accurate knowledge of physical science" with "all the stamp of a Christian apologetic writing".[151] Clifford's attention to such an obscure book is puzzling until it is learned that one of its two anonymous authors was the Victorian physicist Peter Guthrie Tait. It is obvious that Eddington thought quite highly of Clifford's gentle, but withering, criticism of such a fusion of physical science and theism. The essay's title, "The Unseen Universe", taken from the book under review, was later virtually adopted by Eddington for his own widely read lecture to the Yearly Meeting of the Society of Friends in 1929.[152] Both Clifford's review and Eddington's lecture offer similar, and unconditionally negative, assessments of attempts to use science to "prove" the truths of religion.[153] However, the significance of the quotation from Clifford lies not with this, or with Clifford's anticipation of Einstein, but with the pan-psychist metaphysics to which Clifford believed the theory of evolution pointed, that "things-in-themselves" are in fact "elements of feeling".

This view is considerably elaborated in an 1878 essay in which Clifford deliberately pursued a suggestion of Kant "that the *Ding an sich* might be of the nature of mind".[154] The essay also must have been familiar to Eddington, for in it Clifford adopted the name "mind-stuff" for these ultimate constituents of the universe.[155] More generally, Eddington affirmed that "the stuff of the world is mind-stuff", but such a "crude statement" is hardly clarified by an accompanying remark that by "mind" Eddington does not "exactly mean mind", and by "stuff" he does "not at all mean stuff". Now we have seen that this term is employed by Eddington in his Gifford lectures of 1928 to refer to the primitive relations and relata of some world geometry, "the aggregation of relations and relata which form the building material for the physical world".[156] In world building, this "basal stuff of the world" is employed to build a world identical to the known physical world, featuring objects mathematically identical to those occurring in the known fundamental laws, and so betraying the selective activity of mind. However, world building is concerned to reproduce only the aspects of the world that, as metrical, lie within the purview of physics, whereas "mind-stuff" pertains also to nonmetrical aspects of the world. These are glossed in STG by stating that "geometrical notions are only partial aspects of the relation [Clifford] called 'elements of feeling' ".[157] In any case, "mind-stuff" is not to be thought completely identical with consciousness, which is not sharply defined, fading away into inattention, lapses of memory and the subconscious. But it is the protean element, rising "only here and there . . . to the level of consciousness, but from such islands proceeds all knowledge".[158] Beyond mind-stuff is posited something even more indefinite, "yet continuous with our mental nature", termed "world stuff". The latter, a necessary limiting boundary, can be likened to conscious feelings if only because, since "now that we are convinced of the formal and symbolic character the entities of physics there is nothing else to liken it to". Like Clifford, Eddington came to the arresting conclusion that the "substratum of everything is of mental character". Appearances to the contrary, this is not a declaration of subjective idealism. Rather, Eddington, like Clifford, was hankering after a metaphysics that, with a minimum of speculation, was fully consistent with the most recent findings of physical science but nonetheless accorded recognition to the obviously primary reality of the stream of consciousness. These truncated metaphysical glimpses were not at all regarded as definite, but modestly as merely calling attention to what might be reasonably conjectured on the basis of such findings.

> The recent tendencies of science do . . . take us to an eminence from which we can look down into the deep waters of philosophy; and if I rashly plunge into them, it is not because I have confidence in my powers of swimming, but to try to show that the water is really deep.[159]

8

GEOMETRIZING PHYSICS

Eddington's Theory of the Affine Field

> ... physics has in the main contented itself with studying the
> abridged edition of the book of nature.
>
> Eddington (1921, 108)

8.1 Toward a More Perfect Union

Einstein's so-called geometrization of gravitational force in 1915 provided the modern geometrical unification program in physics with its first, albeit partial, triumph as well as its subsequent impetus. Recall that, in general relativity, the fundamental ("metric") tensor $g_{\mu\nu}$ of Riemannian geometry appears in a dual role thoroughly fusing its geometrical and its physical meanings. As is apparent from the "Pythagorean" expression for the space-time interval between neighboring events $ds^2 = g_{\mu\nu}dx^\mu dx^\nu$, the $g_{\mu\nu}$ is at once the geometrical quantity underlying measured lengths and times. But the ten independent magnitudes $g_{\mu\nu}$ are also the "potentials" of the gravitational (or "metrical") field whose values, at any point of space-time, depend, via the Einstein field equations, on surrounding physical magnitudes of mass (energy) and momentum. In the new view, the strength of gravitational "force" is replaced by degree of "curvature" of space-time. Taking a homely example, Earth's mass determines a space-time curvature manifested as a source of gravitational action. By the "tidal force" of Earth's gravitational field, two freely falling bodies, at a given separation a certain height above Earth's surface, will approach one another. The freely falling body is no longer to be regarded as moving through space according to the "pull" of an "attractive force", but simply as tracing out the "laziest" track along the bumps and hollows of space-time. At the same time, the gross mechanical behavior of massive bodies, comprising all the gravitational-inertial phenomena of mechanics, in principle can be derived as solutions of the Einstein field equations. According to these equations, not only are space-time (the metric field) and matter in dynamical interaction, but the metric field itself, as a locus of gravitational energy, is both a source of its own curvature and of a characteristic form of radiation, gravitational waves. An abbreviate way of characterizing this dual role of the $g_{\mu\nu}$ is to say that in the general theory of relativity, gravitation, including mechanics, has become "geometrized", that is, incorporated into the geometry of space-time.

In making space-time curvature dependent on distributions of mass and energy, general relativity is in principle capable of encompassing all matter fields. However, in classical general relativity there is a fundamental asymmetry between gravitational and nongravitational fields, in particular, electromagnetism, the other fundamental interaction definitely known prior to the discovery of nuclear forces in the early 1930s.[1] This shows up visibly in one form of the Einstein field equations where, on the left-hand side, the "Einstein tensor" $G_{\mu\nu}$ represents the curvature of space-time for given sources. A purely geometrical object, it is entirely constructed from the Riemann curvature tensor, itself derivable from the uniquely compatible linear symmetric "Levi-Civita" connection associated with the metric tensor $g_{\mu\nu}$. It is set identical to a nongeometrical phenomenological representation of "matter" (regarded as a neutral "dust of particles"),

$$G_{\mu\nu} = \kappa T_{\mu\nu} \quad \left(G_{\mu\nu} \equiv R_{\mu\nu} - \frac{1}{2}g_{\mu\nu}R\right), \tag{1}$$

where $\kappa = (8\pi G/c^4)$, a coupling constant comprising the Newtonian gravitational constant and the velocity of light. Since $G_{\mu\nu}$ is a tensor, the expression on the right side of the identity must also be a tensor. In Eddington's terms, it is an "*omnium gatherum*", a merely abstract representation of "matter" in the gross form of the stress-energy-momentum of all nongravitational sources of the gravitational field in a region of space-time. In the familiar interpretation, equation (1) states that "space-time tells mass how to move" (reading from left to right) and that "mass tells space-time how to curve" (from right to left); space-time and matter stand in dynamical interaction.[2] But since the geometry of space-time resides solely on the left-hand side of Einstein's equation, an unsatisfying asymmetry was already apparent from the beginning of general relativity. Indeed, Einstein likened his famous equation to a building, one wing of which (the left) was built of "fine marble", the other (the right) of "low grade wood".[3] In its classical form, then, general relativity accords only the gravitational field a direct geometrical significance; other physical fields reside *in* space-time, they are not *of* space-time. Einstein gave a particularly vivid declaration of the need to remedy this asymmetry in his "Nobel lecture" of July 1923:

The mind striving after unification of the theory cannot be satisfied that two fields should exist that, by their nature, are quite independent. A mathematically unified field theory is sought in which the gravitational field and the electromagnetic field are interpreted as only different components or manifestations of the same uniform field. . . . The gravitational theory, considered in terms of mathematical formalism, that is Riemannian geometry, should be generalized so that it includes the laws of the electromagnetic field.[4]

The demanded generalization has turned out to be a tall order, for a dynamically transparent geometrical unification of gravity with the other fundamental fields remains today an open issue, and the posited connection of gravitational and nongravitational matter fields is the paradigmatic problem of quantum gravity and string theory.

Of course, Einstein was not the first to embark on the quest to extend space-time geometry to encompass electromagnetism. Rather, as we have repeatedly seen, the program began in 1918 with Weyl's broadening of the geometry of general relativity into a theory of "gravitation and electricity". Despite a failure to overcome Einstein's objections in the eyes of the relativity community, Weyl's guiding example of a geometrical unification stimulated a variety of similar efforts, all aimed at finding a suitable generalization of the (pseudo-)Riemannian geometry of Einstein's theory to encompass as well nongravitational physics. Minimally, it was widely believed, such a generalization must yield new variables capable of standing for the field quantities of nongravitational fields, in particular, of electromagnetism, then regarded the constituent field of matter. Although widely held at the time, the tacit assumption that formal incorporation of electromagnetism into space-time geometry necessitates a generalization of the Riemannian geometry of general relativity, is not quite correct, as results stemming from Rainich in 1925 and continuing through Wheeler, Misner, and Geroch have shown.[5] Maximally, the generalized geometry should yield field equations possessing solutions (spherically symmetric, static, etc.) corresponding to matter's particulate structure, the "Holy Grail" of the geometrical unified field theory program.

Just as 1921 was a turning point for philosophical interpretations of relativity and for subsequent philosophy of science (chapter 2), so it proved to be "the pivotal year" for the nascent unified field theory program.[6] On 19 February, the *Proceedings of the Royal Society of London* received A. S. Eddington's generalization of Weyl's theory postulating a nonmetrical affine connection as the basis of a "geometry of space and time and things". In Weyl's geometry, the relative magnitudes of vectors at the same point, but pointing in different directions, are immediately comparable; in Eddington's, a direct comparison could be made only for vectors at the same locus pointing in the same direction. The geometry of Eddington's "theory of the affine field" included both Weyl's geometry and the Riemannian geometry of Einstein's theory as special cases. Scant attention was paid, however, to a claim prefacing Eddington's paper, that his objective had not been to "seek unknown laws (of matter)", the *raison d'être* of a unified field theory. Rather, it lay "in consolidating the known (field) laws" wherein "the whole scheme seems simplified, and new light is thrown on the origin of the fundamental laws of physics".[7]

Eddington's theory was received and understood otherwise, as its considerable influence on Einstein demonstrates. The most innovative step, the postulate of an affine connection introduced prior to, and independently of, the metric, prefigured many of Einstein's subsequent efforts in unified field theory. Eddington himself informed Einstein about his novel proposal in a conversation at "a dinner party of quite exceptional nature" held in Einstein's honor on Friday, 10 June 1921, at the London home of Viscount Haldane, Einstein's host during his first visit to England.[8] Einstein's initial impressions were apparently favorable, as can be inferred from a note accompanying an offprint of the paper that Eddington mailed to Einstein in London two days later. In it, Eddington emphasized that his work "is an *extension*, but is not in the slightest degree an *emendation*, of yours", since, in his theory, but not in Weyl's, the "*ds* is actually comparable at a distance (i.e., integrable)".[9]

However, the "favourable expectations" Eddington reported detecting during his conversation with Einstein did not immediately materialize. Prior to his visit to England in June, the *Proceedings of the Berlin Academy* published on 17 March contained Einstein's note exploring Weyl's hypothesis that the interval is only conformally invariant ($ds^2 = 0$), thus lacking the straightforward empirical tie to scale and clock measurements for which he had chastized Weyl.[10] In a letter to Weyl toward the end of the summer, Einstein pronounced on Eddington's theory the dim verdict that it was "beautiful but physically meaningless".[11] On 8 December, having advised T. Kaluza some two and a half years before against publication, Einstein presented to the Berlin Academy Kaluza's novel proposal for unification of gravitation and electromagnetism upon the basis of a five-dimensional Riemannian space-time geometry.[12] Although the theory remained without empirical support, Einstein at once addressed himself to the vital question of whether the Kaluza field equations had singularity-free particlelike solutions. In a paper submitted on 10 January 1922, written with his assistant Jakob Grommer, that question was answered in the negative.[13]

Einstein's assessment of Eddington's theory had not appreciably improved by the summer of 1922. Writing to Weyl on 6 June, Einstein remarked that, like Mie's earlier electromagnetic theory of matter, Eddington's undertaking was "a beautiful framework" but that "one absolutely cannot see how it must be filled up".[14] Then, seemingly suddenly, early in 1923, Einstein underwent an abrupt *volte face*. From aboard ship "near Singapore" en route from Japan on 11 January 1923, Einstein enthusiastically reported to Bohr,

> I believe that I have finally understood the connection between electricity and gravitation. Eddington has come closer to the truth than Weyl.[15]

The new-found excitement lasted through the year. On 23 May, Einstein declared to Weyl that "I absolutely must publish since Eddington's ideas must be thought through to the end".[16] By 1923, in Vizgin's authoritative narrative, Einstein had become the recognized leader of the geometrical unification program, publishing in that year alone four papers on an Eddington-inspired "theory of the affine field".[17] The stimulus Eddington provided spilled over into the so-called Nobel lecture of which we have already had several occasions to mention, delivered to the Nordic Assembly of Naturalists at Gothenburg, on 11 July 1923. There Einstein judged Eddington's idea "to replace Riemannian geometry by the more general theory of the affine (connection)" to be "the most successful" of the varied attempts to find

> a mathematically unified field theory...in which the gravitational and the electromagnetic field are interpreted as only different components or manifestations of the same uniform field.[18]

It will be seen that a reductionist ambition of this kind is at quite some distance from the intent of Eddington's paper, concerned not the discovery of new physical laws, but rather to "the origin" of the existing field laws. This difference in aim from all subsequent "geometrical unified field theories" is obscurely stated in Eddington's several declarations of his purpose. For example, in the penultimate paragraph of his 1921 paper, he observed that "[w]hat we have sought is not the geometry of actual space and time, but the geometry of the world-structure, which is the common basis of space and time and things".[19] A similar statement occurs at the conclusion of the purely mathematical treatment of affine field theory in the last chapter of *The Mathematical Theory of Relativity*: "[W]e have developed a pure geometry, which is intended to be descriptive of the relation-structure of the world".[20]

Without reference to the intended epistemological context of "world building", such statements are still commonly misunderstood as indications that Eddington's affine field theory is a precursor to the "geometrodynamics" of J. A. Wheeler and C. W. Misner, supposing masses and fields entirely built out of space-time geometry.[21] From chapter 7, however, we may surmise that these statements have a rather different sense when placed in the reconstructive nexus of "world building".

Of course, Eddington's account of the origin of the fundamental field laws involved a highly idiosyncratic interpretation of the Einstein field equations as "definitions" that few could fathom, and that Einstein certainly did not endorse.[22] Like the other "geometizers", Eddington was alert to the geometrical asymmetry between the left- and right-hand sides of the Einstein field equations. But far from wanting to eliminate that asymmetry, Eddington idiosyncratically sought to underscore it, exploiting the concept of a "world geometry" he had discovered in Weyl. The latter, distinct from physical theory, nonetheless furnishes a "graphical representation" of it for expressly "heuristic" purposes. The disparity between Einstein's and Eddington's conceptions of an "affine field theory" pointedly emerges in their contrasting attitudes toward the usual field theoretical procedure of deriving field laws from variation of a Lagrangian density according to "Hamilton's principle". In two further papers, Einstein supplied the action principle Eddington had unaccountably omitted from his theory; the last, his final publication on Eddington's theory, appeared in 1925 as an appendix to the German translation of *The Mathematical Theory of Relativity*.[23]

However this may be, throughout the 1920s Einstein alternated in exploring the myriad formal possibilities afforded by Eddington and Kaluza. In 1925 came Einstein's first "homegrown" unified field theory, exploiting Eddington's postulate of introducing an affine connection independently of the metric.[24] Two years later he returned to the five-dimensional scheme of Kaluza.[25] This opening phase of the geometrical unified field theory program essentially ended with Einstein's "distant parallelism" theory of 1928–1931, perhaps his final public sensation.[26] Of course, none of these theories attained the Holy Grail, a classical field-theoretic account of the particulate structure of matter. Lecturing at the University of Vienna on October 14, 1931, Einstein himself laid these efforts to rest. Surveying the failed attempts, each conceived on a different differential geometrical basis, he forlornly referred to a "graveyard of dead hopes".[27] By then, it had become abundantly clear that while the electromagnetic field might be endowed with geometric significance in numerous formally different ways, it was quite another matter to construct a common geometric structure for gravitation and electromagnetism that "formed a harmonic whole, without intrinsically distinguishable parts".[28] Most other theoretical physicists had already concluded that the route to new physics did not lie through geometrical unification, and that the new quantum theory revealed a far more promising route to the problem of matter. As applications of quantum mechanics built up an impressive record of empirical success, the geometrical unification program correspondingly became a dissonant minority tendency already by 1930. When the new physics of nuclear forces emerged in the early 1930s, it could not but also seem greatly premature. Yet Einstein and his various assistants were the exception, doggedly pursuing the path of geometrical unification within the frame of space-time geometry until the bitter end. Indeed, Einstein's very last proposal for geometrical unified field theory, the "Relativistic Theory of the Non-symmetrical Field", rested upon the postulate of a (nonsymmetrical) affine connection, or "infinitesimal displacement field", a concept stemming from Levi-Civita but whose importance Eddington had especially stressed. Writing to Eddington's biographer in 1953, Einstein observed:

> [Eddington] was one of the first to recognize that the displacement field was the most fundamental concept of general relativity theory, for this concept allowed us to do without the inertial system.[29]

8.3 Eddington and Weyl

We saw in chapter 4, and again in chapter 6, that Weyl's second version of his theory, in which he strove to account for the origin of the "natural gauge of the world", did not meet with a widely favorable reception. Eddington proved a crucial exception; as correspondence with Weyl reveals, Eddington was an early and enthusiastic convert. On 16 December 1918, scarcely a month after the end of the World War, Eddington wrote to Weyl requesting a reprint of *"Gravitation und Elektrizität"* (1918c), "a paper which fascinated me very much".[30] On 18 August 1920, apparently having sent Weyl a copy of *Space, Time and Gravitation*, Eddington wrote again, professing delight at Weyl's satisfaction with that book's treatment of Weyl's theory in its penultimate chapter.[31] Responding to points Weyl had raised regarding his new "theory of the affine field", Eddington confessed, in a third letter dated 10 July 1921, that he had initially regarded his generalization to be only a slight reworking, hardly more than a modification, of Weyl's theory. He had been "so directly inspired by" Weyl's theory, Eddington wrote, that he "scarcely considered it [his affine field theory] a rival theory, until I learnt that you stuck to your own theory without my proposed modifications".[32]

The "proposed modifications" were largely directed to Weyl's explanation of the origin of "the natural gauge of the world". Recall from chapter 6, §6.4.2.3, that the entire purpose of Weyl's distinction between *Aethergeometrie* and *Körpergeometrie*, wherein material bodies are in a constant state of "adjustment" to the surrounding field, is to reconcile the epistemological postulate of a "principle of relativity of magnitude" with the actual behavior of measurement bodies. But Eddington assessed Weyl's schematic hypothesis of a compensating dynamical mechanism of "adjustment", the details of which depended upon an unknown theory of matter, as equivocal, unnecessarily obscuring the meaning of what, after all, had been deemed a purely epistemological principle. The ambiguity arose in Weyl's attempt to accomplish the impossible, reconciliation of a scale-invariant metrical theory of a continuum, in essence a purely mathematical conception for which there was not a shred of independently confirming physical evidence, with a physical geometry (*Körpergeometrie*) based upon the empirically attested behavior of rods and clocks. Weyl's difficulty stemmed from the fact that he had not seen fit to explicitly affirm that his *Aethergeometrie* is, and could be, but an ideal "graphical representation" of existing physical theory, not a physical hypothesis, a claim about the structure of a space-time world devoid of matter. Eddington's key modification accordingly sought to remove any source of equivocation resulting from the appearance of two contrasting metrical geometries, and so two conflicting notions of *length*, by building up the *ideal* world geometry from a nonmetrical axiom.

As I showed in chapter 6, §6.3.2 in Weyl's *ideal* "pure infinitesimal geometry", an intrinsic and immediate relative comparison of the magnitudes of differently oriented vectors *at the same point* is secured on the grounds of *Evidenz* presented in a localized eidetic intuition, the phenomenological cornerstone of "the epistemological principle of relativity of magnitude". Such *Evidenz* is to be found in an imaginative rotation of one vector in the infinitesimal homogeneous space about the point until it overlays the other, a phenomenological conception rationally justified by a group-theoretic demonstration of an infinitesimally "Pythagorean" orthogonal rotation group at every point of a Weyl manifold. The constituting ego, "pure, sense-giving consciousness", can reach no more primitive layer or stratum than in such a concept of a *metric* at a point. Phenomenological *Evidenz* is directly constitutive of a *metric* geometry in the imaginative act of a direct relative comparison of vectors at a single point (thus allowing for scale freedom), the epistemological postulate of "pure infinitesimal geometry". To be sure, in the context of general relativity, Weyl's supposition of space as "a form of intuition", metrically homogeneous in the infinitesimally small, required supplementation with an epistemological distinction between the *a priori metrical essence* of space, conceptually represented by infinitesimal orthogonal groups at every point, and the *a posteriori* mutual "orientations" of these groups differing from point to point.

Understandably in the dark regarding the phenomenological requirement of *Evidenz* that point-bound relative comparison of lengths be "given" or "immediately present" to a constituting consciousness, Eddington quite naturally assumed that there must be some tacit reference to material or optical standards of comparison. So, in the letter of 10 July 1921 already cited, Eddington objected,

> You admit an intrinsic comparability of lengths in different directions at the same point; but it seems to me that that can only be done by considering how *material* or *optical metrical standards* behave on changing direction....

Eddington's thought here, that length comparisons at the same point but in different directions necessarily depend on material or optical standards, lends the "principle of relativity of magnitude" a more prosaic meaning. On the one hand, there can be no talk of "length" or "metric" without such means of comparison; on the other hand, the physical behavior of these appliances, as standardized in measurement practice, *fixes the meaning* of all metrical concepts in physical geometry. Metric concepts may be given a wide range of meanings in pure geometry. But in physics, they have necessary reference to the practice of mensuration with material or optical apparatus, whatever it may be. Tying the meaning of metrical concepts to material or optical apparatus in this way clears up the fatal ambiguity plaguing Weyl's account of the origin of the "natural gauge of the world". Weyl's great, and essential, contribution to relativity theory had been point out the essential tension between an *a priori* epistemological principle of relativity of magnitude and the existence of a "natural gauge of the world" testified in actual measurements with rods and clocks. The fundamental flaw of the Einstein theory, in simply assuming the machinery of Riemannian geometry, was that it had swept the problem under the rug. But Weyl's proposed remedy could not convince because it failed to clarify the relation of the ideal, *a priori* and nondynamical, scale-free *Äthergeometrie* to the *Körpergeometrie* of actual measurements; both were *metrical* geometries. Far better, Eddington urged, to adopt at the outset a nonmetrical geometry as the ideal world geometry, suitable for graphically representing physics from the "point of view of no one in particular". Within such a representation, the meaning of physical space and time could then be seen as uniquely fixed by measurements made by material and optical appliances. Hence, the sentence cited immediately above continued, "and if you introduce material standards at all you can scarcely stop short of [a geometry with unique metric corresponding to physical measurements]".

Such a geometry is just the (pseudo-)Riemannian space-time geometry of Einstein's theory, which explains Eddington's above remark to Einstein that it was "an *extension*, but is not in the slightest degree an *emendation*, of yours", since the "*ds* is actually comparable at a distance (i.e., integrable)". What Eddington termed a *modification* of Weyl's account of the reconciliation between the epistemological principle of "relativity of magnitude", and the *de facto* existence of "the natural gauge of the world", is therefore a radical change of perspective. Eddington's reconstruction will require giving "all our (physical) variables ... a suitable graphical representation in some new conceptual space—not actual space". The sharp distinction drawn between an ideal geometry of "the world structure", and "the natural geometry of the world", the geometry of actual rod and clock measurements, is therefore Eddington's fundamental innovation, implemented ruthlessly by an insistence that the former can be but a "graphical representation" for reconstructing known physics, a point underscored by the posit of a nonmetrical and purely mathematical primitive relation of *infinitesimal parallel displacement*.

A minimal set of assumptions are adopted, all based upon the postulate that continuous analysis is sufficient for such a graphical representation. First, there is a four-dimensional continuum of points, with the customary topology for performing differential calculus. Differences in sign between space and time indices were to be put in by hand later on. Next, a fundamental relation of extension termed a "displacement" is assumed between (topologically) neighboring points P and P'. A displacement has orientation but not magnitude (as there is, as yet, no metric or gauge system) and so is not to be initially thought of a vector. Canonically, a displacement is represented as the coordinate difference dx^μ between the neighboring points, P and P'. Finally, there is an "axiom of parallel displacement", by means of which a displacement at P may be compared with one at P' by identification of a direction-preserving clone of it at P' ("parallel transport" to P'), corresponds to a symmetrical "affine connection" for vectors. Weyl had demonstrated in *Raum-Zeit-Materie* that the whole calculus of tensors can be derived from such a connection. As against Weyl's theory, however, Eddington's "axiom of parallel displacement" is introduced *ab initio* without any explicit thought of unique compatibility with a metric (by the condition $g_{\mu\nu;\sigma} = 0$), as is the case in both Riemannian and Weyl geometry. Rather its initial *raison d'être* is to be a most general relation of comparison for the very rudimentary reason that a "structure" or "complex of relations" cannot be "entirely devoid of comparability":

> [F]or if nothing in the world is comparable with anything else, all parts of it are alike in their unlikeness, and there cannot even be the rudiments of a structure.[33]

Hence, the axiom of parallel displacement seemed to be "the minimum degree of comparability which permits of any differentiation of structure".[34] Naturally, this assumption meant that the only physical worlds that can be "built up" from such a "primitive relation structure" of points and displacements will be "relation-structures" of tensors, generated from an affine connection. The fundamental series of scale-free "in-tensors" (as they are called by Eddington) are thus relation-structures derived entirely from a fundamental relation of *local comparability*. Locality here has nothing immediately to do with causation, or prohibitions of action at a distance, ideas belonging to a world of physical things. Rather it is demanded by the strictures of "the point of view of no one in particular": Expansion of the democracy of viewpoints can hardly go further than in the provision of materials for building a physical world presupposing only a principle of local comparison of "displacements". In turn, the ensuing relation structure comprises a purely conceptual space, an abstract perspective, containing the "lumber" from which to build space and time and "things", the measurable quantities of the actual physical world. As shown at the end of chapter 7, the aggregate of relata and relations comprising a world geometry is regarded as "mind stuff", building material for a world identical to the known physical world. But in beginning with a nonmetrical connection, Eddington has also ensured that his world-building materials do not implicit presuppose particular choices of gauge ("scale") since a metric, and so gauge, does not as yet exist. Eddington's world geometry therefore cannot be initially the geometry of physical space and time. Such a geometry, manifesting "the natural gauge of the world", will be constructed only later on, marking the transition from pure geometry to physics. The exclusive emphasis placed upon the metrical quantities of space and time is attributable to the state of physics in 1921. Still, Eddington was one of the first to recognize[35] the idea of local comparison is broader, and with hindsight, it can be said that his affine field theory anticipates the enormous role that linear connections would come to play in the fiber bundle formalism of gauge structures in quantum field theories.

8.4 World Building in the Theory of the Affine Field

In the language of the laity, the "geometry of the world structure" is to furnish all needed resources, or "lumber",[36] for a graphical representation of physical theory from "the point of view of no one in particular". Eddington's novel idea is to construct these mathematical resources from the most minimal assumptions needed to generate the tensors of the relativistic description of the classical gravitational and electromagnetic fields. In 1921, it was still widely believed that the whole of fundamental physics might be characterized through tensor algebra and analysis. For Eddington the use of tensors was mandatory since, by definition, tensors satisfy the relativistic requirement of transformation between all systems of coordinates, a necessary, but not sufficient, condition for attaining "the point of view of no one in particular".

> I do not think it is too extravagant to claim that the method of the tensor calculus, which presents all physical equations in a form independent of the choice of measure-code, is the only possible means of studying the conditions of the world which are at the basis of physical phenomena.[37]

Such a declaration expresses a presumption that fundamental physical laws are exhaustively describable by the methods of (real) analysis on a continuum of space-time points. While Eddington allowed that this assumption may be questioned,[38] he was nonetheless taken aback in 1928 by Dirac's wave equation of the electron, which is Lorentz invariant but is not constructed as a tensor equation. As he was fond of quoting C. G. Darwin, "apparently something has slipped through the net".[39] It has recently been claimed that with the Dirac equation, "Eddington's whole intellectual framework was shattered".[40] This surely is an overstatement. But it can be said that the Dirac equation forced Eddington to modify his mathematical materials. From 1928 on, Eddington developed what he called "wave-tensor calculus", based on the (Clifford) algebra of his "E numbers", an elaborate structure of anti-commuting matrices, and geometries derived from it. There are many statements attesting to the continuity between the affine field theory deemed suitable to macroscopic physics, and these new mathematical methods for epistemological reconstruction of known microscopic physics.[41] Indeed, he sought to show how the geometry of the affine field theory for the macroscopic physics of gravitation and electromagnetism emerges from the algebra.[42]

In accordance with world building's mandatory sharp distinction between mathematical and physical theory, the 1921 paper is divided into two parts, "Geometrical Theory", developing a "world geometry" from the axiom of parallel displacement, and "Physical Theory" in which existing physical theory is reconstructed in world geometric terms.

8.4.1 Geometrical Theory

The purely geometrical part of Eddington's affine field theory develops a fundamental series of "in-tensors" from the indicated postulate basis, and then evaluates these tensors in terms of the Levi-Civita connection of Riemannian geometry. It concludes with the derivation of a relation, called the "gauging equation", identical to the Einstein law of gravitation, supplemented by the cosmological constant, in "empty space". As already discussed in chapter 7, §7.5.4, this is ultimately the reason for Eddington's "passionate attachment" to a nonzero cosmological constant, else "the bottom of space" be "knocked out".

Primitive relation structure. A relation of extension between two "nearby" points P and P' is termed a *displacement* A^μ (an "in-vector" as there is, as yet, no such thing as a metric or a *length*).

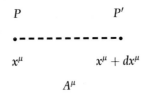

Axiom of parallel displacement. A relation of comparability of displacements "sufficiently close together"; that is, it is possible to identify, among all the displacements at P', one that is "equivalent" to A^μ at P. There is no required "observational test" of this equivalence. As shown in chapter 6, §6.3.1, the most general continuous formula for the change in A^μ is given by

$$dA^\mu = -\Gamma^\mu_{\nu\alpha}A^\alpha dx^\nu \tag{2}$$

where the "affine connection" $\Gamma^\mu_{\nu\alpha}$ is not assumed to be a tensor. Its 64 components, variable from point to point, are reduced to 40 by imposition of the "parallelogrammatical property" that two displacements originating from the same point shall meet in a single point. This is the geometrical meaning of the requirement of symmetry in the lower indices, $\Gamma^\mu_{\nu\alpha} = \Gamma^\mu_{\alpha\nu}$. In familiar fashion, the condition for parallel displacement of A^μ around a small circuit is given by

$$\frac{\partial A^\mu}{\partial x^\nu} = -\Gamma^\mu_{\nu\alpha}A^\alpha dx^\nu \tag{3}$$

and the total change in A^μ when parallel displaced around an infinitesimally small closed circuit $dS^{\mu\nu}$ is

$$\delta A^\mu = -\frac{1}{2}\iint {}^*B^\mu_{\varepsilon\nu\sigma}A^\varepsilon dS^{\nu\sigma}, \tag{4}$$

with the factor $1/2$ indicating that each component of surface appears twice in the integrand, for example, as dS^{12} and as $-dS^{21}$ Writing $\Sigma^{\nu\sigma} = \iint dS^{\nu\sigma}$ for a small circuit, this infinitesimal displacement approaches the limit,

$$\delta A^{\mu} = -\frac{1}{2} {}^{*}B^{\mu}_{\varepsilon v \sigma} A^{\varepsilon} \Sigma^{v\sigma}, \tag{5}$$

showing that ${}^{*}B^{\mu}_{\varepsilon v \sigma}$ is a tensor (since the difference of two in-vectors is an in-vector, and $\Sigma^{v\sigma}$ is a tensor). A metric-independent generalization of the familiar Riemann–Christoffel tensor of Riemannian geometry, this is also the most fundamental such "in-tensor". Using a more familiar arrangement of indices, this in-tensor analogue of the Riemann–Christoffel curvature tensor is defined as

$$ {}^{*}B^{\varepsilon}_{\mu v \sigma} = -\frac{\partial}{\partial x^{\sigma}} \Gamma^{\varepsilon}_{\mu v} + \frac{\partial}{\partial x^{v}} \Gamma^{\varepsilon}_{\mu \sigma} + \Gamma^{\alpha}_{\mu \sigma} \Gamma^{\varepsilon}_{v\alpha} - \Gamma^{\alpha}_{\mu v} \Gamma^{\varepsilon}_{\sigma \alpha}, \tag{6}$$

the asterisk, following Eddington, is adopted to remind of its independence of any gauge or metric. ${}^{*}B^{\varepsilon}_{\mu v \sigma}$ is a measure of world structure at each point; its 256 components are the "bricks of primitive clay" for building a world mathematically identical to the known physical world.[43] But it also contains much "excess lumber", and so an antisymmetrical in-tensor of the second rank is formed by contraction,

$$ {}^{*}B^{\sigma}_{\mu v \sigma} = {}^{*}R_{\mu v} \tag{7}$$

where ${}^{*}R_{\mu v} - {}^{*}R_{v\mu} = 2F_{\mu v}$, so that $F_{\mu v}$ is the antisymmetical part of ${}^{*}R_{\mu v}$, and, it turns out, the "curl" of a covariant vector κ_{σ} already present in the world geometry $(F_{\mu v} = (\partial \kappa_{v} / \partial x^{\mu}) - (\partial \kappa_{\mu} / \partial x^{v}))$, just as the Faraday tensor of the electromagnetic field is the "curl" of the vector potential. These two in-tensors comprise the fundamental material for "world building"; they are "the most fundamental measures of the intrinsic structure of the world".[44] Both "express intrinsic properties of the continuum", the first being "a description as complete as possible of the structure of the continuum (the interconnection of relations) so far as it can be studied by these methods", while the second is "an abbreviated summary of the information contained in the first". Since it will be shown that gravitation and electromagnetism can be reconstructed from the ${}^{*}R_{\mu v}$ alone, Eddington deliciously quipped that "physics has in the main contented itself with studying the abridged edition of the book of nature".[45] Up until this point, there has been no reliance on a metric tensor; a fortiori, there is neither means of raising or lowering indices, nor of forming tensor densities. But a metric is required to assign a length to displacements, $dx^{\mu} = A^{\mu}$. However, there is no requirement that lengths according to it need be consistent with lengths as determined through measurement; after all, matter and so measuring apparatus have not yet been "built". But pursuing the analogy to the space-time interval, Eddington conventionally specifies a tensor, conveniently denoted $g_{\mu v}$, defining the *length l* of the displacement A^{μ} through the relation,

$$l^2 = g_{\mu v} A^{\mu} A^{v} \tag{8}$$

It may be necessary to remind that, despite appearances, this is not yet an expression for the interval of space-time. Nonetheless, it gives a unique quantity, conventionally termed *the distance*, between infinitesimally close points P and P'. Since it is required that this is an invariant, the $g_{\mu v}$ must be a tensor but it is otherwise arbitrary, and without loss of generality, it may be held to be symmetrical. Although purely definitional, it is obviously *analogous to the metric tensor*.

The next step is to specify how the defined length of a displacement changes under parallel transport. Following the rule of parallel displacement, the change in l as A^μ changes under displacement dx^σ between P to P' is given by

$$dl^2 = \left(\frac{\partial g_{\mu\nu}}{\partial x^\sigma} - g_{\alpha\nu} \, \Gamma^\alpha_{\sigma\mu} - g_{\mu\alpha} \, \Gamma^\alpha_{\sigma\nu} \right) A^\mu A^\nu dx^\sigma \tag{9}$$

Since the difference of two invariants is an invariant, the quantity in brackets is a covariant tensor of the third rank, clearly symmetrical in the μ and ν indices. Eddington rewrote it as $2K_{\mu\nu;\sigma}$ the semicolon denoting covariant differentiation, and then showed that Weyl's geometry—*considered as a purely mathematical geometry*—arises in the particularization,

$$K_{\mu\nu;\sigma} = g_{\mu\nu}\kappa_\sigma \tag{10}$$

Hence Eddington's geometry of parallel displacement yields Weyl's as a special case. So far, the theory is purely mathematical.

8.4.2 Physical Theory

As repeatedly noted, in "world building" the field laws must appear as *mathematical identities*. But to interpret these identities as physical laws, a world geometric counterpart must first be found for the *lengths* and *durations of physical measurement*. Then the hitherto purely conventional tensor $g_{\mu\nu}$ is to be chosen so that "the lengths of displacements agree with the lengths determined by measurements made with material and optical appliances.[46] Such appliances, as Einstein's theory demonstrated, display "the natural gauge of the world". With this, the reason is now apparent behind Eddington's contention to Einstein, that his theory "is an *extension*, but is not in the slightest degree an *emendation*, of (Einstein's)" since, in Eddington's theory, but not in Weyl's, the (interval) "ds is actually comparable at a distance (i.e., integrable)".

Recall that in chapter 7, §7.5.4, an account of the existence of this natural gauge has already been given through the derivation of $R_{\mu\nu} = \lambda g_{\mu\nu}$ within a world geometry, fundamentally based, as is Einstein's theory, upon the metric tensor $g_{\mu\nu}$. This led to the "explanation" of Einstein's law of gravitation in empty space as "a gauging equation". Here Eddington followed a similar course. But in affine field theory the account is conceptually cleaner in beginning with an inherently nonmetric world geometry. The argument has the aura of pulling rabbits out of hats, but the logic is clear enough and, so far, impeccable. In the world-building scheme so far, "matter" has still to be "built"; strictly speaking, there are as yet no material appliances to measure lengths and so on. But then neither "matter" nor "space" nor "time" have any physical meaning independently of its respective measures. The physical notions of *length* and *duration* are constituted by their measures according to the practice of mensuration that standardizes the material appliances of measurement. Every measurement, however, is a comparison of the region surveyed (directly, or more likely indirectly, as an in principle possible direct measurement) with the surveying apparatus (the "relativity of field and matter").

In consequence, this apparatus has no *absolute* size: the corrections introduced by measurement practice to counter disturbing influences (to wit, Reichenbach's "differential forces") cannot restore something that does not yet exist.

For this reason, the physical *definitions* of *length* and *duration* must incorporate the "relativity of gauge" or "relativity of magnitude". Since any apparatus used to measure the world is itself part of the world, the natural gauge represents the world as self-gauging".[47] That is, the "natural gauge" must be inherent in the world itself, as its name indeed suggests. This can only mean that the tensor $g_{\mu\nu}$, presupposed in *physical measures* of length and duration by Einstein's theory, must already be contained in the world geometry. Just one fundamental covariant tensor of this rank has been built up within that geometry's primitive relation structure, namely, $*R_{\mu\nu}$. Making the only possible identification, "natural length", that is, length manifesting the "natural gauge", is specified in world geometry as

$$l^2 = {}^*R_{\mu\nu}A^\mu A^\nu \qquad (11)$$

On multiplication, the antisymmetrical part drops out, and the relation reduces to

$$l^2 = R_{\mu\nu}A^\mu A^\nu \qquad (12)$$

where the symmetrical part of $*R_{\mu\nu}$ is denoted $R_{\mu\nu}$. This tensor yields a Riemannian geometry defining an indirect system of measurement. Then, the required agreement between lengths of displacements and lengths determined by measurements with material and optical appliances, depending on $g_{\mu\nu}$ is expressed then by $R_{\mu\nu} = \lambda g_{\mu\nu}$, where λ is now a universal constant introduced "in order to remain free to use the centimeter instead of the natural unit of length whose ratio to familiar standards is unknown". As shown before, the cosmological constant is here simply an expression of the proportionality between the symmetric parts of the universe and the metric tensor, and Einstein's law of gravitation in "empty space" once again emerges as the "gauging equation of the world": "The gauging equation is, in fact, an *alias* of the law of gravitation".[48] After this result establishing a physical metric, the requisite physical identifications can be made, which are simply listed here: (1) The gravitational field is identified with the symmetric part of $*R_{\mu\nu}$, that is, $R_{\mu\nu}$. (2) The electromagnetic field tensor is identified with the antisymmetric tensor $(F_{\mu\nu} = (\partial\kappa_\nu/\partial x^\mu) - (\partial\kappa_\mu/\partial x^\nu))$, whereas the charge and current vector j^μ, from which the other half of the Maxwell equations may be derived, is the covariant divergence $F^{\mu\nu}_{;\nu}$. (3) "Matter" is identified, as before, as $R_{\mu\nu} - (1/2)g_{\mu\nu}R$. The affine field theory thus recapitulates, from the vastly more general world geometry of an affine connection, that the Einstein law of gravitation emerges, in one form ("empty space") as a *"gauging equation"*, but in the presence of matter, as a *definition of matter*. Most important, the fundamental field equations of gravitation and electromagnetism already appear through mathematical identification with tensors of world geometry, and so no appeal need be made to a special action principle to derive them.

8.5 Eddington's Transcendental Idealism

At this point all the essential contours of Eddington's "relativity standpoint" of the early 1920s are in place. Relativity theory, while indeed furnishing new physical phenomena and a new law of gravitation, is, perhaps above all, an epistemological "outlook" in imposing the ideal of objectivity of a completely impersonal world. Adopting this outlook has revealed hitherto unsuspected subjectivity in the perspectives of individual observers. But then relativity theory has also provided a geometrizing template for overcoming these individual particularities under the injunction that physical theory is continuously guided by a completely impersonal conception of the external world. The current state of such a picture of the world has been obtained in several steps, each corresponding to recognition of new observer-dependent features of measurement and representation. These dependencies have been successively purged through a generalization of the geometrical representation of the four dimensional continuum that relativity posits as the basis of all phenomena. Each generalization accommodates a greater degree of symmetry corresponding to a further synthetic representation of possible measurements that might be made by any of a given class of observers. In this development, Weyl's epistemological principle of relativity of length is a further refinement wherein the ideal of an impersonal world becomes a locally symmetric world.[49] From Weyl onward, the "relativistic outlook" mandates infinitesimal world geometry as the canonical means of representation of objectivity in physical theory, a constraint satisfied by identification of the tensor quantities of known physics with purely mathematical counterparts derivable within this geometry. Through such identifications, physics itself (or at least that part of physics dealing with the phenomena of gravitation and electromagnetism) is "geometrized", embedded within the conceptual space of a world geometry. On the other hand, it is only through such infinitesimal geometrical "graphical representation" that an epistemological justification can be fashioned for the *prima facie* inconsistent posit of rigid rods and perfect clocks employed in the empirical tests of general relativity.

The principal epistemological achievement of relativity theory has been to dethrone the "absolute observer", removing the tacit semantic equivalence between "completely impersonal" and "mind independent". The "absolute world" of events and intervals is simply *defined* as the desired "conception of the real world not relative to any particularly circumstanced observer" a conception obviously not mind independent. Its "absolute properties", geometrically represented in four-dimensional space-time, have only been attained or *constituted* via a "synthesis" of all "relative" measures, each presupposing an essentially arbitrary partitioning of this world into the space *and* time of a given observer. Their synthetic origin provides the only route of epistemic access; abstracted from this origin, they are purely mathematical objects. The mutual reciprocity of "absolute" and "relative" in Eddington reminds of Weyl's injunction that the very *sense* of an object of physical theory invokes necessary reference to the arbitrary posit of a coordinate system, the inevitable residuum of "the annihilation of the ego" in "that geometrico-physical world that reason sifts from the given under the norm of 'objectivity'" (see chapter 5, §5.3.4). But whereas Husserl's transcendental-phenomenological idealism underlay Weyl's epistemological reconstruction of general relativity in "pure infinitesimal geometry", Eddington, philosophically untutored but schooled within the sober tradition of Cambridge mathematical physics, recognized only the relativity of measures presented to indifferently situated co-moving observers. In place of the methodological solipsism of phenomenological transcendental constitution of all

objectivities from the "immediate life of a pure, sense-giving Ego", Eddington locates objectivity in the structures of an *a-perspectival* "absolute world", a world itself
constituted by ("a synthesis of") all its "relative" aspects. Yet common, and fundamental, to both is the core idea of transcendental idealism: that cognition is not of
objects given (*gegeben*), as it were from a "God's eye point of view", but is *objective*
because it is "posed as a problem" (*aufgegeben*), constituted *as objective* within the
frame of subjective conditions reflecting the structure and operations of the human
mind. In this sense, representation of an objective ("absolute") physical world can
only signify a representation synthesized from subjective or "relative" conditions,
now generalized to include the hypothetical measures of any and all conceivable
observers. Whereas Weyl followed Husserl in regarding as nonsensical a realist
epistemology of physical objects, processes, and events that "have nothing to do
with consciousness", Eddington similarly distinguished between knowledge that
does not particularize the observer and knowledge that does not postulate an observer at all. The methods of physics are capable of providing only the former, while
various questions surround the latter:

> [W]hether, if such knowledge could be obtained, it would convey any intelligible
> meaning; and whether it could be of any conceivable interest to anybody if it
> could be understood—these questions need not detain us now. The answers are
> not necessarily negative, but they lie outside the normal scope of physics.[50]

While frame dependent measures cannot be taken to stand for, nor in any way
resemble, elements of the geometrically represented absolute world, they nonetheless, as ingredient to the "synthesis" constituting that world, indicate the
structure of that world, the world corresponding to "the point of view of no one in
particular".

This is then the epistemological significance of Eddington's affine field theory.
The field laws of gravitation and electromagnetism are displayed within the most
unrestricted world geometry imaginable in 1921, a graphical representation of the
most general "conditions of possible experience" of a world constructed in accordance with the requirement of meeting "the point of view of no one in particular".
Beginning with the concept of an affine connection, a starting point much less
specialized than that of a metric, a series of fundamental tensors is derived, among
which are tensors mathematically identical to those of the field laws of gravitation
and electromagnetism. Physical space and time and "things" can be "identified"
with their world structural counterparts once a world geometric tensor identical to
the metric tensor has been "found". Appealing to the Einstein–Maxwell equations
(which Eddington simply *assumed*), a "Principle of Identification" is employed to
obtain a physical meaning in terms of measurable quantities for these tensors of
world geometry. In this way the Einstein equations for the gravitational field are
interpreted as "world geometric" definitions of "matter" and of "empty space", the
latter as a "gauging equation", a statement that the world radius of curvature
everywhere supplies the standard of measured lengths with rods and clocks. That,
from such a general basis, these definitions can be obtained, is to realize, at least in
part, an ideal of physical explanation: "[T]o show, not that the laws of nature come
from a special construction of ultimate basis of everything, but that the same laws
of nature would prevail for the widest possible variety of structure of that basis".[51]

Accordingly, while the reconstruction of field laws in the structural terms of world geometry does not imply "any hypothesis as to the nature of the quantities so represented",[52] that is not yet the end of the story. Mathematical counterparts for the tensors of the known field laws comprise but a small proper subset of geometrical objects derivable within this theory of locally symmetric world structure. Given Eddington's premise that a world geometry based on an affine connection affords the most general representation for the conditions of physical objectivity corresponding to "the point of view of no one in particular", these identifications are explained as due to the selective activity of "mind". Represented in this context, they are palpable proof that the order and regularity in appearances encapsulated in the classical field laws have their origin in the nature of mind.

Eddington's often-misleading expressions give rise to the unfortunate, but erroneous, impression of a subjective idealism, to which there has been no shortage of critics. But we have argued that his references to the predominant role of the "mind" in determining the fundamental field laws are better regarded as regarding the transcendental function of consciousness in bestowing an epistemologically responsible meaning on, and thus "constituting", objects of fundamental physical theory. Originating not from a prior philosophical position, but from his own reflections on adopting what he recognized as the standpoint of relativity theory to the physical world, his epistemology of physical science is a species of transcendental idealism shorn of "noumenalism".[53] Just as it was Kant's aim, in limiting spatial and temporal properties to appearances, to distinguish between human standpoints and nonhuman ones of *the same world*, Eddington's operative contrast lies in the difference between any given *relative* "particular point of view" and the *absolute* view of "no one in particular"—any and all conceivable observers— *regarding one and the same world of events*. While thus widening the boundaries of possible experience, Eddington's view of physical knowledge is not merely an instance, projected by a master of the theory of general relativity, of that transcendental idealism recently, and disparagingly, referred to as "a becoming form of epistemological modesty", but is an impressive attempt to carry over Kant's metaphilosophical critique of transcendental realism (and so, scientific realism) to the cognition of relativistic space-times.[54]

9

EPILOGUE

The "Geometrization of Physics"
and Transcendental Idealism

Doch das ist jetzt noch Zukunftsmusik....

Weyl (1919b, 116)[1]

9.1 An Indifferent Reception

The various currents of transcendental idealism underlying and motivating the "world geometries" of Weyl and Eddington were either generally ignored, misunderstood, or went unrecognized, even as their theories were mined for their mathematical ideas. In mathematics itself, Weyl's stepwise construction of a metrical manifold stimulated the development in Princeton of a "school" of differential geometry. Named the "geometry of paths" by Eisenhart and Veblen, its intuitive idea was that "we are dealing not with the empty void of *analysis situs*, but with a manifold in which we find our way around by means of the paths".[2] In Paris, Élie Cartan generalized Weyl's notion of a "affine connection", a basic notion for what became known around 1950 as the fiber bundle formalism of modern differential geometry.[3] Weyl himself was led to a study of the representations and invariants of semisimple Lie groups out of a wish to "understand what is really the mathematical substance behind the formal apparatus of relativity theory".[4] Meanwhile his group-theoretic treatment of the "new problem of space" (see chapter 6, §6.3.2) spurred in him a renewed interest in the value of group theory for investigating the mathematical foundations of physical theories, an interest that bore fruit in a 1928 book introducing the methods of group theory into quantum mechanics.[5] As I have already shown in chapter 8, §8.1, Einstein in 1923 took up Eddington's idea of introducing an affine connection independently of the metric in the attempt to unify gravitation and electromagnetism, exploiting it in various theories over the next thirty years.

Within philosophy, Eddington became controversial, then heretical, mainly on the basis of passages in his best-selling popularizations of science (chapter 7, §§7.1–7.2). With the notable exception of Émile Meyerson's *La Déduction Relativiste* (1925), the richly philosophical context of Eddington and Weyl's works on relativity theory was ignored. Meyerson, however, regarded the philosophical import of Einstein's theory as lying precisely in its geometrizing tendency, and he accordingly paid particular attention to the theories of Weyl and Eddington for extending this tendency even further. Meyerson's book elicited a rare, and substantive book review from Einstein, probably for this reason.[6] Moreover, Meyerson's evaluation of the geometrical unifications of Weyl and Eddington sets out, as clearly as anyone has, a realist interpretation of the significance of geometrical unification, while nonetheless leading directly to, but not crossing, the threshold of comprehending Weyl and Eddington's guiding epistemological motivations. The stumbling block proved to be Meyerson's failure to recognize how transcendental idealism might be supported once detached from the now-impugned doctrine of pure intuition of the Transcendental Aesthetic. Instead, Meyerson offered a realist, if not neo-Platonist, interpretation of this geometrizing tendency, a response that is characteristic of many philosophers of physics to current geometrical unification programs. As, in the early 1920s, the very idea of geometrical unification had laid dormant in physics since the "dream of Descartes" in the 17th century,[7] Meyerson sought to give a comprehensive account of the broad philosophical motivations linking notions of unification, explanation, and truth. In considering this realist response to the theories of Weyl and Eddington, it therefore will be helpful to rehearse Meyerson's general philosophy of science and its curious, by present lights, attention to Hegelian *Naturphilosophie* as the contrasting modern paradigm of a system of global explanation of nature.

9.2 Meyerson's Philosophy of Science

To Meyerson, science itself is above all anti-positivistic because it is "saturated with metaphysics"; the "unexposed bedrock" on which "the whole of science rests" is "belief in a being independent of consciousness".[8] The goal of science exclusively lies in finding causes, and thus in theoretically comprehending the qualitative phenomena of perception in terms of a reality underlying them. "Cause" and "explanation" are synonymous terms.[9] Conclusions about the nature of science are to be reached not on *a priori* grounds but from reflections upon the activity of scientists themselves. As did Cassirer, with whom Meyerson is otherwise in fundamental disagreement, Meyerson regarded himself as an epistemologist of science, concerned to identify and track the patterns of thought revealed by the history of science, particularly of the physical sciences since the birth of the modern age. And like the

Marburg Neo-Kantians in general, he believed that from this mass of material an *a priori* kernel of reason might be extracted. In two previous works, *Identité et Réalité* (1908) and *De l'Explication dans les Sciences* (1921), Meyerson identified this kernel as the *a priori* inclination of thought to impose unity upon a recalcitrant world of phenomenal diversity. The two books portray the growth of science as resulting from continual struggle between the persistent will of the mind to "identify the dissimilar" and the resistance to identification by the "irrational" elements of reality. The very impetus for science lay in surmounting this irrationality, in fashioning explanations rooted in an *a priori* principle of "negation of diversity".

Deductive explanation is, of course, the favored mode for rationalization of nature. While already well established in ancient Greek thought through the paradigm of Euclidean geometry, physical science and philosophy in the modern period presented two opposing models for carrying out such explanations. Descartes deserved the accolade of "the true legislator of modern science"[10] for his proposal of mathematical deduction as the canonical mode of physical explanation. Moreover, in maintaining the identity of matter and extension, Cartesian metaphysics provided the possibility for the fundamental step of physical unification, the bold reduction of matter to the purely kinematical properties of otherwise featureless extension, a purely quantitative representation of a fundamentally spatial reality.

> The goal of explanations and theories is really to replace the infinitely diverse world around us by identity in time and space, which clearly can only be space itself.[11]

With Descartes, *panmathematicism*[12] was established as the *ideal* of physical explanation; indeed, Descartes had actually taken upon himself the "crushing burden" of "the global deduction of physical reality", but this is an impossibility, beyond human capabilities.[13]

To Meyerson, the emergence of Hegelian *Naturphilosophie* in the early 19th century created, for the first time, a rival paradigm of unifying explanation. Deeming mathematics an unsuitable instrument for the purpose of satisfactory metaphysical explanation, Hegel sought a dialectical deduction of natural scientific phenomena (e.g., magnetism) from the categories of his absolute logic; the *panlogism* of the *Naturphilosophie* is the final culmination of the entire Hegelian system of idealism. Despite its utter scientific failure, *Naturphilosophie*, as an alternative system of global explanation, was a subject of considerable interest for Meyerson. Just a few years prior to his examination of the geometrizing aspirations of general relativity, he could still regard Hegelian *Naturphilosophie* as "the most recent of the great attempts at a global explanation of nature", whose value to posterity lay in "the completeness of (its) failure", in the colossal demonstration of an overreaching hubris.[14]

Meyerson's closer diagnosis of the failure of *Naturphilosophie* illuminates his own neo-Platonistic tendencies. Whereas Cartesian dualism posited space (or extension) as a fundamental substance totally distinct from thought, Hegel's idealistic monism madly ignored the "wide and nasty ditch" (quoting Schelling) between thought and reality, logic and natural philosophy. In doing so, it contravened the very "bedrock" realist metaphysics of science. Certainly, the complete rationalization of nature that is the goal of natural science cannot ignore the yawning abyss between thought and nature; somehow, it had to be crossed. Descartes, although not fully cognizant of the difficulty of doing so, had indicated the way forward. Spatiality is the key because "geometrical truths belong to our thought as much as to nature; thought and nature seem to merge".[15] In quasi-Platonic mode, Meyerson declared in *La Déduction Relativiste* that "what is mathematical" is neither completely the same as mind nor entirely alien to it, belonging *"at one and the same time* to our reason and to nature". Responsible for the agreement of thought and reality, and so for linking the twin poles of Cartesian dualism, mathematics is "the true *intermediate substance"*.[16] That Einstein's general relativity succeeded, to the extent that it did, "in the rational deduction of the physical world", is due to the fact that "Einstein followed in the footsteps of Descartes, not Hegel". The result is a partial realization of Descartes' program of reducing the physical to the spatial through geometric deduction.[17]

This difference between a tempered Cartesian *panmathematicism* and an absolutist Hegelian *panlogism* is also fundamental methodologically, for it informed Meyerson's falsificationist understanding of the progress of science. A scientific theory can be "dogmatic only in what it denies", not in what it affirms. If the claims of a new theory seem otherwise, this is only because the more successful replacing theory has "destroyed the old reality", the previous agreement of reason and nature. The new theory, however, is an explanation "accepted only for lack of something better", that is, "provisionally".[18] The necessary change occurs when accepted explanations encounter insurmountable "irrationalities", when reason is confronted with recalcitrant and unyielding phenomena. Inverting Hegel's formula, Meyerson defines the real as that ever resisting the rationalizing impulse of reason. Scientific progress, "the progress our understanding makes in comprehending nature", consists in delimiting the agreement between reason and nature, but it lies in the very nature of science that this agreement can never be known, and so known to be completed.[19]

> We do not know—and shall never know—where the agreement exists, since we can never be sure that there will be no new irrationals to add to the old one. That is why we shall never be able really to *deduce* nature, . . . we shall always need new experiments and these will always pose new problems, causing contradictions between our theories and our observations to leap out at us, as Duhem puts it.[20]

To be sure, Hegelian *Naturphilosophie* altogether lacked any epistemological modesty of this kind and so could only give birth to an extravagant metaphysics. Hegelian *panlogism* is thus unfavorably contrasted with Cartesian *panmathematicism*; as empirical confirmations of Einstein's theory of general relativity testify, the latter method has actually engendered a greater comprehension of nature. In brief, the latter method, but not the former, has proven capable of yielding falsifiable scientific theories. But Cartesian *panmathematicism* is also capable of overreaching itself, of presenting a totalizing scheme of deduction, and when it does, all the mortal sins of Hegelian *panlogism* threaten. In fact, these are the transgressions committed by Weyl and Eddington.

Only a few years before publishing *La Déduction Relativiste*, Meyerson had written that whereas a "truly complete panmathematicism" formed "a basis for many current conceptions among the scientists and thinkers of our time", "a clear and peremptory affirmation of it" was then rarely heard, "outside of certain formulas of the Marburg school".[21] But in *La Déduction Relativiste*, the geometrical unifications of gravitation and electromagnetism of Weyl and Eddington are explicitly recognized as revolutionary scientific affirmations of such a complete *panmathematicism*, or rather of a *pangeometrism* in distinction from the *panalgebrism* of Cassirer.[22] Yet their attempts to deduce all fundamental physics from geometry have led only to a reawakening of the metaphysically disastrous totalizing logic of Hegel. Citing Weyl's impassioned declaration at the end of the third edition of *Raum-Zeit-Materie*,

> it no longer seems daring to believe that we could so completely grasp the nature of the physical world, of matter, and of natural forces, that logical necessity would extract from this insight the univocal laws that underlie the occurrence of physical events.[23]

Meyerson observed: "It can be seen that we did not exaggerate when we spoke of universal deduction in the case of relativity, comparing this bold construct with that of Hegel".[24] Meyerson proceeded to show how closely statements made by the 19th century German *Naturphilosophien* parallel those of Weyl. Weyl's *pangeometrism* is singled out in particular as appearing to be a reversion to Hegelian absolute idealism, and it bears witness of the same fatal flaw. In maintaining that nature is completely intelligible, it has abolished the thing-in-itself and leads to the identity of self and nonself, the colossal error of the *Naturphilosophien*. According to a realism that Meyerson shared with Schlick (chapter 2, §§2.4.1–2.4.2), reason can furnish only "an empty framework"; accordingly, any attempt, like that of *Naturphilosophie* "to introduce idealistic concepts into science" must turn out to be a "chimerical enterprise".[25] The *pangeometrism* of Weyl and Eddington, in Meyerson's assessment, similarly falls victim to this fatal fallacy.

On the other hand, Meyerson openly admitted to puzzlement regarding certain seemingly idealist formulations of Weyl and Eddington, particularly in the face of other philosophical pronouncements, more customary of natural science. Thus, quoting again from the conclusion of the third edition of *Raum-Zeit-Materie*, Meyerson noted Weyl's claim there that, in the last analysis, the merger of physics and geometry has resulted in a conception of "physical reality [that] appears only as a pure and simple form".[26] Similarly, he is perplexed by Eddington's remark from *Space, Time and Gravitation*, noted in chapter 7:

> We did not consciously set out to create a geometrical theory of the world; we were seeking physical reality by approved methods, and this is what has happened.[27]

Meyerson's interpretive hesitation regarding these, and other similar passages in the texts of Weyl and Eddington, might be contrasted with Einstein's opinion, as expressed in his review of Meyerson's book. To Einstein, "the essential point of the theories of Weyl and Eddington" was not that they are *geometrical* unifications but that "they have shown a possible way to represent gravitation and electromagnetism under a unified point of view". Indeed, for Einstein, "the term 'geometrical' used in this context is entirely devoid of meaning", "geometry" denoting simply "the study of the possible positions and displacements of rigid bodies".[28] Of course, this is an expression of Einstein's "official view", beloved by logical empiricism (chapter 3, §3.3). In a different way, Meyerson's own ambivalence regarding the assertions of Weyl and Eddington equally stemmed from an unwillingness to accept them as avowals of a coherent philosophical interpretation of relativity theory. He noted that Weyl, in claiming that physics is only concerned with "the formal constitution of reality" has, like Cassirer, put a " 'Copernican' about-face" upon the whole question, in effect, making reality revolve around thought.[29] Likewise, he recorded Weyl's remarks about "penetrating into the essence of space and time, and matter in so far as they participate in the construction of objective reality". But he cannot accept such transcendental idealist declarations at face value. The whole tenor of relativity theory, providing all space and time determinations with a form equally valid for all observers, is an expression of "the opposite tendency" to Kant. And the new conception of objectivity in relativity theory only arises

> by separating [these determinations] from the subject and giving them an existence independent of the self, that is to say, by putting them back into the thing-in-itself.

Like realists both then and since, Meyerson essentially identifies relativistic "observer independence" with mind-independence, a metaphysical inference expressly contested by the epistemologies of Weyl and Eddington.

Meyerson had no doubt at all that relativity theory, like all science, is essentially realistic, hence that space-time curvature is to be interpreted realistically, as *the Ding an sich*:

> [T]he relativistic space filled with *curvature* or *"puckers"* is objective; it belongs to things ... It alone constitutes all things; it is ... truly and literally *the* thing-in-itself".[30]

Thus he cannot take Weyl and Eddington at their word when they assert that relativity theory demonstrates that space and time are not properties of objects of a mind-independent four-dimensional physical reality. Consequently, he deems these "subjectivist affirmations" as "a sort of *hors-d'oeuvre*"; they are "besides the point", attributable to "simple flights of fancy".[31] While having "all due respect to the writings of such distinguished scientists", Meyerson can only judge these formulations to be misguided attempts "to associate themselves with a philosophical point of view that is in fact quite foreign to the relativistic doctrine".[32] That doctrine of course is transcendental idealism. It is above all "foreign" to relativity theory because Meyerson cannot see how it is possible to "reintegrate the four-dimensional world of relativity theory into the self". After all, Kant's own argument for transcendental idealism proceeded "in a single step", by establishing the subjectivity of the space and time of "our naïve intuition". But this still leaves "the four dimensional universe of relativity independent of the self". Any attempt to "reintegrate" four-dimensional space-time into the self would have to proceed to a "second stage" where, additionally, there would be no "solid foundation" such as the forms of spatial and temporal intuition furnished Kant at the first stage. Perhaps, Meyerson allowed, there is indeed "another intuition, purely mathematical in nature", lying behind spatial and temporal intuition, and capable of "imagining the four-dimensional universe, to which, in turn, it makes reality conform".[33] This would make intuition a "two-stage mechanism". While none of this is "inconceivable", it does appear, nonetheless, "rather complex and difficult if one reflects upon it". In any case, this is likely to be besides the point, for considering the matter "with an open mind", "one would seem to be led to the position of those who believe that relativity theory tends to destroy the concept of Kantian intuition".[34]

As we see, Meyerson had come right up to the threshold of grasping the Weyl-Eddington geometric unification schemes within the epistemological contexts in which they were intended. His failure is instructive, for it is not difficult today to find representatives among both philosophers and physicists of the realist position Meyerson so clearly articulated. The stumbling block for him, and for others, is a conviction that transcendental idealism can be supported only from an argument about the nature of intuition, and intuitive representation as that doctrine appears à la lettre in the Transcendental Aesthetic.

But that doctrine need not be taken literally. The geometric framework for Weyl's constitution of the objective four-dimensional world within transcendental subjectivity is based upon the *Evidenz* available in "essential insight" *locally* within the homogeneous space of intuition of the "pure, sense-giving ego". Mathematically comprehended as linear relations and mappings in the tangent vector space to a point P, the momentary locus of the "*Ich Zentrum*", Weyl's "purely infinitesimal geometry" marked the fundamental transcendental constitutive divide between direct (local) and indirect (global) relations of comparison. The cost of holding the former to be epistemologically privileged was high: a somewhat artificial explanation of the emergence of the integrable relations of metrical comparison, attested by observation. Nonetheless, the "pure infinitesimal" framework for world geometry did not rest simply on the basis of phenomenological "essential insight", rather it was only considered fully justified by a group-theoretic, and so *conceptual* argument. Eddington, on the other hand, without the cultural context of Husserlian phenomenology or indeed of philosophy generally, ignored the explicitly intuitional basis of transcendental idealism altogether, as if unaware of its prominence, while emphasizing, nonetheless, the "relative" nature of particular measures and observations. Then he sought a superior *conceptual* basis for incorporating the "principle of relativity of length" necessary to a fully objective physical world, into an account of the very physical meaning of *lengths* and *durations*. The "world geometry" within which relativity theory is to be reconstructed accordingly is a *conceptual space* set on the basis of a nonmetrical affine connection, the "simplest conceivable relation of comparison in a continuum". He was then free to find his own way to the "practically rigid rods" and "perfect clocks" of Einstein's theory. But in each case, the intent of their respective "geometrizations" was not to formulate a physical hypothesis grasping at the totality of physical reality, but rather to be an epistemological reconstruction illuminating the salient transformation of the notion of physical objectivity by general relativity.

9.4 "Structural Realism"?

Despite the attempts of Weyl and Eddington, it has been routinely assumed that all attempts at the "geometrization of physics" in the early unified field theory program shared something of Einstein's hubris concerning the ability of mathematics to "grasp" the fundamental structure of physical reality. The geometrical unified field theory program thus appears to be inseparably stitched to what has recently been recognized as a distinct offshoot of scientific realism, "structural realism".[35] Structural realism currently exists in two varieties. The first, and older version is *epistemic*, holding that whatever the "nature" of the primitive entities and processes comprising the world, only the "structure" of this fundamental reality can be, and is *known*, insofar as that structure appears and is retained in the basic equations of our best theories. At the end of chapter 7 we saw that Russell indeed maintained such a view in 1927. A more recent variety of structural realism is *ontological*, asserting that the fundamental constituents of the physical world *are* structures. To the adherents of ontological structural realism, it is not surprising that physical knowledge is only knowledge of structure, for *structure is all there is*. In both cases, structural realism sees the ontological continuity across theory change required by scientific realism is to be found solely in a continuity of structure, as the equations of the earlier theory can be derived, say, as limit or special cases, from those of the latter. Geometrical unification theories seems tailored for this kind of realism, as is already clear from general relativity where solutions of the Einstein field equations in the weak field approximation correspond to those of the Newtonian theory of the same phenomena. Moreover, *if* a geometrical theory is taken to provide a true, or approximately true, representation of the reality underlying all physical phenomena, then, on realist assumptions about the nature of representation, what *is* geometrically represented has the structure of the representing geometrical objects and relations.

While a comprehensive criticism of structural realism must be deferred to another venue, we wish simply to conclude this book by underscoring that neither Weyl nor Eddington understood their respective theories of gravitation and electromagnetism in this way. For both, geometrical unification is not, nor can be, a limning of the structure of a mind-independent world, for essentially the reasons so beautifully articulated by Poincaré two decades before.

> Does the harmony the human intelligence thinks it discovers in nature exist outside of this intelligence? No, beyond doubt, a reality completely independent of the mind that conceives it, sees or feels it, is an impossibility. A world as exterior as that, even if it existed, would for us be forever inaccessible. But what we call objective reality is, in the last analysis, what is common to many thinking beings, and could be common to all; this common part ... can only be the harmony expressed by mathematical laws. It is this harmony then which is the sole objective reality. . . . [36]

The harmony of the world displayed in geometrical unification is not, intelligibly, the Pythagorean view that the world is inherently mathematical. Rather, the geometrical unifications of Weyl and Eddington, intertwining geometry, physics, and philosophy in a manner unprecedented since Descartes, were above all explicit attempts to comprehend objectivity in physics, in the light of general relativity, from epistemological standpoints that have the character of transcendental idealism. They were, and remain, illustrations of the harmony of field laws of gravitation and electromagnetism within a geometrically constructed objective reality of the greatest possible indifference to individual points of view, "what is common to many thinking beings, and could be common to all". Their unorthodox arguments, cloaked, by necessity, in the language of differential geometry, have long remained unknown and unexplored by serious philosophical inquiry into just what sense a "geometrized physics" *can* have. While that investigation has still to come, it will certainly profit by taking heed of its precursors within the reign of relativity.

APPENDIX TO CHAPTER 2

Michael Friedman and the "Relativized *A Priori*"

In a number of recent writings, Michael Friedman has returned to Reichenbach's 1920 conception of principles of coordination in a sustained attempt to revive an account of scientific rationality and change based upon a neo-Kantian account of the "relative *a priori*". Initially, in accord with the difference noted in chapter 2, §2.2, between covariance and invariance groups, Friedman has argued that, for Reichenbach, what is "relatively *a priori*" for a theory is determined by the invariance group of the theory. Hence, the distinction between the constitutive and empirical is, in current parlance, that between the *absolute* and the *dynamical objects* of the theory. This is to say that the constitutive function of the "relative *a priori*" (coordination principles) to establish and define the spatiotemporal framework within which the particular dynamical laws operate, a necessary condition for the empirical testing and confirmation of these laws.[1]

However, in several later publications that position has changed somewhat, eliminating the analogy to absolute and dynamical objects. As with Reichenbach, the task is to "coordinate" two independent sources of knowledge, analytically reconstructed as different "parts" of empirically successful physical theories. On the one side of the coordination are purely abstract conceptual (mathematical) structures; on the other, the concrete empirical phenomena to which these representations are intended to apply.[2] The application of the former to the latter is only attained through theory-specific "coordination principles"; in the case of relativity theory, these are the principle of the constancy and source independence of the velocity of light, as well as the principle of equivalence. The former constrains events that can be considered as causally related; the latter serves to define accelerations in gravitational fields in lieu of the notion of a global inertial frame. The general role of such principles is to provide empirical meaning to the mathematically expressed precise laws of nature, furnishing "the necessary framework within which the testing of properly empirical laws is then possible". Without these two fundamental "mathematical-physical presuppositions", Einstein's field equations have no empirical content.[3]

Friedman's revival of Reichenbach's account of the "relative a priori" is largely articulated as a critical response to Quinean holism and the Kuhnian theory of scientific revolutions. Against Quine, Friedman maintains that his "relatively *a priori*" constitutive principles are *not* simply empirical principles deeply embedded in our current overall system of belief. Rather, as first making possible an empirical knowledge structured and framed by such principles, they have a unique and necessary "meta-empirical" standing that Quine's holist account of knowledge cannot recognize without surrendering its fulcrum point, the attack on the analytic/synthetic distinction. With regard to Kuhn, Friedman holds, much as with Reichenbach's "method of successive approximation", that in the development of modern science these constitutive principles comprise a convergent series, successively refined in the direction of greater generality and adequacy. In the particular instance of the transition from Newtonian gravity to general relativity, the generalization emerges perspicuously through a generally covariant formulation of Newtonian theory in a four-dimensional space-time with a particular inertial frame singled out as "absolute" ("Newtonian space-time").[4] In contrast to a Kuhnian account of the discontinuous and a rational character of scientific change, Friedman views fundamental conceptual change or paradigm shifts as eminently rational, evolving through continuous change, a later paradigm or framework arising by "natural transformation" from an earlier one.[5]

This brief overview of Friedman's significantly more complex position must suffice for our purposes; a detailed examination goes beyond the scope of this book. Here I raise just three observations about this resuscitation of Reichenbach's "relative a priori".

(1) Regarding Friedman's original explication of the "relative *a priori*" in terms of the invariance group of the theory, as shown in §2.2 for general relativity, the invariance group is the diffeomorphism group leaving invariant the totality of "point-coincidences" of the theory. In particular, according to "active general covariance", for any general relativistic space-time, there is a (nonempty) class of diffeomorphically equivalent models. However, recall from §2.2, and as earlier discussed in chapters 3 and 4, that Einstein considered his gravitational theory not as a finished product but as a template for a "total field theory". The requirement of general covariance is above all a heuristic guide in the search for such a field theory, encompassing gravitation and all matter fields but from which all reference to a nondynamical background space-time has been eliminated. But even with respect to general relativity or, rather, Einstein's intentions regarding general relativity, the distinction between absolute and dynamical objects is problematic. It is often said that M^4 is the only "absolute object" in general relativity, but that it is a trivial one since it has no nontopological properties. From this point of view,

general covariance appears as a principle of "general invariance", a generalization of the principle of relativity from the Lorentz invariance of special relativity.[6]

However, while reference to successive generalizations of the invariance group of the theory (from the Galilean group, to the Lorentz group, to $Diff(M)$) emphasizes continuity with Newtonian theory and special relativity, the presence of M^4 as an "absolute object" is not a minor difference with Einstein's conception. For, in holding that space has existence "only as a structural quality of the field", Einstein was underscoring his heuristic postulate that, according to Stachel, "spatio-temporal individuation of the points of the manifold in a general-relativistic model is possible only *after* the specification of a particular metric field, that is, only after the field equations of the theory (which constitutes its dynamical problem) have been solved".[7] In this respect, the (now standard) formulation of the class of diffeomorphically equivalent models of general relativistic space-times $\{\langle M, g_{\mu\nu}, T_{\mu\nu}\rangle, \langle M, D^*g_{\mu\nu}, D^*T_{\mu\nu}\rangle\}$ already violates the spirit, if not the letter, of this injunction, for apparently suggesting a manifold M^4 of points as the space-time background for treating gravitational dynamics. In addition, as already mentioned above, there are unsolved internal difficulties with the proposal for it has proven to be rather difficult to cleanly distinguish between "absolute" and "dynamical" objects by means of an invariance group, without in addition making controversial physical assumptions to guide the distinction.[8]

(2) In general relativity, the applicability of the Riemannian theory of manifolds to physics is indeed made possible in particular by two "coordinations" or "identifications" (in the reconstructive mode of Weyl and Eddington; see chapters 6–8). However, the considerably more mathematically sophisticated approach of Weyl, rather than any notions stemming from Reichenbach, provides the basis of modern treatments. The affine (or projective) structure of a four-dimensional manifold is coordinated to the paths of freely falling particles (of negligible mass) and the conformal structure of the manifold to the paths of light rays. As Weyl first showed (see Chapter 4, §4.2.3), the affine and conformal structures yield the metrical structure of space-time together with its $(3 + 1)$ signature [obviating any need for a "coordinative definition" of the metric via rods and clocks in the absence of "universal forces", as in Reichenbach (1928); see chapter 4]. It should be observed that the geometrical notions are linked to objective coordinate-independent physical processes, which, of course, is not the case of measurements made with scales and clocks. In general, the geometrical properties of space-time cannot be empirically ascertained independently of other physical laws, as Schlick and Reichenbach, post-1921, claimed. In particular, the law of motion of freely falling bodies (the "geodesic hypothesis") may be derived from the generally covariant field equations of gravitation.[9] Unlike in Reichenbach's neo-conventionalist treatment of the metric of space-time, such "coordinations" are not at all "arbitrary" but abide a reasonable requirement of consistency, that "the physical interpretation of the mathematical notions occurring in a physical theory must be compatible with the equations of the theory".[10]

Once these coordinations are laid down, other mechanical phenomena may receive an interpretation within space-time geometry. For example, the paths of accelerating bodies are definable with respect to free-falling neutral mass particles, giving the local affine structure (or, with Weyl, *Führungsfeld*) that is the gravitational substitute for inertial structure while a physical definition of rotation is afforded by the conformal structure. More familiarly, the "bending of light rays" passing close by the surface of the sun can be interpreted as due to the "curvature" of space-time, that is to say, to the nonvanishing of a certain curvature tensor in that region. In any case, the geometrical representation of these and other physical processes is guided by Einstein's requirement of general covariance. It is a contention of *transcendental philosophy* that this constraint on coordination yields a new conception of possible object within physics, a claim made not only by Reichenbach in 1920, but independently and in various formulations by Weyl, Cassirer, Hilbert, and Eddington, who also make clear that, in this regard, it has precisely the function of a regulative idea.

(3) Friedman rightly maintains that the fundamental coordinations listed under (2) above are meta-empirical statements that indeed fulfill the constitutive function of "relativized" Kantian *a priori* principles, fallible but in a distinctively different manner from other empirical statements of relativity physics. What, then, of Quinean naturalism where the meta-empirical can only be regarded as a body of high-level empirical generalizations about lower level empirical claims? Consistent with Quinean precepts, this "definition" of the meta-empirical must be regarded an empirical assertion, not an analytic statement.

Quine does not entirely eschew a "normative domain within epistemology", countenancing methodological norms such as simplicity and a "maxim of minimum mutilation".[11] However, these norms are but poor resources for enlightenment regarding the sense or meaning of abstract geometrical statements in relativistic physics. Of course, Quinean naturalism can scarcely dignify this to be a specifically *philosophical* problem. But regarded as a *scientific* problem about the physical meaning of such statements, it is addressed only in vaguely picturesque generalities about linkages between more and less peripheral elements within the web of belief that is our "total theory of the world". Inevitably, this deflects the issue into the platitudes of underdetermination of theory by evidence on which benighted skeptics feast. But then the prior question, for example, as to how the geometry of Riemannian manifolds has found application in mechanics at all, must be, for the Quinean, merely a matter of "psychology of discovery". And so for Hilbert spaces, C* algebras, local symmetry groups, and all the other abstract mathematical apparatus in fundamental physical theories. Even so, this is an entirely unredeemed, and probably unredeemable, promissory note that willfully ignores considerable historical evidence of an internal dynamics of conceptual change in physics constrained by posits of "*a priori* reasonableness", for example, by symmetry and conservation principles—often in the face of dissenting empirical evidence.

Friedman has further argued that Quinean naturalism similarly must fail to adequately describe episodes of revolutionary conceptual transformation in science, also an empirical issue of scientific historiography, and here again we may let the chips fall where they may. Still, in view of the difficulties already posed to Quinean naturalism, one can hardly be sanguine about the prospects of an informative history of science under its auspices. Nor is it clear to me that Quineans might wish to endorse the "naturalism" of various constructivist or sociological programs in science studies. The burden then lies with Quinean naturalism. It must show that the very principles that enable fundamental physical theories to be brought to "the bar of experience" are themselves just further statements within the web of belief. More pointedly, the question of whether relativity theory has demonstrated the existence and role of meta-empirical "constitutive" principles of knowledge can, presumably, only be addressed by close attention to the principles and statements of that theory. But then why should we give any credence *at all* to an epistemological holism that has said nothing (or nothing of interest) specifically about relativity theory? Until Quinean naturalists deign to provide *epistemological analyses* (however this be understood) *of particular physical theories* in something like the same degree of detail as Reichenbach, or Friedman, or any of those listed above, I see no reason *not* to regard its naturalistic injunctions against the "relativised *a priori*" as the armchair bluster of a rather naïve "first philosophy".

NOTES

Chapter 1

1. Clark (1971), 232.
2. Russell (1926), 331.
3. Russell (1927). See the brief discussion in chapter 7, §7.6.1.
4. Schlick (1922b); Engl. trans. (1979), 351. Herbert Feigl was present on this occasion to witness the scene. See Feigl (1937–1938), xix.
5. Von Laue (1921), 42.
6. An unpublished remark of Husserl ca. 1914–1915, as cited in Bernet et al. (1993), 204.
7. Cassirer (1921), 78; Engl. trans. (1953), 418.
8. Chandrasekhar (1982), 93.
9. Hempel (2000).
10. Dirac (1931), 60.

Chapter 2

1. The quotation continues: "This obviously means: There is such a thing as a conceptual construction for the grasping of the interpersonal, the authority of which lies purely in its validation. This conceptual construction refers precisely to the 'real' (by definition), and every further question concerning the 'nature of the real' appears empty".
2. The term is Sommerfeld's (1949), 102.
3. See Friedman (2000a, 2000b, 2001, 2002), and the appendix in this volume.

4. Although Cassirer's book went to press in the summer of 1920, before Reichenbach's, the latter was published (by Julius Springer, Berlin) in November 1920, and Cassirer's (Verlag Bruno Cassirer, Berlin) in January 1921. According to Cassirer's *Vorwort*, dated 9 August 1920, his book was read in manuscript by Einstein. We know from marginalia in his copy that Einstein, to whom Reichenbach dedicated his book, read at least parts of Reichenbach's book. (I am grateful to Thomas Oberdan for providing me with photocopies of Einstein's marginalia.) Here and throughout the book, I have generally made my own translations from the original works; citations to the translations are made for ease of reference and purposes of comparison. I have used a reprinted edition of Cassirer's monograph; page references to the 1921 German text are to this edition.

5. Schlick (1921a); see discussion in ch. 3.

6. Rovelli (2001), 108.

7. Einstein (1916a); as repr. in (1996), 283–339; 287 and 291 (original emphasis in both quotations). "In the general theory of relativity, space and time cannot be defined in such a way that differences of the spatial coordinates can be directly measured by the unit measuring-rod, or differences in the time coordinate by a standard clock. . . . This comes to requiring that:—*The general laws of nature are to be expressed by equations which are valid for all coordinate systems, that is, are covariant with respect to arbitrary substitutions (generally covariant)*. It is clear that a physics which satisfies this postulate will be suitable for the postulate of general relativity. . . . That this requirement of general covariance, which takes away from space and time the last remnant of physical objectivity [*den letzten Rest physikalischer Gegenständlichkeit*], is a natural requirement, will be seen from the following reflection. All our space-time verifications [*Konstatierungen*] invariably amount to a determination of space-time coincidences. If, for example, events consisted merely in the motion of material points, then ultimately nothing would be observable but the meetings of two or more of these points. Also, the results of our measurements are nothing other than verifications of such meetings of the material points of our measuring rods with other material points (respectively, observed coincidences between the hands of a clock and points on the clock dial)—point-events happening at the same place and at the same time. The introduction of a reference system serves no other purpose than to facilitate the description of the totality of such coincidences. Since all our physical experience can be ultimately reduced to such coincidences, there is no immediate reason for preferring certain coordinate systems to others, that is to say, we arrive at the requirement of general covariance".

8. Norton (1993).

9. Barbour (2000), 68, expresses the conception in its purest form: "What is the reality of the universe? It is that in any instant the objects in it have some relative arrangement".

10. See, e.g., Earman (1989), ch. 5.

11. The customary definition of a reference frame in general relativity is as a congruence of timelike curves (each of which is the world line of a reference point of the frame); an equivalence class of coordinate systems can then be "adapted" to the reference frame in several ways.

12. Einstein (1911), 899, as repr. in (1993), 487; Engl. transl. (1952), 100.

13. Cf. Ohanian and Ruffini (1994), 53, 377. Earman (1989), 102, details the difficulties of finding a space-time structure that is recognizably relativistic for which the existence of rotation is not absolute.

14. The field equations of general relativity permit generically nonflat space-times in which no global Galilean (inertial) coordinate systems can exist; hence, general relativity, unlike Newtonian or neo-Newtonian gravitational theory, *must be* a generally covariant theory. Interestingly, Einstein did not give this reason as a justification of general covariance.

15. Kretschmann (1917).

16. See Earman (2003) for an examination of this claim in the context of space-time theories.

17. The positive part of Kretschmann's paper involved an attempt to identify the relativity principle fulfilled by the general theory of relativity as a symmetry principle associated with a group of transformations preserving the geodesic structure of all space-times permitted by general relativity. For an analysis, see Rynasiewicz (1999).

18. Einstein (1918), 242, repr. in (2002), 39. Presumably, Einstein had in mind such features of generally covariant space-time theories as are described in note 22, below. Four-dimensional generally covariant formulations of Newtonian theory were produced by Cartan and others in the early 1920s.

19. For example, Earman and Glymour (1978); Pais (1982), 222–223, 243–244; Earman (1989), 105.

20. Hoffmann (1972), 127. The footnote to this passage reads: "This belief was based on a confusion between, among other things, coordinate systems and reference frames".

21. Friedman (1983), 212. Friedman's analysis is based upon Anderson (1967).

22. This theorem, a special case of Noether's second theorem, was first stated by Hilbert (1915), in the more general setting of Einstein's gravitational theory coupled to Mie's electromagnetic theory of matter, adapted to general covariance. Thus, Hilbert's variational principal varies the total action of the combined gravitational and electromagnetic fields with respect to the 14 independent gravitational and electromagnetic potentials. See Brading and Ryckman (forthcoming). Cf. Pauli (1921); Engl. trans. (1958), §56. The usual modern view of this result pertains only to the gravitational field. Because of the twice-contracted Bianchi identities (i.e., the vanishing of the covariant divergence of the Einstein tensor, linking the left-hand side of the Einstein field equations with the right-hand side, and so coupling the metric field to its matter sources), there are only six independent field equations for the particular values of the $g_{\mu\nu}$. There are, however, 10 independent $g_{\mu\nu}$. So there are four additional conditions that must be specified in addition to the field equations to fully determine the $g_{\mu\nu}$. E.g., Davis (1970), 266: "In any generally covariant field theory in space-time, by definition, one must have the freedom to make four independent general transformations on the coordinates. Once the coordinate frame has been fixed (e.g., four noncovariant conditions specified) the six field equations are sufficient to determine the form of the ten $g_{\mu\nu}$ with the components of the stress-energy tensor [i.e., the right-hand side of the Einstein field equations for matter sources] given".

23. Brown and Brading (2002) cite results of Lovelock and Gregore showing that this constraint (in particular, that the Lagrangian density entering into the gravitational action is constructed out of the $g_{\mu\nu}$ field and its derivatives alone) must lead to the Hilbert–Einstein gravitational action of general relativity.

24. Stachel (1986b, 1989a, 1992a, 1993a).

25. Einstein (1952), 155. See also Einstein (1950), 348: "According to general relativity, the concept of space detached from any physical content does not exist".

26. Rovelli (1997), 209.

27. Stachel (1989a) and Norton (1984).

28. Stachel (1996), xix: "It is only through Einstein's correspondence that we know that he came to reject [the "Hole Argument"] when he realized that not the metric field, but only the totality of space-time coincidences has physical meaning".

29. See Stachel (1989a), 65ff; Norton (1984), 120ff.

30. Einstein (1914), 1067; repr. in (1996), 110.

31. This bald statement ignores subtler issues about "observables" in general relativity; for a classic discussion, see Bergmann (1961).

32. To say that the metric field has arbitrarily small, or even zero values, at a point is of course distinct from the claim that the metric field does not exist.

33. Stachel (1989a), 78.

34. A complicating consideration is that Einstein expressed his condition of general covariance in terms presupposing that the underlying differential manifold was a "number manifold", as was common practice in the differential geometry of the time. Such a manifold contains additional superfluous structure that leads to further obscurities in interpreting Einstein's intentions; see Norton (1992b).

35. See the discussion in Ryckman (1992).

36. Subsequently, Einstein frequently stressed that in the transition from the special to the general theory of relativity, coordinates lose their immediate physical significance [e.g., Einstein (1925d), 794–797, a passage not found in the original 1915 version of this article]: "In the general theory of relativity . . . there is no geometry and kinematics independent of the remainder of physics since the behavior of measuring rods and clocks is conditioned by the gravitational field. It is connected with this circumstance that the general theory of relativity brings with it a much deeper modification of the doctrine of Space and Time than in the special theory. According to the latter, namely, space and time coordinates have an immediate physical significance. . . . To the coordinates in the general theory of relativity such an immediate physical significance cannot be ascribed. To be sure, one arranges the system of occurrences, that is, the point-events, also here in a four-dimensional continuum (space-time), however the behavior of rods and clocks (the geometry, that is, in general the metric) is determined in the continuum by the gravitational field; the latter is therefore a physical condition of space which simultaneously determines gravitation, inertia and the metric. Herein lies the deepening and the unification which the foundations of physics have experienced through the general theory of relativity".

37. Anderson (1967); Trautman (1966); Friedman (1983).

38. Of course, this is not really a group. Diffeomorphisms (point transformations) are automorphisms globally defined on the manifold and form a (non-abelian) group, whereas coordinate transformations are not, in general, globally defined and do not form a group. Cf. Göckeler and Schücker (1987), 48.

39. Although each model of general relativity is diffeomorphically invariant in the sense of §2.2.2, it is not true that any two randomly chosen models are diffeomorphically related, as is readily seen by contrasting models of an always-expanding Friedmann world, and an expanding-then-contracting Friedmann world. Thanks to Roberto Torretti for the example, one of several he adduced.

40. Brown and Sypel (1995); Brown and Brading (2002).

41. Ohanian and Ruffini (1994), 378.

42. Utiyama (1956).

43. Einstein (1924), 87–88; Engl. trans. in Saunders and Brown (1991), 15–17; instead of "ether", Einstein noted that one could equally speak of "physical qualities of space".

44. In fact, Wald (1984), 57, terms this a statement of "the principle of general covariance". Maidens (1998) argues that a clean distinction between "absolute" and "dynamical" objects by means of an invariance group cannot be made without in addition making controversial physical assumptions to guide the distinction.

45. Cf. Rovelli (1997), 190.

46. Smolin (2003), 12.

47. As was first discussed in Hilbert (1915, 1917); see note 22 above. In brief, the initial data required for satisfaction of the Cauchy (initial-value) problem (and so for a unique solution for given sources) must include specification of an intrinsic geometry on the initial-value spatial surface, together with its intrinsic curvature describing its embedding into the space-time manifold. For details, see Misner, Thorne, and Wheeler (1973), ch. 21, or the concise discussion in Earman (1989), 107.

48. "I see the most essential thing in the overcoming of the inertial system, a thing that acts upon all processes, but undergoes no reaction. This concept is in principle no better than that of the center of the universe in Aristotelian physics" [Einstein to Georg Jaffe, 19 January 1954; cited in Stachel (1986b), 1858]. See also the discussion in Stachel (1993b), 277–278.

49. Einstein to Besso (15 April 1950) in Speziali (1972), 438–439, as cited in Stachel (1992a), 25.

50. E.g., Einstein (1949), 81. Earman and Eisenstaedt (1999) show that Einstein used singularities only as a calculational device for treating the problem of motion within the context of the pure gravitational field.

51. Bergmann (1949), 680: "All theories which are covariant with respect to general coordinate transformations share the property that the existence and form of the equations of motion is a direct consequence of the covariant character of the equations".

52. Bergmann (1979), 63.

53. As emphasized by Howard (1990).

54. This is, however, a prohibition violated in the customary operator formalism of quantum field theory because it is necessary to specify commutation relations, in addition to the field equations, in order to define the theory. As the commutativity of two observables is based on their causal independence, the usual (nonalgebraic) formulation of quantum field theory makes use of global knowledge of the causal structure of space-time, available in special relativity ("Minkowski space-time") but not in the generic variably curved space-times of general relativity. For discussion and details, see Haag (1992).

55. E.g., Rovelli (1997, 2001); Smolin (1992).

56. Kant (1781/1787), A50/B74. As is customary, "A" refers to pagination of the 1781 edition, "B" to that of 1787.

57. For the issues involved, see Allison (1973), 73ff.

58. Allison (2004) terms this the "discursivity thesis".

59. See Moynahan (2003), 50ff.

60. Beiser (1987), 41; see also 171, 292–293.

61. Cassirer (1931), 17.

62. Postcard from Cassirer to Reichenbach (7 July 1920), Hans Reichenbach papers in the Archive for Scientific Philosophy, University of Pittsburgh, Hillman Library, HR 015-50-09. Quoted by permission. I am grateful for Marlene Bagdikian's assistance in deciphering Cassirer's handwriting.

63. Cassirer (1907), 716.

64. Cassirer (1913, 1927).

65. See Ryckman (1991).

66. Henrich (1994), 47. The Marburg position is forcefully stated in Cohen (1902).

67. Reichenbach (1920), 46; Engl. trans. (1965), 48–49.

68. Reichenbach (1920), 47–48; Engl. trans. (1965), 50.

69. To be sure, already in his Ph.D. dissertation, Reichenbach (1916–1917), regarded cognition as involving a coordination (*Zuordnung*) of conceptual and nonconceptual elements. E.g., (1916; pt. 3), 230, "the proper task of physics" is the coordination [*Zuordnung*] of mathematical propositions [of a given class] to objects of empirical intuition [*Gegenständen der empirischen Anschauung*]".

70. Reichenbach (1920), 83; Engl. trans. (1965), 86. This generality must be regarded as *pro tem* since, "*There are no most general concepts*".

71. Schlick (1918); Engl. trans. (1985) of the 2nd (1925) German ed.

72. Schlick (1918), §7; Engl. trans. (1985), §7.

73. Schlick (1918), 36; Engl. trans. (1985), 38.

74. Schlick (1918), 20, 74, and 326; Engl. trans. (1985), 23, 89, and 383.

75. Schlick (1918), 55–65; Engl. trans. (1985), 59–72.

76. See Schlick (1918), 74; Engl. trans. (1985), 89. "Empiricist critical realism" and "empirical realism" are the designations Herbert Feigl gave to Schlick's (1918, 1925) epistemology; see Feigl (1937–1938), xx; also (1956, 1969).

77. Schlick (1918), 307ff; Engl. trans. (1985), 361ff.

78. Schlick (1918), 306; Engl. trans. (1985), 360.

79. Schlick (1918), §§1, 3; (1985), §§1, 3.

80. Reichenbach (1920), 34; Engl. trans. (1965), 36–37. Obviously, Reichenbach wrote before the limitative results of Gödel and Tarski.

81. Reichenbach (1920), 38, original *Sperrdruck* emphasis; Engl. trans. (1965), 40.

82. Reichenbach (1920), 40; Engl. trans. (1965), 42–43.

83. Reichenbach (1920), 109–110, n. 27; Engl. trans. (1965), 115–116.

84. Reichenbach (1920), 81; Engl. trans. (1965), 85. Throughout his career, Reichenbach consistently held that every coordination of a physical body to a mathematical theory involves the notion of approximation, and so the concept of probability.

85. Reichenbach (1920), 39: "*Aber eine Definition des Wirklichen leistet die Wahrnehmung nicht*". Engl. trans. (1965), 42.

86. Reichenbach (1920), 50; Engl. trans. (1965), 53.

87. Coffa's comment (1991), 203, is instructive for it reveals that the "dual origin" thesis itself comprises, for Coffa, "a Kantian framework": "Reichenbach shared Schlick's realist instincts, but he did not find the expression of these instincts within a Kantian framework any easier than Schlick had years earlier".

88. In Schlick's view, Reichenbach's account of knowledge as coordination comes to close to the fatal delusion entertained by Marburg Neo-Kantians: "I believe that only the undefined side determines—through the mediation of perception—the conceptual side, and not vice-versa. Your theory seems to me to emerge from the fact that it is so easy to confuse the *concept* of reality [*Begriff der Wirklichkeit*] with reality itself... an illusion to which the Marburg Neo-Kantians have fallen prey". Schlick to Reichenbach (26 November 1920) [HR 015-63-22], as quoted and translated in Coffa (1991), 204.

89. Reichenbach (1920), 29; Engl. trans. (1965), 31.

90. Reichenbach (1920), 85; Engl. trans. (1965), 89.

91. Reichenbach (1920), 84; Engl. trans. (1965), 87.

92. Reichenbach (1920), 67–68; Engl. trans. (1965), 70.

93. Reichenbach (1920), 43; Engl. trans. (1965), 45.

94. Reichenbach (1920), 66; Engl. trans. (1965), 69.

95. Reichenbach (1920), 78; Engl. trans. (1965), 81.

96. Reichenbach (1920), 83; Engl. trans. (1965), 87.

97. Reichenbach (1920), 84; Engl. trans. (1965), 88.

98. Reichenbach (1920) 22–23; Engl. transl. (1965), 23–24.

99. Norton (1985), 29–30.

100. Einstein to Schlick 21 March 1917: "*im Unendlichkleinen ist jede stetige Linie eine Gerade*". Einstein (1998), Part A, Doc. 314. See the discussion in Norton (1985), 35.

101. Reichenbach (1922c); Engl. trans. (1978), 44 n. 20.

102. See Dingler (1920). In his book Reichenbach's response to this objection is a classic example of "bootstrapping" [(1920), 62; Engl. trans. (1965), 64]: "[O]ne formulates a theory by means of which the empirical data are interpreted and then checks for univocality [*Eindeutigkeit*]. If univocality is not obtained, the theory is abandoned. The same procedure can be used for coordinating principles. It does not matter that the principle to be tested is already presupposed in the totality of experiences used for the inductive inferences. It is not inconsistent to assert a contradiction between the system of coordination and experience".

103. Reichenbach (1921), 382–384.

104. Reichenbach (1920), 85; Engl. trans. (1965), 88.

105. Reichenbach (1920), 86; Engl. trans. (1965), 89–90.

106. In the Engl. trans. of Reichenbach (1920), this term is translated as "the method of logical analysis", presumably in view of later developments in logical empiricism.

107. Reichenbach (1922c); Engl. trans. (1978), 39.

108. Reichenbach (1920), 86–87; Engl. trans. (1965), 90–91.

109. Reichenbach (1920), 87–88; Engl. trans. (1965), 91–92.

110. Reichenbach (1920), 86–87; Engl. trans. (1965), 90–91.

111. The "weak equivalence principle", sometimes called the "Galilean equivalence principle", states, in the formulation of Ciufolini and Wheeler (1995), 13–14: "[T]he motion of any freely falling test particle is independent of its composition and structure". They explicitly note (91) that it "is at the base of most viable theories of gravitation". Ohanian (1977), 904, gives an alternative formulation of the weak principle of equivalence, called "Newton's principle": "The gravitational mass of any system equals its inertial mass". Normally, these are not distinguished, though the Newtonian principle imposes more severe restrictions on possible gravitational theories.

112. Reichenbach (1920), 86; (1965), 90.

113. Such definitions invoke equivalence classes of coordinate atlases, and so automatically only chart-independent statements are held to be meaningful.

114. Norton (1992b), 290–291.

115. See, e.g., Anderson (1967), 98–101; Stachel (1992b), and note 22 above.

116. Reichenbach (1920), 83–84; Engl. trans. (1965), 87.

117. With one significance exception: Carnap (1928). However, Carnap's Konstitutionstheorie initiates the program of assimilating the problem of coordination to a logical systematization of scientific theories within a conventionally chosen logico-linguistic framework, where the distinction between form and content is rendered as that between analytic and synthetic statements. See Friedman (2001), 82, n. 14, and Ryckman (forthcoming b).

118. Cassirer (1910); Engl. trans. (1953). For discussion of Zuordnung as an epistemological term in early 20th century epistemology, see Ryckman (1991).

119. Cassirer (1910), 21, original emphasis; Engl. trans. (1953), 17, has "law of arrangement".

120. Cassirer (1910), 412; Engl. trans. (1953), 311.

121. Cassirer (1910), 47; Engl. trans. (1953), 37.

122. Cassirer (1910), 403; Engl. trans. (1953), 303.

123. Prologeomena §19: "Es sind daher objektive Gultigkeit und notwendige Allgemeingultigkeit [vor jedermann] Wechselbegriffe".

124. For discussion, see Ihmig (2001), 171ff.

125. Cassirer (1910), 356–357; Engl. trans. (1953), 268–269.

126. Cassirer (1910), 248–249; Engl. trans. (1953), 187.

127. Cassirer (1910), 395; Engl. trans. (1953), 297. Cf. Cassirer (1910), 403; Engl. trans. (1953), 303: "Thus we do not know 'objects' as if they were already independently determined and given as objects,—but we know objectively, in that we create certain limitations and fix certain enduring elements and modes of connection. The concept of object in this sense is no more the ultimate limit of knowing, but rather the basic means by which all that has become its permanent possession is established and expressed".

128. Cassirer (1910), 453; Engl. trans. (1953), 341.

129. Cassirer (1910), 408; Engl. trans. (1953), 307.

130. Planck (1910), 25.

131. Cassirer (1921), 49; Engl. trans. (1953), 392: "In general epistemological consideration, [the theory of relativity] is precisely distinguished in that it carries out, more consciously and clearly than ever before, the progress from the copy theory [Abbildtheorie] of knowledge to the functional theory [Funktionstheorie]".

132. Cassirer (1929), 552; Engl. trans. (1957), 473.

133. Cassirer (1921), 8; Engl. trans. (1953), 356.

134. Cassirer (1921), 10; Engl. trans. (1953), 357.

135. Cassirer (1921), 37; Engl. trans. (1953), 381.

136. Cassirer (1921), 38; Engl. trans. (1953), 383.

137. Cassirer (1921), 76; Engl. trans. (1953), 416. Schlick (1921a) found this formulation vacuous; see chapter 3, §1.

138. Cassirer (1921), 56; Engl. trans. (1953), 398.

139. Cassirer (1921), 89; Engl. trans. (1953), 428.

140. Cassirer (1921), 39; Engl. trans. (1953), 384.

141. Cassirer (1921), 33; Engl. trans. (1953), 377.

142. Cassirer (1921), 50–51; Engl. trans. (1953), 393.

143. Cassirer (1999), 118: "*Daß es sich hierbei nicht um die einfache Feststellung eines Faktums, sondern um eine methodische Maxime, um ein 'regulatives Prinzip' für die Naturbetrachtung handelt, hat Einstein selbst betont*".

144. Cassirer (1921), 78; Engl. trans. (1953), 418. In these passages the Marburg neo-Kantian rejection of Kant's account of pure intuition as an independent nonconceptual faculty of knowledge is readily apparent. See Cassirer (1902) and (1907), 684, as translated in Friedman (2000a), 90: "The pure intuitions of space and time, like the concepts of pure understanding, are just different aspects and manifestations of the basic form of the synthetic unifying function".

145. Cassirer (1921), 71; Engl. trans. (1953), 413.

146. Cassirer (1921), 80; Engl. trans. (1953), 421.

147. Cassirer (1921), 80–81, 108; Engl. trans. (1953), 420–421, 448. To be sure, Cassirer does not explicitly affirm the non-existence of absolute objects, but such a ban is certainly in the spirit of his discussion of the significance of general covariance, and indeed of his general account of an epistemological development toward fully functional and relational concepts in physical science.

148. Hilbert, in his unpublished summer semester 1921 lectures, titled *Grundgedanken der Relativitätstheorie*, explicitly interprets general covariance as a principle of physical objectivity: "Hitherto, the objectification of our view of the processes in nature took place by emancipation from the subjectivity of human sensations. But a more far reaching objectification is necessary, to be obtained by emancipating ourselves from the *subjective* moments of human *intuition* with respect to space and time. This emancipation, which is at the same time the high-point of scientific objectification, is achieved in Einstein's theory, it means a radical elimination of *anthropomorphic* slag, and leads us to that kind of description of nature which is *independent* of our senses and intuition and is directed purely to the goals of objectivity and systematic unity". As cited by, and translated in, Majer (1995), 283.

149. As has been emphasized in Friedman (2000a), 117.

Chapter 3

1. HR 015-49-26; quoted with permission of the Archive for Scientific Philosophy, Hillman Library, University of Pittsburgh.

2. Schlick (1915), 163; Engl. trans. (1979a), 178.

3. Cassirer (1937), 132; (1939), 201–202.

4. Cassirer (1921), 71; Engl. trans. (1953), 412.

5. Details of Cassirer's life, see Krois (1987).

6. Cassirer (1906, 1907, 1920).

7. Cassirer (1918).

8. See Holzhey (1992) and Orth and Holzhey (1995) for discussion and further references.

9. Cassirer (1921), 2: "Albert Einstein has read the following essay in manuscript and encouraged [*gefördert*] it through some critical comments that he attached to the reading. I cannot let this book go out without expressing my heartfelt thanks to him again in this place".

10. A. Einstein to M. Schlick (17 October 1919); as cited and translated in Howard (1984), 625.

11. A. Einstein to E. Cassirer (5 June 1920); EC 8-386, Einstein Archive, Jerusalem.

12. Schlick (1904).

13. Herneck (1970), 9.

14. Schlick (1917).

15. See Howard (1984) and Hentschel (1986).

16. Letter of Einstein to Born (9 December 1919), in Born (1969), 38.

17. A. Einstein to M. Schlick (17 October 1919), as cited and translated in Howard (1984), 620.

18. Einstein (1921a), 5, 6, 14; (2002), 387, 388, 396.

19. Coffa (1991), 199.

20. Von Laue (1921), 42.

21. Schlick (1921a), 98; Engl. trans. (1979a), 323.

22. Friedman (2000a), 115.

23. Schlick to Reichenbach (26 November 1920), HR 01563-22, as translated in Coffa (1991), 201–202.

24. Schlick (1921b), 340.

25. Neurath (1929), 89: "The scientific world conception knows no unconditionally valid cognition from pure reason, no synthetic judgments *a priori*, as lie at the foundation of the Kant's epistemology, and really all pre- and post-Kantian ontology and metaphysics".

26. Schlick (1921a), 98; Engl. trans. (1979a), 324.

27. Coffa (1991), 204.

28. Carnap (1928).

29. Schlick (1918), 12; Engl. trans. (1985), 13.

30. Schlick (1918), 36; Engl. trans. (1985), 38.

31. Schlick (1918), 39, 42; Engl. trans. (1985), 42, 44.

32. Schlick (1918), 326–327; Engl. trans. (1985), 384.

33. Schlick (1918), 309; Engl. trans. (1985), 363.

34. Schlick (1918), 46; Engl. trans. (1985), 49.

35. Schlick (1918), 63; this text is replaced by new material in the second edition (1925), 64–69; Engl. trans. (1985), 69–75.

36. Schlick (1918), 44, also 46; Engl. trans. (1985), 46, also 49.

37. Schlick (1918), 47; Engl. trans. (1985), 50.

38. Schlick (1918), 58; Engl. trans. (1985), 62.

39. Schlick (1915), 153; Engl. trans. (1979a), 171.

40. Schlick (1915), 150–151; Engl. trans. (1979a), 168–169.

41. Schlick (1917), 183; Engl. trans. (1979a), 244–245.

42. Schlick (1925), 64–72; Engl. trans. (1985), 69–78.

43. See Howard (1994), 68–70. As Howard points out, this is the decisive step toward the modern empiricist dichotomy between analytic and synthetic statements attacked in 1950 in Quine's "Two Dogmas of Empiricism".

44. Schlick (1925), 66; Engl. trans. (1985), 71.

45. Friedman (2002), 227, n. 63, rightly insists that Schlick distinguished between concrete definitions and conventions.

46. Howard (1994), 71.

47. Einstein (1921a), 124; cf. Engl. trans. (1983), 28. Excerpts in Feigl and Brodbeck (1953), 189–194.

48. Einstein (1921a), 125–126; cf. Engl. trans. (1983), 31–33.

49. Einstein (1921a), 126; cf. Engl. trans. (1983), 35.

50. Stachel (1989b).

51. Here, and throughout, when pertaining to extension between events in space-time, I have translated the German word *Strecke* as "tract", although in the context of foundations of geometry, it normally translates as "segment".

52. Einstein (1921a), 129; cf. Engl. trans. (1983), 38.

53. Pauli to Eddington (20 September 1923), in Pauli (1979), 118: "Certainly, the most beautiful achievement of the theory of relativity was to have brought the measurement results of measuring rods and clocks, the paths of freely-falling mass particles and those of light rays, into a determinate inner connection [*Verbindung*]. Logically, or epistemologically, this postulate does not admit of proof. However, I am persuaded of its correctness".

54. Poincaré (1902) and Engl. trans. (1929), ch. 4. See Ben-Menahem (2001) for an illuminating discussion.

55. Einstein (1921a), 127; cf. Engl. trans. (1983), 36.

56. Einstein (1914), 1079–1080.

57. Weyl (1918d), 385; repr. in Weyl (1968), vol. 2, 2.

58. This is now a commonplace, nowhere better and more simply motivated than in the remarkable Geroch (1978).

59. Einstein (1923e), 483–484.

60. Einstein (1925a), 17 and 18–19.

61. Einstein (1925a), 20.

62. Poincaré (1902), 73; Engl. trans. (1929), 63.

63. Helmholtz (1921); Engl. trans. (1977).

64. Koenigsberger (1906), 312.

65. Helmholtz (1866). See Torretti (1978), 162.

66. Helmholtz (1868), repr. in Helmholtz (1921), 38; Engl. trans. (1977), 39.

67. Torretti (1978), 391, n. 8. Hilbert (1917) refers to "measure threads" (*Massfaden*).

68. Lie (1890), 466–471.

69. Helmholtz (1868), repr. in Helmholtz (1921), 43; Engl. trans. (1977), 44–45.

70. Lie (1890), 454ff. See the discussion in Torretti (1978), 161.

71. P. Hertz in Helmholtz (1921), 62 (n. 15 to the text of 43); Hertz observed: "One is not allowed to apply to the infinitely small the axioms which were laid down for the finite case, as S. Lie has shown"; Engl. trans. (1977), 65.

72. Helmholtz (1870), repr. in Helmholtz (1921), 15; Engl. trans. (1977), 15.

73. Schlick in Helmholtz (1921), 30; Engl. trans. (1977), 31.

74. Helmholtz (1870), repr. in Helmholtz (1921), 18; Engl. trans. (1977), 19.

75. Schlick in Helmholtz (1921), 33; Engl. trans. (1977), 34.

76. Helmholtz (1868), as repr. in Helmholtz (1921), 38; Engl. trans. (1977), 39.

77. Friedman (1997).

78. Friedman (1997), 33.

79. Helmholtz (1878), as repr. in Helmholtz (1921), 117–118; (1977), 124.

80. Helmholtz (1878), as repr. in Helmholtz (1921), 140; (1977), 149.

81. This is what I take Torretti (1978), 168, to have meant in stating, "The notion of a rigid body must therefore be regarded, if Helmholtz is right, as a concept constitutive of physical experience, that is, as a transcendental concept in the proper Kantian sense". Unlike DiSalle (1993, 513–514; and Dingler, whom DiSalle cites), I do not see a conflict with Helmholtz's stated aim of providing a *factual* foundation for geometry, such facts being "constituted" and not simply (and naively, in the light of Riemann) "given".

82. Helmholtz (1870), as repr. in Helmholtz (1921), 24; Engl. trans. (1977), 25.

83. Torretti (1978), 170.

84. Helmholtz (1870), as repr. in Helmholtz (1921), 23–24; Engl. trans. (1977), 24–25 and Helmholtz (1878), as repr. in Helmholtz (1921), 144; Engl. trans. (1977), 154.

85. Helmholtz (1878), as repr. in Helmholtz (1921), 144; Engl. trans. (1977), 154.

86. Helmholtz (1870), as repr. in Helmholtz (1921), 24; Engl. trans. (1977), 25; (1878) in Helmholtz (1921), Engl. trans. (1977), 157.

87. Helmholtz (1878), as repr. in Helmholtz (1921), 145–146; Engl. trans. (1977), 155–156.

88. Schlick in Helmholtz (1921), 157–158; Engl. trans. (1977), 168.

89. Schlick (1922b); Engl. trans. (1979a), et al.

Chapter 4

1. A brief classic account is Bondi (1960); see also Kerszberg (1989).

2. A recent overview is O'Raifeartaigh and Straumann (2000); translations of original papers and instructive commentary are in O'Raifeartaigh (1997).

3. Weyl (1949a), 288.

4. Because such theories are provably "renormalizable"; see, e.g., Hoddeson et al. (1997).

5. "Neo-conventionalist" in view of his "mature" view that, "[a]ccording to the theory of relativity, the choice of a geometry is arbitrary: but it is *no longer* arbitrary once congruence has been defined by means of rigid bodies". Reichenbach (1922c); Engl. trans. (1978), 38.

6. This theme is taken up in a somewhat different way in Friedman (2002).

7. See Einstein (1998), part B; doc. s 472, 498, and 507; the relevant passages from this exchange are quoted in Straumann (1987).

8. For a typical account, see Pais (1982), 341; for further details, see Yang (1985) and Straumann (1987).

9. Einstein (1933), 113: "*Höret nicht auf ihre Worte, sondern haltet euch an ihre Taten!*"; Engl. trans. (1954), 270.

10. Struik (1989), 104.

11. Weyl (1918c), 466; (1918d), 385; cf. (1919a), 91.

12. Weyl (1923b), 9.

13. Weyl (1918c), 479; as repr. in Weyl (1968), vol. 2, 41; Engl. trans. in O'Raifeartaigh (1997), 35.

14. Weyl (1923a), 124. Also Weyl (1921a), 112; Engl. trans. (1953), 124.

15. Physically, a manifold with an affine connection is interpreted as a space-time with a gravitational field, given the suggestive name "guiding field" (*Führungsfeld*), in view of "the undeniable fact that a body released in a given world-direction fulfills a uniquely determined natural motion, out of which it can be disturbed only by outside forces". Weyl (1923b), 13.

16. Weyl (1919b), 244.

17. Weyl (1919b), 251.

18. Weyl (1919b), 251–252; (1923a), 308, 314–315; Weyl argued that just as conservation of energy-momentum corresponds to coordinate invariance, "conservation of electricity" (both the electric potential and the current satisfy the boundary condition $(\partial\varphi_\mu/\partial x^\mu) = 0$) corresponds to gauge invariance, as is reflected in the fact that the Maxwell equations are already invariant under conformal transformations, i.e., gauge invariant in Weyl's sense. For insightful analysis and discussion, see Brading (2002).

19. Weyl (1919b), 253.

20. Weyl (1918d), 385.

21. Reichenbach (1920), 73; Engl. trans. (1965), 76.

22. In a letter (2 February 1921) to Reichenbach thanking him for sending a copy of his book, apparently the first extant correspondence between them, Weyl ridiculed Reichenbach's criticism: "[Y]ou say that I hold that mathematics (e.g., the theory of the [Riemann] ζ-function !!) and physics are but one discipline. I maintain only that the concepts of *geometry* and of field physics have come together". Weyl's letter, in the Reichenbach archive at Pittsburgh [HR 015-68-04], is quoted by permission.

23. Weyl (1918d), 385: "The sole distinction between geometry and physics is this: that geometry investigates generally what lies in the essence of metric concepts, while physics determines the law through which the actual world is singled out from among all possible four-dimensional metric spaces of geometry and explores its consequences".

24. Cf. Minkowski (1908), 55: "*Die Mannigfaltigkeit aller denkbaren Wertsystem x, y, z, t soll die Welt heißen*".

25. Weyl (1922a), 114–115; (1923b), 47: "*Die Metrik hängt am Begriffe der **Kongruenz**, der jedoch rein infinitesimal gefaßt werden muß*".

26. Weyl (1921b), 473.

27. Weyl (1921a), 206–207, 285–286; Engl. trans. (1953), 228–229; 313–314.

28. Weyl (1921c).

29. Weyl (1921c), 100; repr. in Weyl (1968), vol. 2, 196: "In the theory of relativity, projective and conformal properties have an immediately intuitive significance.... The tendency of persistence [*Beharrungstendenz*] of world direction of a moving material particle, forces upon it, once it is set loose in a determinate world direction, a determinate "natural" motion, is a unity of inertia and gravitation. However, the infinitesimal (light) cone realizes in the neighborhood of a world point the distinction between past and future; the conformal property is the action-connection [*Wirkungszusammenhang*] of the world, through which is determined which world points stand in possible causal connection with one another. Hence it is also a meaningful fact for physics which comes to expression in the following theorem: *Theorem I. The projective and conformal properties of a metric space univocally determine its metric*...[the proof follows]".

30. See, e.g., Norton (1993), 832. Ehlers et al. (1972) carry out a constructive method illustrating Weyl's claim that the conformal and projective properties determine metrical structure (up to a constant factor) and a uniquely compatible affine connection; see also Ehlers (1988a).

31. E.g., Grünbaum (1973), 733ff; Sklar (1985), 139.

32. Eddington (1923a), 207. After making some order of magnitude assumptions that set $F_{\mu\nu}$ comparable to the force at the surface of an electron, Eddington concludes, "Thus dl/l [the ratio of the change in length of an infinitesimally displaced vector to its original length at P] would be far below the limits of experimental detection".

33. For example, a standard text reports, "The strength of Einstein's objection seems not as powerful now as at the time when it was raised, since we know the classical physics does not describe atomic phenomena without certain quantum-theoretical modifications". Adler et al. (1975), 506.

34. Pauli (1921), 202. Pauli (1919a) shows that the two theories also agree on the prediction of gravitational redshift. In another early paper (1919b), Pauli took up the issue of the action function, seeking in particular, a function yielding static, spherically symmetric singularity-free solutions (corresponding to the atomic composition of matter). Weyl found solutions containing singularities, i.e., corresponding to an electrical particle in that the electrostatic potential has a singularity at the radial center, briefly leading him (only in the 4th ed. of *Raum-Zeit-Materie*) to proclaim, "Matter is accordingly a real [*wirkliche*] singularity of the field". (1921a), 273; Engl. trans. (1953), 300. Given Weyl's skepticism, evident already in 1919–1920, of the possibility of a field theoretic account of matter, such remarks seem puzzling unless placed into the context of his "agent-theory" of matter (e.g., 1920a, 1921d, 1924a). Here, matter particles are

literally nonextensive statistical elements of a multiply connected topology, living "beyond the field" as what John Wheeler later termed "wormholes". In any case, the fact that Weyl's fourth-order field equations were almost impossible to integrate was certainly a stumbling block to a more favorable assessment of his theory.

35. Taking the simple case of a static gravitational field, and so letting the electrostatic potential φ_μ be a function of *time*, Pauli used the "proper time" equation $\tau = \tau_0^{e\alpha\varphi(t)}$ where α is a factor of proportionality. Pauli explained: "Let two identical clocks C_1, C_2, going at the same rate, be placed at first at the point P_1, at an electrostatic potential φ_1. Let the clock C_2, then be taken to point P_2, at potential φ_2 for t seconds, and then finally returned to P_1. The result will be that the rate of clock C_2, compared with that of clock C_1, will be increased or decreased, respectively, by a factor $\exp[-\alpha(\varphi_2 - \varphi_1)t]$ (depending on the sign of α and of $\varphi_2 - \varphi_1$). In particular, this effect should be noticeable in the spectral lines of a given substance, and spectral lines of definite frequencies could not exist at all. For, however small α is chosen, the differences would increase indefinitely in the course of time, according to [this equation]". Pauli (1921); Engl. trans. (1958), 196.

36. A more complete account of the two versions of Weyl's theory is given in chapter 6, §§6.4.1 and 6.4.2.

37. Weyl (1919b), 113.

38. Eddington (1923a), §83 ("Natural geometry and World geometry"): "The new view entirely alters the status of Weyl's theory. Indeed it is no longer a hypothesis, but a graphical representation of the facts, and its value lies in the insight suggested by this graphical representation". Pauli subsequently also finds this interpretation to characterize the situation very well; see Pauli (1926), 273–274, and also his letter to Eddington of 20 September 1923, quoted below in §4.3.

39. As Coffa (1979), 283–284, forcefully put this objection.

40. Weyl's book *Raum-Zeit-Materie* exhibits a shift in view between the 3rd and 4th eds., from a "pure field theory of matter" to an "agent-theory of matter" ("giving birth to the field"); for the former, see Weyl (1919b) §34; for the latter, (1921a), §36, and (1923a), §38.

41. Weyl (1919a), 113; repr. in (1968), vol. 2, 67: "If then these instruments [i.e., measuring rods and clocks] also play an unavoidable role as indicators of the metric field then it is apparently perverse [*verkehrt*] to *define* the metric field through indications taken directly from them".

42. Weyl (1923a), 298.

43. Einstein (1923d), 448.

44. This distinction seems to have been made in print first in Weyl (1920b), 649; the distinction is discussed more fully in (1921a, 1921b, 1921d, 1921e).

45. It is remarkable that his distinction between "adjustment" and "persistence" is still retained some thirty years later, long after Weyl abandoned his theory, in the form of an objection raised from the point of view of a (now future) "systematic theory" to the rigid rods and clocks of Einstein's theory; the objection occurs in a new appendix to Weyl (1949a), 288.

46. See the extended discussion in chapter 6, §6.4.

47. The motivation is briefly hinted at in Einstein (1921b), 474. In his (1919), Einstein attempts to reinterpret the cosmological term $(-\lambda g_{\mu\nu})$ appended to his field equations in 1917 in an effort to satisfy what he would come to call "Mach's principle", as a constant of integration, rather than as hitherto, "a peculiar universal constant". To do so, Einstein slightly modifies his field equations without the cosmological term in a manner that explicitly accords gravitational force a role in holding atoms together. Now the proportionality factor on the left-hand side is changed (from 1/2 to 1/4). But this modification makes the scalar R of the left-hand side (i.e., of $R_{\mu\nu} - (1/4)g_{\mu\nu}R$) vanish identically as does the scalar of the right-hand side $(-\chi T_{\mu\nu})$ when the stress-energy

tensor is written in terms of the Maxwell–Lorentz components of the electromagnetic field and the divergence taken. The modification enabled him to derive the result that the scalar of curvature R is constant (1) in all domains in which the current density of electricity vanishes ("empty space") and (2) on every world line of the motion of electricity (i.e., regarding electricity as a moving charge density). Einstein then gives the following "intuitive" (*anschaulich*) interpretation (352): "The curvature scalar R plays the role of a negative pressure which outside of the electrical corpuscles has a constant value R_0. In the interior of each corpuscle there subsists a negative pressure $(R - R_0)$, whose drop maintains the equilibrium of the electrodynamic forces. The minimum of pressure (respectively, the maximum of the curvature scalar) *does not alter with time* in the interior of the corpuscle" (emphasis added). Finally, Einstein shows that in regions where gravitational and electrical fields are present, $\lambda = (1/4)R_0$. Pais (1982), 257, refers to this paper as "Einstein's first attempt at a unified field theory".

48. See Weyl (1921a), 121, for the definition of F (the curvature scalar of Weyl's generalized metric) in terms of the Riemann curvature scalar R; on the basis of F, Weyl develops the notion of a "natural gauge": it is to this "natural gauge" that measuring rods and the frequencies of atomic clocks "adjust". See chapter 6, §6.4.2.1, and also the discussion in (1923a), §40, esp. 303: "[M]easuring rod lengths and frequencies of atomic clocks are conserved on the basis of the natural gauge [*natürliche Eichung*], thus in fact are determined through adjustment [*Einstellung*] to the radius of curvature".

49. The physicist Arnold Sommerfeld seems to have thought, at this point, that there was only a minimal difference between Einstein and Weyl, and he urged Einstein to meet Weyl halfway; see his letter to Einstein of 10 August 1921: "I have the feeling as if between you and [Weyl] there is only a really small distinction [*ein ganz kleiner Unterschied*]. [Weyl] would overcome the practical effect of his measuring rod alterations through his [concept of] "adjustment" [*Einstellung*] and you would restrict the indeterminacy of the world function [*Weltfunktion*] if you would take the $g_{\mu\nu}$ as relational magnitudes". Hermann (1969), 87.

50. With Pauli, Einstein viewed the failure of Weyl's theory to yield solutions corresponding to electrons to be a fundamental flaw; see the letter to Ehrenfest of 4 December 1919, cited in Seelig (1960) 280, and the letter to Besso of 12 December 1919 in Speziali (1972), 148. However, this failure would attend all of Einstein's attempts at a unified field theory as well.

51. Einstein (1920), 662.

52. Einstein (1921a), 127–128: "*Die Existenz scharfer Spektrallinien bildet einen überzeugenden Erfahrungsbeweis für den gennanten Grundsatz der praktischen Geometrie*".

53. Einstein (1921a); translation altered in accord with Stachel (1989a), 94 n. 38.

54. Einstein (1925a), 19.

55. Einstein (1925a), 20: "*Wie das Ergebnis jener Bestrebungen auch sein möge, jedenfalls*".

56. Einstein (1928), 254–255; the bracketed term is in Einstein's handwritten original article but not in the published version.

57. Einstein (1949), 685–686.

58. E.g., Einstein (1916a), 775; repr. in (1996), 290. Engl. trans. (1952), 117.

59. The argument may be briefly summarized [see Stachel (1989b)]: Rods placed radially along the disk exhibit Euclidean relations, whereas rods placed in the tangential direction of motion on the circumference of the disk do not, because of Lorentz contractions. Therefore, the ratio of the circumference to the diameter is greater than π. Stachel quotes from a letter of 1951 in which Einstein states that this example was of "decisive importance" to him in setting up general relativity because "it showed that a gravitational field (here equivalent to the centrifugal field) causes non-Euclidean arrangements of measuring rods, and thus compelled a generalization of Euclidean space" (55).

60. Einstein to Weyl (15 April 1918): "*Lässt man den Zussamenhang des ds mit Maßstab- und Uhr-Messungen fallen, so verliert die Rel. Theorie überhaupt ihre empirische Basis*"; in Einstein (1998), pt. B, doc. 507; also quoted in Straumann (1987), 416.

61. Letter to Besso of 12 July 1920 in Speziali (1972), 153.

62. Einstein (1920), 651.

63. Herneck (1976), 103–104.

64. Excerpts from letters in 1918 and 1919 are quoted in Straumann (1987). In particular, in a letter of 16 December 1918, Einstein, referring to a possible visit to Zürich the following February, remarks: "You will see that I am not stubborn [*eigensinnig*] but rather am gladly prepared to enter into any line of thought". Hoffman (1972), 223, also speaks of "Einstein's official argument against Weyl's theory".

65. Einstein (1921b), 262.

66. Einstein (1923a), 32.

67. "Letter from Einstein to Bohr", translated in French (1979), 274.

68. Einstein (1933), 117; Engl. trans. (1983), 274: "Experience may suggest the appropriate mathematical concepts, but they most certainly cannot be deduced from it. Experience remains, of course, the sole criterion of the physical utility of a mathematical construction. But the genuinely creative principle resides in mathematics. In a certain sense, therefore, I hold it true that pure thought can grasp reality [*Wirklichkeit*], as the ancients dreamed".

69. Einstein (1923d), 448.

70. Weyl letter of 19 May 1952 to Carl Seelig, in Seelig (1960). 274–275; cf. Sigurdsson (1991), 253.

71. See Einstein (1919, 1921b, 1923a, 1923c, 1923d); the conclusion expressed in (1925b), 371, is typical: "For me, the end result of this consideration unfortunately consists of the impression that the Weyl-Eddington deepening of geometric foundations is incapable of bringing us progress in physical knowledge; hopefully, future development will show that this pessimistic opinion has been unjustified".

72. Letter to Besso, 25 December 1925, in Speziali (1972), 215.

73. Pauli to Eddington, 20 September 1923, in Pauli (1979), 115–119.

74. Pauli (1919b); as cited and translated in Mehra and Rechenberg (1982), 278: "For a physicist this [field strength] is only defined as a force on a test-body and since there are no smaller test bodies than the electron itself, the concept seems to be an empty, meaningless fiction. One should stick to introducing in physics only those quantities which are observable in principle".

75. Letter to Born, 27 January 1920, in Born (1969), 42; Engl. trans. (1971), 21.

76. Pauli (1921); Engl. trans. (1958), 206. See the comments in Bargmann (1960), 189–190.

77. Einstein (1949), 684.

78. See Fine (1996), 94, quoting from a letter to Schrödinger of 17 June 1935.

79. Reichenbach (1920), 50; Engl. trans. (1965), 53.

80. Reichenbach's review of the Hertz–Schlick edition of Helmholtz (1921) clearly indicates the source of the transformation in his views. After an initial sentence describing the contents of the book, Reichenbach ([1922a], 421) continued: "It is surprising [*überraschend*] with what certainty here is recognized the connection of the congruence axioms with the behavior of rigid bodies; even Poincaré has not expressed conventionalism more clearly". He concluded: "Helmholtz' epistemological lectures must therefore be regarded as the source of modern philosophical knowledge of space".

81. Reichenbach (1922b), 34ff.

82. Reichenbach (1924), 68; Engl. trans. (1969), 88.

83. Reichenbach (1928), and Engl. trans. (1958), §5.

84. See Ryckman (1994) for further details. On Schlick's role in effecting this change of terminology, see chapter 2, §2.4.1, and Coffa (1991), 201–204.

85. See also the discussion in Torretti (1983), 232ff, from whom I have borrowed the term.

86. For example, in contrast to some, notably Grünbaum, who viewed Reichenbach's coordinative definitions as only a form of "trivial semantic conventionalism", in that "the meaning we give to words is arbitrary", Putnam recognized, 30 years ago, that Reichenbach's use of coordinative definitions "assert(s) a quite special epistemological thesis" that is not trivial. According to Putnam (1963), 121: "That the words [occurring in an empirical law] must *first* be given a meaning by the laying down of *definitions* is not trivial, and indeed, in the opinion of most philosophers of science today, is not true. Yet it is just this that Reichenbach is concerned to assert and in no uncertain terms. He asserts *both* that *before* we can discuss the truth or falsity of any physical law all the relevant theoretical terms must have been *defined* by means of 'coordinating definitions' *and* that the definitions must be *unique*, i.e. must uniquely determine the extensions of the theoretical terms. Such views were quite common when Reichenbach wrote [in 1928]".

87. Schlick (1922a), 100–101; Engl. trans. (1979a), 265.

88. Reichenbach (1924), and Engl. trans. (1969), §4; for discussion, see Ryckman (1992).

89. Kamlah (1979), 433; the term "rational reconstruction" however seems to have first been used with a related sense in Carnap (1928).

90. Reichenbach (1925), 38.

91. Reichenbach (1927), 130 n. and 133. This paper is dated "Juli 1925".

92. Reichenbach (1921), 684.

93. Reichenbach (1921) erroneously claimed that the Lorentz transformations can be deduced from the light geometry (685) and that light geometry, especially axiom V, renders the metric arbitrary only up to a linear function (686); for discussion; see §4.4.3.

94. Reichenbach (1921), 686: "*Wesentlich ist, dass mit der Massbestimmung der speziellen Theorie auch die der allgemeinen festgelegt ist*".

95. Reichenbach, in (1924), was apparently unaware of the earlier attempts of the Cambridge mathematician A.A. Robb (1914) to axiomatize special relativity using as the only basis concept the signal relation "<" ("after"). Weyl (1923b) had already referred to Robb in this context.

96. Reichenbach (1924), 58; Engl. trans. (1969), 76. The meaning is that the Michelson-Morley experiment concerning the nondetection of an ether drift is taken to show that the behavior of rods and clocks is in accord with the Lorentz, not Galilean, transformations of the coordinates; cf. Weyl (1918a), 136; Engl. trans., 173–174: "[W]e must regard *the Michelson-Morley experiment as a proof that the mechanics of rigid bodies must, strictly speaking, be in accordance not with Galileo's Principle of Relativity but with that of Einstein*". By the 3rd ed. of Weyl's book, the following remark is added (1919b), 149–150: "Since the behavior of rods and clocks remain somewhat problematic for the formation of physical laws, it is of theoretical interest to note that in principle much simpler measuring instruments suffice for fixing the space-time coordinates in an arbitrary reference system, namely light signals and the motions of force-free mass points".

97. Reichenbach (1925), 37. It is therefore somewhat surprising that Reichenbach, responding to the sensational news that Dayton C. Miller claimed to have detected an ether drift, suggested in (1925), 48, that one should not expect anyway the rigorous validity of the matter axioms (they are likened to the ideal gas laws). Einstein had an indifferent response to Miller's alleged findings [Pais (1982), 113–114], which were probably due to temperature differentials; see Shankland (1964).

98. Reichenbach (1927), 143; see Hentschel (1990), 189. Reichenbach made the claim in correspondence with a Finnish critic of his axiomatization.

99. Reichenbach (1924), 132; Engl. trans. (1969), 167.

100. Reichenbach (1925), 37.

101. Reichenbach (1924), 2; Engl. trans. (1969), 4–5: "We might think of the content of a theory as summarized in a variational principle; this principle can never be the direct object of an experiment, and yet, depending on the confirmation of its consequences, it may be called true or false with a certain degree of *probability*. In order to avoid this difficulty, it is advantageous to approach the axiomatization in a different fashion. It is possible to start with the observable facts and to end with the abstract conceptualization.... Such a *constructive* axiomatization is more in line with physics than is a *deductive* one, because it serves to carry out the primary aim of physics, the description of the physical world" emphasis in original German. Pauli (1921), 201, expressed skepticism toward what he saw as the too little empirically oriented Göttingen approach of seeking such a *Weltfunktion*: "[I]t is not at all self-evident from a physical point of view, that physical laws should be derivable from an action-principle. It would, on the contrary, seem far more natural to derive the physical laws from purely physical requirements, as was done in Einstein's theory".

102. Reichenbach (1924), 4; Engl. trans. (1969), 6; and (1925), 36: "*Jedes einzelne Axiom bedeutet einen anschaulich vorstellbaren Tatbestand, in dem weiter gar nichts Geheimnisvolles oder Unvorstellbares steckt*". The term "*Elementartatbestand*" is bit misleading since these "facts", pertaining to the behavior of light rays and rods and clocks, are not elementary in the sense of being theory-independent, but only in the sense of independence of relativity theory; thus, they rely on optics and the standard lore of mensuration.

103. Reichenbach (1924), 68; Engl. trans. (1969), 88.

104. See below; Carrier (1990) has given an illuminating discussion of the "ideal" character of facts underlying all "constructive" axiomatizations.

105. A stationary spatial system centered about a point A consists of all those points $P, P', P'', \ldots$, wherein the time of a light signal from A to P and back does not alter with time. (1924), 34; Engl. trans. (1969), 44. Static systems are those stationary systems in which light signals sent simultaneously from A in opposite directions about a closed path simultaneously return to A. (1924), 39; Engl. trans. (1969), 49.

106. At this point there is a mistake: the proof offered of the uniqueness of the definition of inertial systems contains an error corrected in Reichenbach (1925), 34; the English translation (1969) includes the relevant passage in a footnote at 58–59.

107. In the complex Euclidean plane with the addition of a point at infinity ($z = 0$; $1/z = \infty$) to avoid singularities, these are circular transformations which are of the form $z' = az + b/cz + d$, where a, b, c, and d are complex numbers, and $ad - bc \neq 0$. For details, see Pedoe (1988), ch. 6. For the more involved formulation of Möbius transformations in n-dimensional spaces, see Fock (1964), appx. A.

108. This hit-or-miss procedure in abandoned in (1928), 201; Engl. trans. (1958), 173, where it is recognized that the determination of the class of inertial systems "is not possible unless we avail ourselves of some physical means other than light signals".

109. Reichenbach (1924), 10 and 63; Engl. trans. (1969), 13–14 and 82. Weyl's (1923b), lecture 1; his (1924b) review of Reichenbach's book discussion brings clarity to the problem. Beginning from full projective space (because the group of projective transformations is singularity-free), he shows that the Lorentz group, which is the quotient of the Möbius group and the projective group, is the sole singularity-free subgroup of the full Möbius group. He then identifies the trajectories of force-free point masses with the straight lines invariant under projective mappings. A complete treatment is given in Weyl (1930).

110. Reichenbach (1924), 72; Engl. trans. (1969), 93.

111. Reichenbach (1924), 69; Engl. trans. (1969), 89.

112. Reichenbach correspondingly assigned matter axioms a different status in special relativity and in general relativity. There is "an essential difference" between the two relativity theories in that the light geometry suffices (with the restrictions noted above) for the construction of the metric in the former, but not in the latter where "material structures" (*materielle Gebilde*) are required to determine the absolute values of the $g_{\mu\nu}$. (1924), 119, 122, and 125; Engl. trans. (1969), 151, 155, and 158: "The significance of material structures (*materiellen Gebilde*) becomes clear: they bring about a comparison of the units at different points. This comparison cannot be achieved by means of light signals".

113. Reichenbach (1924), 63 n., 122 n.; Engl. trans. (1969), 82 n. 27, 154 n. 46.

114. Weyl (1921a), 285–286; Engl. trans. (1953), 313–314.

115. Reichenbach (1928) contains a 42-page appendix (*Anhang*; 331–373) developing this argument at length, explicitly arguing that Weyl's attempt to find a "geometrical significance for electricity" rests also upon the choice of an arbitrary coordination definition specifying the object (the mass point, or neutral test particle) of infinitesimal parallel transport of a vector. The appendix does not appear in the English edition (1958).

116. Reichenbach (1928), 24; Engl. trans. (1958), 15.

117. Weyl (1921a), 200; Engl. trans. (1953), 221. Cf. Ehlers (1973), 3.

118. Wald (1984), 73; for a modern treatment, see Geroch and Jang (1975), although Prugovečki (1995), 40, observes that "a *rigorous* derivation of the geodesic postulate from Einstein's equations is still pending". Thanks to Roberto Torretti for emphasizing that "the truly significant and novel thing" about general relativity is that the nonlinearity of the Einstein gravitational field equations straightforwardly precludes assuming the geodesic law as an independent hypothesis, as Einstein had originally done. For in a closed material system, there is every reason to suppose that the behavior of test particles (and all elements) will be severely constrained by their nonlinear gravitational interaction under the field equations. On the history of the problem of motion in general relativity, see Havas (1989).

119. Cf. Ehlers (1973), 34.

120. Reichenbach (1924), 71; Engl. trans. (1969), 91. This reply appears nearly verbatim both in his response to critics (including Weyl) of his axiomatization (1925), 46–47, and in (1928), 233; Engl. trans. (1958), 201.

121. Torretti (1983), 241.

122. Weinberg (1972), vii.

123. Reichenbach (1924), 111–112; Engl. trans. (1969), 142–143.

124. Emphasized in Norton (1985).

125. Reichenbach (1924), 115; Engl. trans. (1969), 146.

126. Einstein explicitly invokes it in his 1921 Princeton lectures, where, in a discussion rotating systems, he states that the behavior of rods and clocks are assumed to depend only upon velocities, not accelerations, "or at least that the influence of acceleration does not counteract velocity". (1922), 62 n.; Engl. trans. (1956), 60 n.

127. Torretti continues: "That is why rods ultimately cannot hold their own as instruments of fundamental measurement in General Relativity" (1983), 315, n. 25. Already in 1910, von Laue, among others, raised fundamental objections to the concept of a rigid body in the theory of special relativity on grounds that rigidity entailed the possibility of transluminal propagation of causal effects; for a recent discussion, see Norton (1992a). The objection is raised again by Weyl four decades later in (1951), 75, by noting the fundamental inappropriateness of the notion of a rigid body already in the special theory of relativity.

128. Reichenbach (1924), 115; Engl. trans. (1969), 147.

129. E.g., French (1968), 27.

130. Reichenbach (1924), 116; Engl. trans. (1969), 148.

131. Reichenbach (1924), 125; Engl trans. (1969), 157–158; 157 of the translation omits part of the German text.

132. For Weyl (1924b), 2127, "the inappropriateness [*Unsachgemässe*] of Reichenbach's *Ansatz* of rods and clocks here becomes fully evident: the axiomatic analysis of the metric field is not based on the world point but on a three-parameter congruence [*Schar*] of world lines" whereas "the nature of the field would be brought home to us only through the permanence of material bodies and their elementary particles". On Weyl's justification of the "infinitesimally Pythagorean" nature of the metric field, see chapter 6, §6.3.2.

133. For example, Ehlers (1973), 34.

134. Reichenbach (1924), 121–122; Engl. trans. (1969), 153–154. Reichenbach made use of a theorem of Schouten and Struik (1921), generalizing Kasner's (1921) result, stating that a necessary and sufficient condition for a conformally flat manifold to be mapped upon a Euclidean manifold is that $R_{\mu\nu} = 0$, i.e., the vanishing of the Ricci tensor.

135. Reichenbach (1924), 141; Engl. trans. (1969), 178.

136. These coordinate restrictions were originally formulated in Hilbert (1917) and were given the name "reality relations" (*Realitätsverhältniße*) in Pauli (1921), 613; Engl. trans. (1958), 62. See Brading and Ryckman (forthcoming) for discussion. Reichenbach, as Hilbert, pointed out that in such systems there can be no closed timelike worldlines (1924), 148; Engl. trans. (1969), 187; see also (1928), 313; Engl. trans. (1958), 273.

137. Reichenbach (1924), 152; Engl. trans. (1969), 192.

138. Reichenbach (1924), 155; Engl. trans. (1969), 195.

139. Here we see clearly the difference between Reichenbach's sense of "topology" and the standard one. The manifold topology of space-time (which is such that an open neighborhood of each space-time point is homeomorphic with R^4), obviously is not derivable from empirical facts about time order. On the other hand, other topologies have been proposed that do reflect, at least ideally, such facts, e.g., Zeeman (1967). I am grateful to Roberto Torretti for pushing for clarification here, and for the reference.

140. I am once again indebted to Roberto Torretti for this last observation. See Ryckman (forthcoming a) for further discussion of Reichenbach's subordination of geometrical structure to the causal order of time.

141. Reichenbach (1928), 308, 326; Engl. trans. (1958), 268–269 and 285: "The fact that an ordering of all events is possible within the three dimensions of space and the one dimension of time is the most fundamental aspect of the physical theory of space and time. In comparison, the possibility of a metric seems to be of subordinate importance. It is only the metric, however, which, in the general theory of relativity, has been recognized as an effect of the gravitational field. The essence of space-time order, its topology, remains an ultimate fact of nature, unaffected by these considerations". The talk of "destruction of the metric" might lead one to think that Reichenbach is here anticipating later results concerning singularities, but actually these results cut against his causal ordering story. As matters are understood today, general relativistic laws hence, the metric break down only at singularities where space-time curvature is infinite; however, according to the Penrose–Hawking singularity theorems, there space-time itself breaks down taking along the simply connected topology the manifold normally supports. However, surely Reichenbach certainly had no such far-reaching anticipations in mind; rather, his location in (1928), 308; cf. Engl . trans. (1958), 269, of the "epistemological foundation" (*erkenntnistheoretische Grundlegung*; not "philosophical result") of the theory of relativity in the causal theory of space and time stems from his narrowly epistemological conception of the metric in general relativity.

Chapter 5

1. "*Wenn wirklich die Naturwissenschaft spricht, hören wir gerne und als Jünger. Aber nicht immer spricht die Naturwissenschaft, wenn die Naturforscher sprechen;...*" See Engl. trans. (1983), 39.

2. "*Vorwort*" (1918a), iii; Engl. trans. (1994), 1. The recent assessment is that of Feferman (1988).

3. Weyl (1918b); Einstein to Weyl (8 March 1918), doc. 476 in Einstein (1998).

4. Weyl to Einstein (1 March 1918), doc. 472 in Einstein (1998).

5. Weyl (1918c); for details of Einstein's negotiation, see the letters between Einstein and Weyl in April 1918; docs. 512, 525, and 526 in Einstein (1998).

6. See Yang (1977), Mielke and Hehl (1988); Cao (1997); Vizgin (1994); Scholz (1994, 1995, 1999, 2001); O'Raifeartaigh (1997); O'Raifeartaigh and Straumann (2000); and Straumann (2001).

7. For details, see Sigurdsson (1991), 7, and Peckhaus (1990).

8. "Introduction", in Smith and Smith (eds.) (1995), 5–7.

9. For details, see Mancosu and Ryckman (2002), and Schuhmann (1977), 113.

10. Weyl (1955); as repr. in Weyl (1968), vol. 4, 637.

11. Weyl (1948), 380.

12. Husserl (1910–1911); Engl. trans. (1965).

13. Weyl (1948), 381; Lange (1996).

14. A.S. Eddington, *Dehnt sich das Weltall aus?* Stuttgart, 1933; trans. by Hella Weyl of Eddington (1933).

15. Courant (1948), "Hella Weyl, Words of Remembrance. Spoken on Thursday, September 5, 1948", *Nachlaß Heinz Hopf*, HS 621:1506, *ETH Archiv*, Zürich; as cited in S. Sigurdsson (1991), 68.

16. The extant correspondence between Weyl and Husserl is published in Schuhmann (1996), 287–295, and in van Dalen (1984); excerpts are translated in Tonietti (1988).

17. Husserl wrote a doctoral thesis on the calculus of variations at the University of Vienna in 1883, serving that summer as Weierstrass's *Assistant* in Berlin.

18. Husserl (1921), "*Vorwort*", vi; Engl. trans. (1970), 663.

19. Schlick (1918), 121.

20. Weyl (1923c), 60.

21. See chapter 3, §3.1 and 3.3.

22. Schlick (1925), 128 and viii; Engl. trans. (1985), 139 and viii.

23. Schlick (1925), 136; Engl. trans. (1985), 148.

24. Schlick (1934); Engl. trans. (1979b), 381 and 382. For Neurath's criticism and a masterful discussion of this debate, see Uebel (1992), 183ff.

25. See Mancosu (1998), 65–85, and Mancosu and Ryckman (2002).

26. Weyl (1928), 88; as repr. in Weyl (1968), vol. 3, 149; Engl. trans. (1967), 484.

27. Weyl (1985), 13.

28. Weyl (1940).

29. Scholz (1995), 1597, remarks on Weyl's "strong fascination" with Fichte's work in the period 1916–1922; indeed, Weyl, at the end of his life, termed Fichte a "constructivist of the purest water (*ein Konstuktivist reinsten Wassers*)". Weyl (1955), 163. At that time, he opposed constructivism to phenomenology, stating that his sympathies lay on the side of constructivism, while deeming Fichte's own execution of the constructive program as "outrageous (*als hanebüchen*)" (165). But, I cannot agree with Scholz's assessment that in the years 1916–1922, "Husserl's phenomenology...shifted into the background" for reasons that will be evident in this chapter. Rather, as Weyl himself stated, his interest in Fichte during the period in question concerned the latter's "metaphysical speculations concerning God, the Ego, and the world". Of these, he wrote in 1955, his memory no longer contained "a trace" (168).

30. Weyl (1949b), 397.

31. Weyl (1955), 170.

32. Eddington (1922), 634. The review is signed "A.S.E".

33. Weyl (1918b, 1919b, 1921a, 1923a), 3–4; cf. Engl. trans. (1952), 4–5.

34. Weyl (1918b), 228; (1919b), 264; (1921a), 289; (1923a), 331; cf. Engl. trans. (1952), 319.

35. Weyl (1919b), 123; (1921a), 124; (1923a), 136; cf. Engl. trans. (1952), 77.

36. Weyl (1918b, 1919b, 1921a, 1923a), 2; cf. Engl. trans. (1952), 2.

37. Weyl (1918b, 1919b, 1921a, 1923a, 9; cf. Engl. trans. (1952), 10.

38. Weyl (1918b, 1919b, 1921a, 1923a), 2; cf. Engl. trans. (1952), 2.

39. The *Husserliana* III edition (1950) of this work includes at this point the following remark added by Husserl to a copy of the (1913) original text: "as a being experienceable in subjects of consciousness in virtue of appearances, and possibly becomes confirmed *ad infinitum* as a verificational unity of appearances".

40. Weyl (1921a), 4; cf. Engl. trans. (1953), 5.

41. Husserl (1931), 30.

42. Cassirer to Husserl (10 April 1925), in Schuhmann (1994), 6.

43. Sigurdsson (1991), 7.

44. "Categorial intuition" must be understood in the extended sense given to perception by Husserl, as a type of super-sensuous (i.e., "raised above sense, or categorial") perception of ideal entities, in particular of "states of affairs" (*Sachverhalten*), such as "S is P". Whereas straightforward acts of perception give an object in one or another manner, categorial intuitions are *founded* acts in which objects are given "in themselves", in such a way that assertive thought finds fulfillment, e.g., *seeing that* "S is P". See Husserl (1921); Engl. trans. (1970), §§44–48.

45. Spiegelberg (1982), 120. According to Walter Biemel (see Husserl, 1950, vii–viii) a former student of Husserl and later editor of several volumes of *Husserliana*, the continuing series of posthumously issued collected works, this was the result of a "crisis" having to do, in part, with personal matters, but also with doubts about the phenomenological method of the *Investigations*, characterized there as a presuppositionless, purely descriptive, psychology.

46. Husserl (1907).

47. Husserl (1910–1911); Engl. trans. (1965).

48. Husserl (1907), 22–23; Engl. trans. (1964), 17–18.

49. Husserl (1910–1911), 290; Engl. trans. (1965), 74.

50. Husserl (1907), 17; Engl. trans. (1964), 1.

51. Husserl (1910–1911), 298; Engl. trans. (1965), 85.

52. Husserl (1910–1911), 299; Engl. trans. (1965), 87.

53. Husserl (1910–1911), 298; Engl. trans. (1965), 85.

54. Husserl (1910–1911), 293; Engl. trans. (1965), 78.

55. Husserl (1913) and Engl. trans. (1983), §55.

56. Husserl (1910–1911), 301; Engl. trans. (1965), 90.

57. Husserl (1910–1911), 294; Engl. trans. (1965), 79.

58. Husserl (1910–1911), 295; Engl. trans. (1965), 80.

59. Husserl (1910–1911), 294; Engl. trans. (1965), 78.

60. Husserl (ca. 1908), 382; this remark, dated approximately 1908, appears in one of many *Beilagen* published together with Husserl's lectures of 1923–1924 on "*Kritsiche Ideengeschichte*". Husserl was a prodigious writer; Bell (1990), xi, reports some 7,000 pages have been published in German, whereas over 40,000 pages of manuscript in shorthand are in the Husserl Archive in Louvain.

61. Husserl (1913) and Engl. trans. (1983), §33.

62. Husserl (1913) and Engl. trans. (1983), §§33, 49, 50.

63. Husserl (1913) and Engl. trans. (1983), §55.

64. Spiegelberg (1982), 168. For discussion, see Mancosu and Ryckman (2002).

65. E.g., Philpse (1995), 249ff.

66. See Welton (2000).

67. Weyl (1955); repr. in Weyl (1968), vol. 4, 637.

68. The reconstruction is based upon Philpse (1995), without sharing a number of his assessments of the argument.

69. Husserl (1913) and Engl. trans. (1983), §46.

70. See the discussion in Welton (2000), 104–111.

71. Husserl (1913) and Engl. trans. (1983), §51.

72. Husserl (1913), §47: "[W]hatever *physical things are*—the only physical things about which we can make statements, the only ones about the being or non-being, the being-thus or being—otherwise of which we can disagree and make rational decisions—*they are as experienceable physical things*".

73. Husserl (1913) and Engl. trans. (1983), §47.

74. Husserl (1913) and Engl. trans. (1983), §47, original emphasis.

75. Husserl (1913) and Engl. trans. (1983), §20.

76. Husserl (1913) and Engl. trans. (1983), §48.

77. Husserl (1913) and Engl. trans. (1983), §52.

78. Husserl (1913) and Engl. trans. (1983), §52.

79. Husserl (1913) and Engl. trans. (1983), §55.

80. Husserl (1913) and Engl. trans. (1983), §55.

81. Husserl (1913) and Engl. trans. (1983), §41.

82. Husserl (1913) and Engl. trans. (1983), §52.

83. Husserl (1907), 55; Engl. trans. (1964), 43.

84. Husserl (1913) and Engl. trans. (1983), §49.

85. Husserl (1907), 55; Engl. trans. (1964), 43; also (1910–11), 299–300; Engl. trans. (1965), 87.

86. Husserl (1936–1937), 193: "Es gilt nicht, Objektivität zu sichern, sondern sie zu verstehen". Engl. trans. (1970), 189.

87. Becker (1923).

88. Details in Sluga (1993), 219–220; and in Mancosu and Ryckman (2002).

89. Shuhmann (1994), 293–294. See the discussion in Mancosu and Ryckman (2002).

90. The letter is reproduced and discussed in Mancosu and Ryckman (2004).

91. Weyl (1918a), 65–74: "Anschauliches und mathematisches Kontinuum"; Engl. trans. (1994), 87–97.

92. Weyl (1918a), 17; Engl. trans. (1994), 25. Two "simple facts" pertain to this relation: (1) that for every number x there is a unique number y that is its successor; (2) that there is a unique number 1 that is not the successor of any number.

93. Weyl (1918a), 39; Engl. trans. (1994), 48.

94. Feferman (1988); Mancosu (1998), 70–74.

95. Weyl (1918a), 71; Engl. trans. (1994), 94. For a sustained defense of Weyl's perspective, see Feferman (1998).

96. Weyl (1918a), 67; Engl. trans. (1994), 88.

97. Weyl (1918a), 70; Engl. trans. (1994), 92.

98. Weyl (1918a), 72; Engl. trans. (1994), 93.

99. Husserl (1913) and Engl. trans. (1983), §50; cf. §§49, 55.

100. Weyl (1926), 98; cf. (1949a), 135.

101. Weyl (1918b), 3; cf. Engl. trans. (1952), 3.

102. Weyl (1918b), 181–182; cf. Engl. trans. (1952), 227.

103. Weyl (1921a), 8; cf. Engl. trans. (1952), 8.

104. Weyl (1921a), 8; cf. Engl. trans. (1952), 8–9.

105. Weyl (1921a), 9; cf. Engl. trans. (1952), 10.

106. Weyl (1926), 57; cf. (1949a), 75.

107. Weyl (1926), 82–83; cf. (1949a), 116.

108. Carnap (1928), 198–200; Engl. trans. (1969), 227–230.

109. Husserl (1922), 190; Eng. trans. (1970), 194.

110. Weyl (1918a), 11 n.; Engl. trans. (1994), 119 n. 19.

111. Husserl (1913) and Engl. trans. (1983), §59.

112. Weyl (1918b), 69; (1923a), 71; Engl. trans. (1952), 77.

113. Weyl (1923c), 60.

114. Weyl (1925a), 23–24; Engl. trans. (1998), 136.

115. Weyl (1928), 88; repr. in (1968), vol. 3, 149; Engl. trans. (1967), 484.

116. Weyl (1918b), 60–61; (1923a), 60; cf. Engl. trans. (1952), 67–68.

117. Husserl (1913) and Engl. trans. (1983), §3.

118. Husserl (1913) and Engl. trans. (1983), §2.

119. Husserl (1913) and Engl. trans. (1983), §83.

120. Husserl (1913) and Engl. trans. (1983), §6.

121. Husserl (1913) and Engl. trans. (1983), §2; emphasis in original.

122. Husserl (1913) and Engl. trans. (1983), §79.

123. Husserl (1913) and Engl. trans. (1983) §1: "*Die Welt ist der Gesamtinbefiff von Gegenständen möglicher Erfahrung und Erfahrungserkenntnis, von Gegenständen, die auf Grund aktueller Erfahrungen in richigen theoretishen Denken erkennbar sind*".

124. Weyl (1918b), 3–4; cf. Engl. trans., 4–5.

125. Weyl (1918b), 4; cf. Engl. trans. (1952), 5.

126. Husserl (1913) and Engl. trans. (1983), §79.

127. Husserl (1913) and Engl. trans. (1983), §4.

128. Husserl (1913) and Engl. trans. (1983), §§52, 55; Carnap (1928) and Engl. trans. (1969), §§3, 64, 65.

129. Cited and translated from a Husserl manuscript of the early 1920s in Bernet et al. (1993), 79.

130. Husserl (1913) and Engl. trans. (1983), §7: "But for the *geometer* who explores not actualities but "ideal possibilities", not predicatively formed actuality-complexes but predicatively formed eidetic affair-complexes, *the ultimate grounding act* is not experience but rather *the seeing of essences*".

131. Husserl (1913) and Engl. trans. (1983), §72.

132. Husserl (1913) and Engl. trans. (1983), §70. "Presentiate", although rare, is an English word, attested since 1659 in the *Oxford English Dictionary*, where it is given the meaning *to make or render present in place or time; to cause to be perceived or realized as present*. Hence, "presentiation" or "presentiated" is *the act of rendering present*. Here again, I am indebted to Roberto Torretti.

133. Husserl (1913) and Engl. trans. (1983), §79.

134. Husserl (1913) and Engl. trans. (1983), §7.

135. Husserl (1913) and Engl. trans. (1983), §16. In a private communication, Roberto Torretti has pointed out the qualified character of Husserl's claim. For Kant synthetic cognitions *a priori* were never justified by "evidence" or "intuition" but required vindication through their Transcendental Deduction.

136. Husserl (1913) and Engl. trans. (1983), §9.

Chapter 6

1. Weyl (1920), 738; repr. in (1968), vol. 2, 113. See also Ryckman (2003a).

2. Weyl (1918b) and (1923a), 3. See also (1918b), 172; (1923a), 218. Engl. trans. (1952), 217.

3. Weyl (1918b) and (1923a), 5 and 7. Engl. trans. (1952), 6 and 8. Also (1923b), 24, 44ff.

4. Weyl (1926), 93; cf. (1949a), 130: "The penetration of the Here-Now and the Thus is the general form of consciousness; something *is* only in the insoluble unity of intuition and sensation, in which continuous extension and continuous quality overlap. Phenomenologically, one cannot get beyond this".

5. Weyl (1920a), 738; repr. in (1968), vol. 2, 114: *"Die Welt kommt uns nur zum Bewußtsein in der allgemeinen Form des Bewußtseins, welche da ist: eine Durchdringung des Seins und Wesens, des 'Dies' und 'So'. (Das innige Verständnis dieser Durchdringung ist, nebenbei bemerkt, meiner Ueberzeugung nach der Schlüssel zu aller Philosophie.) In Akten der Reflexion sind wir imstande, des* Wesen, das *So-Sein der Phänomene zur Abhebung zu bringen, für sich zu bemerken, ohne es doch von dem einzelnen Sein des jeweils anschaulich Gegebenen, in dem es erschient, de facto lösen zu können. Hier der Ursprung der* Begriffe!" (Emphasis in original.)

6. Weyl (1921f), 57–58; repr. in (1968), vol. 2, 159. Cf. Engl. trans. (1998), 100. "It may be said that, through the mathematical treatment of actuality [*Wirklichkeit*], the attempt is made to represent, in the absoluteness of pure being [*reinen Seins*], the world that is given to consciousness in its more general form of a penetration of being [*Sein*] and essence [*Wesen*] [of the 'this' ['*dies*'] and the 'thus' ['*so*']]".

7. Weyl (1924), 81: "That for the purpose of its theoretical description we must set the actual [*das Wirkliche*] upon the background of the possible [*des Möglichen*] (of the space-time continuum with its field structure) signifies, when all is said and done, the appearance of geometry in physics". This passage occurs in a section of *Erläuterungen und Zusätze* that appeared first in the monographic publication. See also Weyl (1926), 94; cf. Eng. trans. (1949a), 131: "It is rooted in the double nature of the actual [*des Wirklichen*] that we can only design a theoretical image [*Bild*] of the existing [*des Seienden*] upon the background of the possible. Thus the four-dimensional continuum of space and time is, above all else, the field of *a priori* subsisting possibilities of coincidences. Accordingly, Leibniz named the "abstract space the order of all positions [*Stellen*] assumed possible" and adds: "consequently it is something ideal (*etwas Ideales*)" (citing Leibniz's Fifth Letter to Clarke, in Alexander, 1956, 89).

8. Weyl (1931), 49 and 52. The quotation continues: "That is mirrored in theoretical construction in the relation between the curved surface and its tangent plane at the point *P*: both cover the immediate surroundings of the center *P*, but the further one proceeds from *P*, the more arbitrary becomes the continuation of an unambiguous correspondence of the covering relation between surface and plane". Also, Weyl (1926), 98; Engl. trans. (1949a), 135: "[A] space of intuition whose metrical structure on essential grounds [*aus Wesensgründen*] fulfills the Euclidean laws does not contradict physics in so far as it clings to the Euclidean character of the infinitely small region of a point O (at which I momentarily find myself)".

9. Weyl (1926), 61; Engl. trans. (1949a), 86.

10. Weyl (1918a), 72; Engl. trans. (1949a), 93: "The coordinate system is the unavoidable residuum of the ego's annihilation [*das unvermeidliche Residuum der Ich-Vernichtung*] in that geometrico-physical world which reason sifts from the given under the norm of 'objectivity'—a final scanty token in this objective sphere that existence [*Dasein*] is only given and *can* only be given as the intentional content of the conscious experience of a pure, sense-giving ego". Also Weyl (1921a), 8; cf. Engl. trans. (1952), 8.

11. Weyl (1918b), 82; (1923a), 86. Engl. trans. (1952), 66. Also Weyl (1926), 61; Engl. trans. (1949a), 86.

12. Weyl (1918c), 480; repr. in (1968), vol. 2, 42. Engl. trans. (1997), 36.

13. Cf. Weyl (1955), 162; repr. in (1968), vol. 4, 640.

14. About *Wesenschau* Weyl wrote near the end of his life (1954), 629: "At the basis of all knowledge there lies: (1) *Intuition*, mind's originary act of 'seeing' what is given to him [sic]; limited in science to the *Aufweisbare*, but in fact extending far beyond these boundaries. How far one should go in including here the *Wesenschau* of Husserl's

phenomenology, I prefer to leave in the dark". *Aufweisbare* may be translated as "evident" or "what is recognizable with certainty".

15. Weyl (1918d).

16. *"Vorwort zur dritten Auflage"*, in Weyl (1919b), vi.

17. Weyl (1918d), 385; repr. in (1968), vol. 2, 2.

18. See Weitzenböck (1920).

19. Weyl (1918d), 386; repr. in (1968), vol. 2, 3.

20. Levi-Civita (1917), Hessenberg (1917), Schouten (1919).

21. The terms "transport" and "displacement" are metaphorical. What one does is to continuously *define* at each infinitesimally nearby point along a path from P a vector that is parallel to the given vector at the preceding point.

22. Weyl (1923b), 11.

23. Following Cartan (1923), such a connection is called "without torsion"; see §6.5.

24. Weyl (1923b), 17. A metric tensor is needed only to raise or lower indices.

25. Eddington (1921a); see chapter 8.

26. Weyl (1923a), 113, and (1921d), 542; repr. in (1968), vol. 2, 238. See also Scholz (1994).

27. Weyl (1918c), 466; repr. in (1968), vol. 2, 30. Emphasis in original.

28. Weyl (1923b), 47.

29. Pauli (1921); Engl. trans. (1958), 195–196.

30. Weyl (1923a), 124: *"ein metrischer Raum trägt von Natur einen affinen Zusammenhang"*. Laugwitz (1958) proved Weyl's conjecture, showing that this condition distinguishes infinitesimal Euclidean metrics from the wider class of Finsler metrics.

31. Weyl (1921b), 497; repr. in (1968), vol. 2, 235; Also, Weyl (1922a), 120; repr. in (1968), vol. 2, 269. The full proof only appears as an appendix in Weyl (1923b).

32. See Torretti (1978).

33. See Chern (1996).

34. Coolidge (1940), 410.

35. Weyl (1921g), 221; repr. in (1968), vol. 2, 344.

36. Weyl (1922a).

37. Weyl (1923b).

38. Weyl (1922a), 120; repr. in (1968), vol. 2, 269. See also Borel (1986), 54.

39. Weyl (1923b), 46.

40. Weyl (1923b), 24.

41. Weyl (1923b), 44–45.

42. Weyl (1923b), 47ff. For details, see Scholz (2001), 85–95; Coleman and Korte 215–250; and more briefly, Hawkins (2000), 435–436.

43. Weyl (1923b), 47.

44. Weyl (1923b), 26.

45. Scheibe (1988), 68ff.

46. See Laugwitz (1965), 185ff, and Laugwitz (1958).

47. Weyl (1923a), 103.

48. Weyl (1921b), 479; repr. in (1968), vol. 2, 235.

49. Schuhmann (1994), 291.

50. Weyl (1921a), 133; cf. Engl. trans. (1952), 148.

51. As Friedman (1999), 52–53 has observed, this charge certainly fits Carnap's understanding of phenomenological intuition in his early work *Der Raum* (1922), which however appeared after these lines were written. In a mainly expository review of Carnap's work, Weyl's principal criticism is that it "lacked a deeper epistemological analysis...of the relations of intuitive space and physical space" (1922c), 632; nonetheless, it is selected for citation from among the "the rich German literature" on relativity theory, in the 5th edition of *Raum-Zeit-Materie* (1923a), 334.

52. Weyl (1955), 161; repr. in (1968), vol. 4, 639.

53. Weyl (1918c, 1919a, 1919b).

54. Weyl (1919b), 244.

55. Weyl (1919b), 251.

56. Weyl (1919b), 251–252; (1923a), 314–315; for discussion, see Brading (2002).

57. Weyl (1919b), 253.

58. Weyl (1918c), as repr. in (1968), vol. 2, 41.

59. Pauli (1919a).

60. Eddington to Weyl (16 December 1918), *Weyl Nachlaß*, ETH Zürich, HS 91:522.

61. *Raum-Zeit-Materie*, 4th ed. (1921a), 260, n. 35; Engl. trans. (1953), 286, n. 35.

62. Weyl (1918d), 402; repr. in (1968), vol. 2, 19: "*sachliche Bedeutung schreiben wir nur den Tensoren vom Gewichte o zu*".

63. Weyl (1918c), 474; repr. in (1968), vol. 2, 37. Original emphasis. Engl. trans. (1997), 31.

64. Weyl (1918c), 474; repr. in (1968), vol. 2, 37. Engl. trans. (1997), 31.

65. Weyl (1918d), 410; repr. in (1968), vol. 2, 27

66. Weyl (1918c), 475; repr. in (1968), vol. 2, 38. Engl. trans. (1997), 32. Also, Weyl (1918d), 28 as repr. in (1968), vol. 2, 28. On Weyl's tie of gauge invariance to conservation of charge, see Brading (2002).

67. Weyl (1923a), 308.

68. Weyl (1929a); for discussion, see O'Raifeartaigh (1997).

69. Weyl to Einstein (16 November 1918), doc. 657 in Einstein (1998), Part B.

70. Einstein to Weyl (29 November 1918), doc. 661 in Einstein (1998), Part B.

71. Weyl (1918c), 480; repr. in (1968), vol. 2, 42. Cf. Engl. trans. (1997), 36. "It is to be observed that the mathematically ideal process of vector displacement, on which the construction of the geometry is based, has nothing to do with the real [*realen*] occurrence of motion of a clock, a course determined by the laws of nature".

72. Weyl (1919a), 113: "*Das Funktionieren dieser Meßinstrumente ist aber ein physikalischer Vorgang, dessen Verlauf durch die Naturgesetze bestimmt ist, und hat als solcher nichts zu tun mit dem ideellen Prozeß der 'kongruenten Verpflanzung von Weltstrecken', dessen wir uns zum mathematschen Aufbau der Weltgeometrie bedienen*".

73. Pauli (1921); Engl. trans. (1958), 196: "This relinquishment seems to have very serious consequences. While there no longer exists a direct contradiction with experiment, the theory appears nevertheless to have been robbed of its inherent convincing power, from a physical point of view". For Eddington's response, see chapter 8.

74. Weyl (1926), 95; cf. Engl. trans. (1949a), 132.

75. Weyl (1919b), 260.

76. Weyl (1921b), 475; repr. in (1968), vol. 2, 231.

77. Weyl (1922b), 52; repr. in (1968), vol. 2, 316: "*Meine theorie von Elektrizität und Gravitation, war von den Physikern meist dahin mißverstanden worden, als wolle ich an diese Tatsache rütteln. . . . Ich akzeptiere jene Grundtatsache so gut wie Einstein; wir weichen voneinander ab in ihrer theoretischen Deutung*".

78. Einstein (1917b). In this paper, Einstein attempted to reconcile his field equations with what he would call, in 1918, "Mach's principle", for "in a consistent relativity theory there should be no inertia *with respect to* "space", but only an inertia of masses *to each other*". (1996), 544. He argues this is to be done by abolishing boundary conditions at infinity for his 1915 field equations, and so constructing a model of a finite universe with three-dimensional volume described by a spherical geometry of positive constant curvature. However, he realized that any solution of his 1915 equations in which three-dimensional space has constant positive curvature at each instant of time is necessarily nonstatic, with time dependence of both curvature and volume. It is to remedy this non-static character nature of the universe that the cosmological constant $(-\lambda g_{\mu\nu})$ is introduced. I am indebted to Roberto Torretti here for clarification of Einstein's "order of reasons".

79. Pauli (1921); Engl. trans. (1958), 201.

80. Einstein (1919); 352; repr. in (2002), 134. Engl. trans. (1953), 195.

81. Eddington (1933), 21. Birkhoff (1923) later proved that the requirement of (spatial) spherical symmetry is sufficient to yield the vacuum Schwarzschild solution of the Einstein field equations, and alone entails that the metric field is static and flat at spatial infinity. This result can be interpreted as showing that there is no Machian necessity to Einstein's abolition of boundary conditions at infinity, for flatness at infinity can be seen as a mere consequence of having a single central source of gravity in an otherwise empty universe. For discussion, see Anderson (1967), 386, or Joshi (1993), 68.

82. Weyl (1921a), 251; cf. Engl. trans. (1952), 277.

83. Weyl (1921a), 252–254; cf. Engl. trans. (1952), 277–279; see also the discussion in Kerszberg (1989), ch. 4.

84. It is the arithmetical mean of Riemannian curvatures at a point P of an n-dimensional manifold obtainable from n mutually orthogonal (vector) directions; see e.g., Laugwitz (1965), 126–127.

85. For a general proof, see Weyl (1921a), 287–288; Engl. trans. (1953), 315–316. Hilbert merely assumed this, relying on some local Göttingen "folklore" about differential geometry; see Rowe (2001), 416–418. It is because the second derivatives appear only linearly that the gravitational action based on R yields second-order field equations. See Landau and Lifshitz (1975), 268–269.

86. Einstein (1916b) 412 n.; repr. in (1996), 412. Engl. trans. (1953), 170 n.

87. Weyl (1919b) and (1921a), §17; Engl. trans. (1952), §17. Also (1923a) §18.

88. Weyl (1923a), 129. Weyl's subsequently commented that in principle, the appearance of a distinguished gauge in his geometry is no more surprising than the fact that it is convenient to introduce adapted coordinate systems in Riemannian manifolds.

89. Weyl (1923a), 305.

90. Responding, with some notable irritation to a letter of Einstein (29 November 1918), in which Einstein reiterated his prehistory objection (and added a new objection: if standard units of mass and length are adopted, Weyl's theory gives rise to a new universal constant having the dimensions of an electrical quantity of enormous order of magnitude, over 10^{13}), Weyl writes, on 10 December: "Your repeat, without any backing, your assertions about the geodetic line and the dependence on the prehistory; I think I have given reasons refuting them and cannot say more than that on it;.". Einstein (1998), Part B, doc. 669.

91. Weyl (1919b), 259–260.

92. Weyl (1923a), 298 and 271.

93. Weyl (1923a), 298 and 271.

94. As do Coleman and Korté (2001), 229, in my opinion, wrongly.

95. Weyl (1923a), 298: "*Daraus geht mit aller Evidenz hervor, daß sich die Atommassen, Uhrperioden und Maßstablängen nicht durch irgendeine Beharrunstendenz erhalten; sondern es handelt sich da um einen durch die Konstitution des Gebildes bestimmten Gleichgewichtszustand, auf den es sich sozusagen in jedem Augenblick neu einstellt*" and, 303: "*die Tatsache, daß sich Maßstablängen und Frequenzen der Atomuhren bei Zugrundelegung der näturlichen Eichung erhalten, sich also in der Tat durch Einstellung auf den Krümmungsradius bestimmen*". Original emphasis indicated in underlining.

96. Weyl (1918b), 37; (1921b), 480.

97. Weyl (1923a), 304: "*Die durch das Prinzip der Eichinvarianz erzwungene Erweiterung der Weltgeometrie führt, bei Zugrundelegung eines in einfacher rationaler Weise aus den Zustandgrößen des metrischen Felldes aufgebauten Wirkungsprinzips, zu Folgerungen, die mit der Erfahrung im Einklang stehen, und macht ein bis dahin neben der metrik angenommenes physikalisches Zustandsfeld wie das elektromagnetische überflüssig*". Original emphasis.

98. Weyl (1923a), 308.

99. Weyl (1921b), 480.

100. Weyl (1934).

101. Weyl (1949a), 288: "The rigid rods and the clocks by which Einstein measures the fundamental quantity ds^2 of his metric theory of the gravitational field preserve their length and period in the last instance because charge e and mass m of the composing elementary particles are preserved. The systematic theory, however, proceeds in the opposite direction; it starts with a metric ground form and thus introduces a primitive field quantity to which the Compton wave length m^{-1} of the particle adjusts itself in a definite proportion. (Again, we employ the natural units in which c and h equal unity.)"

102. Cartan (1923); Engl. trans. (1986), 26. See also Cartan (1931), 23. One must be careful here; as "the tidal field is no less a local quantity than, say, the electric field", Cartan's assertion therefore pertains only to experiments not sensitive to tidal effects. Ohanian (1977), 907.

103. Bergmann (1979), 70.

104. Sharpe (1997), x.

105. Weyl (1988), 40.

106. Weyl (1929b), 719; repr. in (1968), vol. 3, 210.

107. Weyl (1929a), 331; repr. in (1968), vol. 3, 246. Engl. trans. (1997), 122.

108. Weyl (1938), and (1949b), 538–539; repr. in (1968), vol. 4, 397–398.

109. See Trautman (1980a) and Sharpe (1997).

110. A "principal connection" maps the tangent space $T_P(M)$ of the point $P \in M$ into the (higher dimensional) tangent space $T_\phi(E)$ over the point $\phi \in \Sigma_P$, the fiber above P. With such a "connection", the tangent space $T_\phi(E)$ is split into a "horizontal subspace" (isomorphic to $T_P(M)$) and a "vertical subspace"; this allows any vector in the bundle space E to be decomposed into horizontal components along M and vertical components along Σ_P. In this way, parallel displacement of fibers can be defined along any timelike curve γ in M by "horizontally lifting" γ into a curve $\bar{\gamma}$ in the bundle space E (of all the associated spaces Σ_P) such that all vectors tangent to $\bar{\gamma}$ are horizontal. For a concise account, see Progovečki (1995), 41–42, and, for further elaboration, see Frankel (1998), ch. 16.

111. A much simplified version of the account in Weinberg (1995), §8.1, and Kaku (1993), 102–103.

112. Cf. Yang (1980); Mills (1989); Auyang (1995), 55–58.

113. See Yang (1977), Mielke and Hehl (1988), Cao (1997), Vizgin (1994), Scholz (1994, 1995, 1999, 2001), O'Raifeartaigh (1997), O'Raifeartaigh and Straumann (2000), and Straumann (2001).

114. Moriyasu (1983), 5.

115. Weyl (1929a), 331; (1968), vol. 3, 246. Engl. trans. (1997), 122.

116. Quoted from Redhead (2002), 299.

117. Gross (1999), 57.

118. Husserl (1937), 48–49; Engl. trans. (1970), 48–49.

119. See Brading and Castellani (2003).

Chapter 7

1. Eddington (1928).

2. When passing a bookshop window in Cambridge prominently displaying one of Eddington's popular books (several were best sellers) on science; see Wood (1957), 202.

3. Lovejoy (1930), 266; Eddington's remark, misquoted by Lovejoy, is in Eddington (1928), 276.

4. Russell (1931), 112, 121; Frank (1934), 130.

5. Campbell (1931), 181–182.

6. Dingle (1937), 786. Since mathematical speculation plays little role in Aristotle's natural philosophy, one can only wonder at Dingle's choice of terms.

7. E.g., Eddington (1936), 327.

8. Beck et al. (1931).

9. Schrödinger (1937, 1938, 1939).

10. Whittaker (1942).

11. Jeffreys (1941).

12. Bastin and Kilmister (1952); Kilmister (1994).

13. Stebbing (1937), 55, x.

14. Ritchie (1948), 36; for similar appraisals, Emmet (1945), 69ff; Dingle (1954), 39; Kilmister (1994), 37, 49–50, 58.

15. Eddington (1928), vi.

16. E.g., French (2003).

17. Feigl (1975), 17 and 25, refers to the thesis that "our knowledge of the 'external world', achieved with the help of physical concepts and theories, is '*structural*', as "Schlick-Russell-Eddington structuralism". Feigl particularly had in mind such mid-1920s texts as Eddington (1928), the *second* edition of Schlick's *Allgemeine Erkenntnislehre* (1925) and Russell's (1927) detailed examination of relativity theory, as well as his last major philosophical work (Russell 1948).

18. See especially Slater (1957) and Kilmister (1994).

19. Sommerfeld to Einstein (31 October 1926), in Hermann (1969), 109.

20. Eddington (1916). To be sure, papers of De Sitter on Einstein's cosmology appeared already in *The Monthly Notices of the Royal Astronomical Society* 9 no. 9 (1916) and in the October 1916 issue of *The Observatory*, the organ of the Greenwich Observatory.

21. For details, see Douglas (1956b).

22. Chandrasekhar (1982), 112.

23. Douglas (1956b), 40–41. Einstein's predicted value of 1.74 arcseconds (3,600 arcseconds = 1°) was only confirmed by taking the mean value of Eddington's photograph with results from photographs taken by an auxiliary telescope at Sobral; see Pais (1982), 305, and Einstein (2002), xxx.

24. Whitehead (1925), 10.

25. See Eddington to Einstein (1 December 1919), EA 9–260.

26. Dirac (1977), 110, 115.

27. Dirac (1982), 82.

28. Russell (1927), 395: "The theory of relativity, to my mind, is most remarkable when considered as a logical deductive system. That is the reason, or one of the reasons, why I have found occasion to allude so constantly to Eddington. He, more than Einstein or Weyl, has expounded the theory in the form most apt for the purposes of the philosopher".

29. Stebbing (1937), 7.

30. Eddington (1939a), 188–189: "But if it were necessary to choose a leader from among the older philosophers, there can be no doubt that our choice would be Kant. We do not accept the Kantian label, but as a matter of acknowledgment, it is right to say that Kant anticipated to a remarkable extent the ideas to which we are now being impelled by the modern developments of physics".

31. See Eddington (1936, 1946). "Heretical" is not too strong a term; witness the infamous controversy with Chandrasekhar (Eddington, 1935b) concerning Eddington's denial of the gravitational collapse of massive white dwarf stars, the controversy (Eddington, 1939b, 1942) with Dirac, Peierls and Pryce over the meaning of Lorentz invariance in quantum theory, or Eddington's derivations of the fundamental constants as

pure numbers (Whittaker, 1945). Kilmister (1994) is a valiant attempt to unravel the thread of argument in Eddington (1936).

32. "Preface", dated August, 1928, to Eddington (1928).

33. Eddington (1921b), 29.

34. What is probably the first notice of Weyl's theory in English was almost certainly written by Eddington in an unsigned report in the November 1918 issue of *The Observatory*, the organ of the Greenwich observatory. There is no mention of Weyl's influence on Eddington's views of relativity theory in the standard works on Eddington, i.e., Witt-Hansen (1958), Yolton (1960), Merleau-Ponty (1965), Dingle (1954), Stebbing (1937), or, more recently in Kilmister (1994).

35. Eddington (1921b), 30, 32.

36. Eddington (1921b), 183.

37. Eddington (1921b), 31.

38. Cassirer (1921), 37. "*Eben diese Unabhängigkeit vom zufälligen Standort des Beobachters meinen wir, wenn wir von dem in sich bestimmten Gegenstand der 'Natur' und von in sich bestimmten 'Naturgesetzen' sprechen*".

39. Eddington (1921b), 36, 32.

40. Eddington (1921b), 183.

41. Stebbing (1937), 99; Dingle (1954), 19.

42. Eddington (1925a), 196.

43. Eddington (1921b), 182.

44. Eddington (1925a), 193–194.

45. Eddington (1928), 286.

46. Eddington (1921b), 31.

47. Eddington (1925a),196; (1928), 284–285.

48. See chapter 2.

49. E.g., Hanna (2001), 39.

50. Eddington (1921b), 87: "In so far as our knowledge of nature is a knowledge of intersections of world lines, it is absolute knowledge independent of the observer".

51. Frank (1917); see the discussion in Ryckman (1992).

52. Reichenbach (1928), 327; Engl. trans. (1958), 285–286.

53. Eddington (1921b), 34: "*Length* and *duration* are not things inherent in the external world; they are relations of things in the external world to some specified observer".

54. Eddington (1921b), 33.

55. Eddington (1921b), 187 and 46.

56. Eddington (1921b), 181, 183.

57. Eddington (1920c), 153.

58. Eddington (1921b), 186, 197.

59. Eddington (1921b), 189.

60. Eddington (1921b), 189. In the passage elided from this quotation, Eddington notes that scalars (invariants) are an exception to the stated limitation. Of course, even scalars are not independent of choice of scale, but Eddington's concern here is only with the Riemannian geometry of orthodox relativity theory.

61. Eddington (1921b), 189.

62. Eddington (1921b), 191.

63. Eddington (1920c), 152. Eddington's notation varies, here writing, following the early papers of Einstein, $G_{\mu\nu}$ for the Ricci tensor, rather than the customary $R_{\mu\nu}$. We shall follow custom throughout.

64. Eddington (1921b), 192.

65. Eddington (1920c), 150.

66. Eddington (1921c), 151.

67. Eddington (1921b), 190.

68. Eddington (1921b), 190.

69. Eddington (1920c), 152.

70. Stachel (1986a), 235.

71. Eddington (1920d), 420.

72. The developmental story is significantly more complicated; see Norton (1984), 253–316.

73. Eddington (1921b), 198.

74. Eddington (1921b), 197.

75. Eddington (1921b), 198.

76. Eddington (1921b), 197.

77. Eddington (1921b), 200.

78. Eddington (1928), 278; and (1921b), 196, 192.

79. Eddington (1920c), 155.

80. Eddington (1921b), 201.

81. Douglas (1956a), 100.

82. McCrae (1991), 94.

83. Eddington (1939a), 28; cf. (1937), 4; (1935), 215; (1933), 98.

84. Eddington (1923a), 1.

85. Eddington (1923a), 2.

86. Eddington (1923a), 6–7. Contrast Bridgman (1927), 21–22, on precisely the same example: "What is the possible meaning of the statement that the diameter of an electron is 10^{-13} cm? Again, the only answer is found by examining the operations by which the number 10^{-13} was obtained".

87. Bridgman (1936), 82–83. Bridgman further asserted, antithetically to Eddington's epistemology: "There is no getting away from preferred operations and a unique standpoint in physics".

88. Eddington (1923a), 3 and 4. Cf. 47: "In using the phrase 'condition of the world' I intend to be as non-committal as possible; whatever in the external world determines the value of the physical quantities which we observe, will be included in the phrase".

89. Eddington (1928), 12, 19.

90. Eddington (1921b).

91. Eddington (1923a), 5.

92. Eddington (1923a), 5–6.

93. Eddington (1933), 17.

94. Cited by Weinberg (1993), 101. Weinberg notes the "half-serious" maxim is attributed to Eddington, but does not give the source; it is Eddington (1935a), 211.

95. Eddington (1923a), 119–120.

96. Eddington (1923a), 196.

97. Eddington (1923a), 154.

98. Eddington (1923a), 10 (italics in original).

99. Eddington (1923a), 41.

100. Eddington (1923a), 37.

101. Eddington (1923a), 41.

102. Eddington (1923a), 81–82. Eddington's grounds for this conclusion are not explicitly given but, presumably, are those that guided Einstein. What is wanted is an equation yielding the Newtonian approximation connecting metrical quantities and matter such that when the energy-momentum tensor vanishes (i.e., a classically defined vacuum, $T_{\mu\nu} = 0$), the metrical quantities do *not* become zero as must not happen in a metrical manifold. This is the case only if the connection to matter is made via the Ricci tensor, $R_{\mu\nu}$, and not the metric tensor $g_{\mu\nu}$ or the Riemann tensor $R_{\mu\nu\sigma\tau}$.

103. Eddington (1923a), 82.

104. Eddington (1923a), 82.

105. See also Eddington (1923b). This heretical view of the Einstein field equations is also found in Schrödinger (1950), 99; for discussion, see Zahar (1989), 45.

106. We overlook that the equation ignores the energy of the gravitational field, the contribution of a so-called pseudo-tensor, and so expresses the conservation of energy only in a Pickwickian sense. See Hoefer (2000) for a clear discussion. Moreover, a tensor divergence equation cannot really express a conservation law because it is not possible to integrate the covariant divergence over a three-dimensional volume in space-time (to add up the quantities $T_{\mu\nu}$ at different points) since doing so presupposes a covariantly constant timelike vector field (and so fixing a definite system of co-ordinates). The usual reading of this equation in general relativity is that it is approximately valid for sufficiently small regions of space-time.

107. Eddington (1923a), 119.

108. In a bibliographic comment added to the second edition of *MTR* (1924), 265, Eddington cites the paper of Harward (1922), appearing while the first edition was in press, as giving a "more elegant" direct proof than that in *MTR*. Harward, however, rederived the identities while noting that "I can scarcely believe that ['the general theorem'', i.e., the Bianchi identities] has not been discovered before" (584). Finally, the history of the Bianchi identities is pointed out in a note on Harward's paper by Schouten and Struik (1924), who trace their discovery to G. Ricci and their first publication to a paper of E. Padova in 1889. For discussion, see Rowe (2002).

109. See, e.g., Ciufolini and Wheeler (1995), 57.

110. Omitting electromagnetism, such "matter" is defined by the macroscopic energy-momentum tensor $T^{\mu\nu} = \rho V^\mu V^\nu$, where ρ is the density of matter and V^μ is the macroscopic velocity field of matter.

111. Eddington (1923a), 120; "The [Einstein] law of gravitation is not a law in the sense that it restricts the possible behaviour of the substratum of the world"; See also Eddington (1920b), 190.

112. Eddington (1923a), 146.

113. Eddington (1923a), 119–120.

114. Braithwaite (1929), 427.

115. Eddington (1925), 213–214; also (1923a), 147; (1928), 240–241.

116. Eddington (1922), 636.

117. Eddington (1923a), 138 and 139.

118. Schrödinger (1950), 92.

119. Eddington (1926), 908.

120. Eddington (1923a), 147.

121. Eddington (1923a), 147; also (1928), 241: "The Hamiltonian derivative has just that kind of quality which makes it stand out in our minds as an active agent against a passive extension of space and time.... Hamiltonian derivatives are virtually the symbol for creation of an active world out of a formless background".

122. Eddington (1923a), §63; see also the extended discussion in Schrödinger (1950), 93–97.

123. Eddington (1923a), 146: "In order that this theory may not be merely an exercise in pure mathematics, but may be applicable to the actual world, the quantities appearing in the theory must at some point be tied on to the things of experience. In the earlier chapters this was done by identifying the mathematical interval with a quantity which is the result of practical measurement with scales and clocks. In the chapter presently discussed [i.e., chapter 4], this point of contact of theory and experience has passed into the background, and attention has been focussed on another opportunity of making the connection. The quantity $R_{\mu\nu} - (1/2)g_{\mu\nu}R$ appearing in the theory is, on account of its property of conservation, now identified with matter, or rather with the mechanical abstraction of matter which comprises the measurable properties of mass, momentum and stress sufficing for all mechanical phenomena. By

making the connection between mathematical theory and the actual world at this point, we obtain a great lift forward".

124. Eddington (1923a), 146.

125. Eddington (1928), 145.

126. Eddington (1923a), 151–152.

127. Hence, Eddington allows, in accordance with Einstein (1917b), *only* for constant curvature. As Torretti has observed (private communication), today's astronomers are "hard put to find even the mass required to secure zero spatial curvature".

128. Eddington (1923a), 153.

129. Eddington (1923a), 153. Original emphasis.

130. Eddington (1928), 153.

131. Eddington (1923a), 155.

132. Eddington (1923a), 153.

133. Douglas (1956b), 191–192. Writing in the third person, Eddington noted: "Eddington's *Mathematical Theory of Relativity* contains numerous independent developments connected with the logical presentations of Einstein's theory as a connected whole. Besides introducing alternative methods of obtaining the results, simplifying certain points, and bringing out the implications more clearly, his particular development is a generalization of Weyl's theory given in ch. 7, pt. II of the book. Connected with this is his explanation of the law of gravitation $\left(R_{\mu\nu} = \lambda g_{\mu\nu}\right)$. This means that the radius of curvature of a section of space-time at any point and in any direction is a constant. Eddington pointed out that this is the same thing as saying that our practical unit of length at any point and in any direction is a definite fraction of the radius of curvature for that point and direction; so that the law of gravitation is simply the statement of the fact that the world-radius of curvature everywhere supplies the standard with which our measure lengths are compared".

134. Stachel (1986a), 232.

135. Eddington (1933), 147.

136. Eddington (1935a), 214–215.

137. Eddington (1933), 104.

138. Earman (2001), 204.

139. Cf. Bastin and Kilmister (1952), 559: "Eddington's work embodies an unorthodox attitude to the interpretation of physical measurement. This attitude regards physical theories as a formulation of conditions presupposed by our experimental procedures, rather than as determined, empirically, by those procedures".

140. Eddington (1920b), 197.

141. Carnap (1928), §16; Schlick (1935), 359 n.

142. Later Russell (1927, 9) asserted that the causal theory of perception has always been "the common sense view", namely, that "all our perceptions are causally related to antecedents which may not be perceptions". For discussion, see Yolton (1960), 64.

143. Russell's (1927) version of neutral monism in brief (383): There is but one kind of "neutral stuff"—events—and one kind of law, causal relations between events. A continuous causal chain connects a distal external event with the event in a percipient's brain, a *percept*. Knowledge of distal events is inferential, grounded on the causal theory of perception, and based on the causal maxim of "different effects, different causes". Of events that are percepts (occurring in regions of space-time where there is a brain), it is possible to know their *intrinsic nature*. But of events occurring in brainless regions of space-time, only their structure is knowable, through inferences warranted by the supposition that differences in structure among percepts correlates with differences in structure among the events that are causes of percepts. This theory underlies Russell's notorious "under-the-hat" view of perception: "I should say that what the

physiologist sees when he looks at a brain is part of his own brain, not part of the brain he is examining".

144. Newman (1928), 144.

145. In a letter to Newman of 24 April 1928, reproduced in Demopoulos and Friedman (1985), 631–632, Russell admits that he had always also presupposed that there is knowledge of the spatiotemporal continuity of nonpercept events with percepts, such as to enable passage in "a finite number of steps from one event to another compresent with it, from one end of the universe to the other".

146. Eddington (1935a), 255 (epigram to ch. 12, "The Theory of Groups"); (1939a), 152; the Russell passage is briefly paraphrased in Eddington (1928), 277 n.

147. Eddington (1923a), 106.

148. For a recent discussion, see French (2003).

149. Eddington (1920c), 145.

150. Perhaps first by Weyl (1921a), 87; Engl. trans. (1953), 97. Since reference to Clifford does not occur in the earlier editions (1918, 1919) of Weyl's book, it is plausible that Weyl added it after encountering the Clifford quote in *Space, Time and Gravitation*, a book Weyl warmly commends to the reader: "*eine ausgezeichnete populär-anschauliche und ausführliche Darstellung der allgemeinen Relativitätstheorie einschl. der hier in §§35,36 besprochenen Erweiterung*"(1921a), 291.

151. Clifford (1875), quotation on 776–777.

152. Eddington (1930).

153. Compare Eddington (1930), 89: "The spirit of seeking which animates us refuses to regard any kind of creed as its goal", with Clifford (1875), 792: "It is clear that the good old gods of our race, sun sky, thunder, and beauty, are to be replaced by philosophic abstractions, substance, energy, and life, under the patronage respectively of the persons of the Christian trinity. But why are we to stay there? ... If there is room in the unseen universe for the harmless pantheistic deities which our authors have put there, room may also be found for the goddess Kali, with her obscene rites and human sacrifices, or for any intermediate between these. Here is the clay; make you images to your heart's desire".

154. Clifford (1878), 87.

155. Clifford (1878), 85: "A moving molecule of inorganic matter does not possess mind or consciousness; but it possesses a small piece of mind-stuff. When molecules are so combined together as to form the film on the underside of a jelly-fish, the elements of mind-stuff which go along with them are so combined as to form the faint beginnings of Sentience. When the molecules are so combined as to form the brain and nervous system of a vertebrate, the corresponding elements of mind-stuff are so combined as to form some kind of consciousness".

156. Eddington (1928), 276–282, 278.

157. Eddington (1920b), 192.

158. Eddington (1928), 277.

159. Eddington (1928), 276.

Chapter 8

1. Weyl (1921e), 800: "Modern physics renders it probable that the only fundamental forces in Nature are those which have their origin in gravitation and in the electromagnetic field".

2. E.g., Wheeler (1990), 12.

3. Einstein (1936), 335; Engl. trans. (1983), 311.

4. Einstein (1923e), 489.

5. See Rainich (1925); Misner and Wheeler (1962); and Geroch (1966). In the "already unified field theory" the space-time geometry of general relativity, constrained by the imposition of four algebraic ("Rainich") conditions and one additional differential equation, arises from a solution of the Maxwell equations while conversely, the electromagnetic field tensor emerges as nearly completely determined from the metric tensor. The qualification stems from the null field case of electromagnetism, i.e., pure radiation, a situation treated exhaustively in Geroch (1966). This "already unified theory" thus shows that nothing from "outside" need be imported into the Riemannian geometry of general-relativistic space-times in order to incorporate the electromagnetic field. According to Geroch, 183: "[T]he already unified theory expresses the content of the Einstein-Maxwell equations in relations involving only the geometry. It brings out the fact that the imprint of the electromagnetic field on the geometry is a very distinctive one—so much so that the electromagnetic field may be determined from the geometry. It does unify electromagnetism and gravity".

6. Vizgin (1994), ch. 4. Vizgin's book is the most thorough history of the first decade of the geometrical unified field theory program.

7. Eddington (1921a), 105.

8. Clark (1971), 273ff. Haldane was an ex-diplomat and statesman well read in German philosophy. His book *The Reign of Relativity* appeared soon after this visit.

9. A.S. Eddington to A. Einstein (12 June 1921) EA 9-277. Eddington began: "Here is the paper we were speaking about on Friday night. Later on I shall be interested to hear whether it fulfills the favourable expectations you formed in our conversations about it".

10. Einstein (1921b).

11. Einstein to Weyl (5 September 1921), Einstein-Sammlung ETH, Zürich, no. 551; as cited and translated in Stachel (1986a), 240.

12. Kaluza (1921). A translation of Kaluza's paper and of the Kaluza-Einstein correspondence, both by C. Hoenselaeus, is in De Sabbata and Schmutzer (1983), 427–433, and 447–457.

13. Einstein and Grommer (1923). For discussion, see Vizgin (1994), 177, and Pais (1982), 333.

14. A. Einstein to H. Weyl 6 June 1922 (EA 24-071): "*Bei den Eddington'schen Ausführungen geht es mir wie bei der Mie'schen Theorie; es ist ein schöner Rahmen, bei dem man absolute nicht sieht, wie er ausgefüllt werden muss*".

15. As translated in French (1979), 274.

16. Einstein to Weyl (23 May 1923), EA 24-080, as cited and translated in Stachel (1986a), 240.

17. Vizgin (1994), 265. The papers are Einstein (1923a, 1923b, 1923c, 1923d). The latter, rendered into English by R.W. Lawson, appeared in the 22 September 1923 issue of *Nature*.

18. Einstein (1923e); Engl. trans. (1967), 489.

19. Eddington (1921a), 121.

20. Eddington (1923a), 222. See also Ryckman (2003b).

21. E.g., Stachel (1986a), 235; Cao (1997), 336; Vizgin (1994), 140. Stachel states that Eddington "sought to explain those properties ordinarily interpreted as material in terms of the metrical structure of space-time" and concludes that Eddington's approach to matter was developed into "a fairly full anticipation of what Wheeler was later to call geometrodynamics". Cao cites Eddington's "idea about 'the geometry of the world structure, which is the common basis of space and time and things' "as containing "elements correlated both with matter and with interaction mechanisms of material systems". Vizgin, whose treatment of Eddington is by far the best in the literature,

speaks ambiguously of Eddington's "support for the program of unified geometrized field theories".

22. Douglas (1956b), 56. As already noted, Schrödinger, who developed his own affine field theory in the 1940s, took over this view from Eddington; see, e.g., (1950), 99.

23. Einstein (1925b).

24. Einstein (1925c); Pais (1982), 343.

25. Einstein (1927).

26. Einstein (1929). In a display of excessive journalistic zeal, on the day of its publication this paper was telegraphed from Berlin to New York, where a translation, the equations suitably transcribed by Columbia University physicists, appeared on 1 February 1929 in the *New York Herald Tribune*. Fölsing (1997), 605.

27. Einstein (1932).

28. Bergmann (1983), 2.

29. Douglas (1956b), 56. As Einstein explained to Michele Besso in a letter at about the same time (10 August 1954), in an inertial system vectors at distinct points P and Q separated by an arbitrary distance but possessing the same components have an immediate invariant relation: they are equal and parallel. Hence in an inertial system, the differentiation of a tensor with respect to the coordinates always yields another tensor. In an arbitrary (hence, noninertial) coordinate system, the affine connection, while not itself an invariant (it is not a tensor), can be differentiated to construct tensors at different points. In this way, it is "an invariant substitute for the inertial system—and thereby the foundation of every relativistic theory". As translated in French (1979), 268.

30. Eddington to Weyl (16 December 1918), ETH Weyl Archives, Zürich (HS 91: 522).

31. Eddington to Weyl (18 August 1920), ETH Weyl Archives, Zürich (HS 91: 523): "I am particularly glad that you are pleased with the chapter on Electricity and Gravitation".

32. Eddington to Weyl (10 July 1921), ETH Weyl Archives, Zürich (HS 91: 525).

33. Eddington (1923a), 224.

34. Eddington (1921a), 121.

35. Together with Schouten (1922), and Cartan (1923).

36. Eddington (1928), ch. 11, "World Building".

37. Eddington (1923a), 49.

38. Eddington (1923a), 225–226.

39. Eddington (1936), 1–2. The spin matrices of the Dirac equation transform as vectors under the spinor representation of the Lorentz group.

40. Kilmister (1994), 101.

41. E.g., Eddington (1936), 323–329.

42. For details, see Kilmister (1994).

43. Eddington (1928), 235.

44. Eddington (1923a), 215.

45. Eddington (1921a), 108.

46. Eddington (1923a), 219.

47. Eddington (1923a), 219.

48. Eddington (1921a), 111; (1923a), 221.

49. Eddington retained this assessment of Weyl's principle of "relativity of length" until the end of his life; see (1937) and (1946).

50. Eddington (1921b), 31; see also 82 n.

51. Eddington (1923a), 106.

52. Eddington (1923a), 197.

53. E.g., for the genus of this species of transcendental idealism, see Bird (1962), Matthews (1969), and Allison (1983).

54. See Guyer (1989), 140, for the remark about "epistemological modesty." I have followed Allison (1983) and (2004) in viewing transcendental idealism as a *metaphilosophical* standpoint, beyond realism and idealism (or anti-realism).

Chapter 9

1. "Yet this is now still future music ...".
2. Eisenhart and Veblen (1922); Veblen (1923), 136.
3. A "Cartan" or "Ehresmann" connection, enabling the study of global properties of manifolds. Sharpe (1997) is a comprehensive treatment.
4. Weyl (1949b), 541; repr. in (1968), vol. 4, 400. See the discussion in Hawkins (2000), ch. 11.
5. Weyl (1928b).
6. Einstein (1928).
7. Of his own theory of "gravitation and electromagnetism", Weyl wrote (1921a), 258: "The dream of Descartes of a pure geometrical physics appears to be fulfilled in wonderful ways, certainly entirely unsuspected by him". Cf. Engl. trans. (1952), 284.
8. Meyerson (1921); Engl. trans. (1991), 27.
9. Meyerson (1921); Engl. trans. (1991).
10. Meyerson (1921); Engl. trans. (1991), 135.
11. Meyerson (1921); Engl. trans. (1991), 137.
12. Meyerson borrows the term "panmathematicism" from Brunschvicg, who coined it to characterize Plato's theories.
13. Meyerson (1921); Engl. trans. (1991), 388.
14. Meyerson (1921); Engl. trans. (1991), 270.
15. Meyerson (1921); Engl. trans. (1991), 426.
16. Meyerson (1925), Engl. trans. (1985), 152.
17. Meyerson (1925), 188–189; Engl. trans. (1985), 129.
18. Meyerson (1925), 354–355; Engl. trans. (1985), 231–232.
19. Meyerson (1921); Engl. trans. (1991), 459: "[T]he true progress of science, which is the progress our understanding makes towards comprehending nature, must ultimately consist in determining the limits and the modalities of the agreement between nature and reason".
20. Meyerson (1921); Engl. trans. (1991), 172.
21. Meyerson (1921); Engl. trans. (1991), 421.
22. Meyerson (1925), 221–222; Engl. trans. (1985), 150–151.
23. Weyl (1919b), 262.
24. Meyerson (1925), 131; Engl. trans. (1985), 92.
25. Meyerson (1921); Engl. trans. (1991), 423.
26. Meyerson (1925), 192; Engl. trans. (1985), 134.
27. Eddington (1920a), 183.
28. Einstein (1928); Engl. trans., 255, and 254–255.
29. Meyerson (1925), 214–215; Engl. trans. (1985), 146.
30. Meyerson (1925), 212–213; Engl. trans. (1985), 145. "Puckers" is Eddington's term, quoted in English.
31. Meyerson (1925), 195; Engl. trans. (1985), 134.
32. Meyerson (1925), 214–215; Engl. trans. (1985), 145–146.
33. Meyerson (1925), 216; Engl. trans. (1985), 146–147.
34. Meyerson (1925), 217; Engl. trans. (1985), 147.

35. Worrall (1989) initiated the recent discussion of "structural realism". The distinction between the "epistemic" and "ontological" varieties of structural realism was first made by Ladyman (1998).

36. Poincaré (1906); Engl. trans. (1958), 14.

Appendix to Chapter 2

1. Friedman (1999, first published in 1994), 66: "The idea is as follows. Each of the theories in question (Newtonian physics, special relativity, general relativity) is associated with an *invariance group of transformations* that presents us with a range of possible descriptions of nature—a range of admissible reference frames or coordinate systems—that are equivalent according to the theory. The choice of one such system over another is therefore arbitrary, and Reichenbach's thought is that those elements left invariant by the transformations in question ... are precisely the constitutive elements of the theory.... [I]n special relativity the relevant group of transformations is the Lorentz group, and so ... the underlying structure of space-time of Minkowski space-time is constitutively *a priori*; particular fields defined within this structure [...] do not count as constitutive. Finally, in general relativity the relevant group includes all one-one bidifferentiable transformations (diffeomorphisms), and so only the underlying topology and manifold structure remain constitutively a priori".

2. Friedman (2001), 45; (2000b), 382; (2002).

3. Friedman (2001), 83 and 79.

4. Friedman (1983), 95–104, is a classic account.

5. Friedman (2001), 63.

6. See, e.g., Auyang (1995), 39.

7. Stachel (1989a), 78.

8. See Maidens (1998).

9. See Havas (1989) for history and discussion.

10. Trautman (1980b), 4.

11. E.g., Quine (1995), 49.

REFERENCES

Adler, R., M. Bazin, and M. Schiffer. 1975. *Introduction to General Relativity*, 2nd ed. New York: McGraw-Hill.

Alexander, H. G. 1956. *The Leibniz-Clarke Correspondence*. Edited and with an introduction by H. G. Alexander. Manchester: Manchester University Press.

Allison, H. E. 1973. *The Kant-Eberhard Controversy*. Baltimore: Johns Hopkins University Press.

———. 1983. *Kant's Transcendental Idealism*. New Haven, CT: Yale University Press.

———. 2004. *Kant's Transcendental Idealism*. Rev. and enl. ed. New Haven, CT: Yale University Press.

Anderson, J. L. 1967. *Principles of Relativity Physics*. New York: Academic Press.

Auyang, S. 1995. *How Is Quantum Field Theory Possible?* New York: Oxford University Press.

Barbour, J. 2000. *The End of Time: The Next Revolution in Physics*. New York: Oxford University Press.

Bargmann, V. 1960. "Relativity," in M. Fierz and V. Weisskopf (eds.), *Theoretical Physics in the Twentieth Century: A Memorial Volume to Wolfgang Pauli*. New York: Interscience, 187–198.

Bastin, E. W., and C. W. Kilmister. 1952. "The Analysis of Observations," *Proceedings of the Royal Society of London* A212, 559–576.

Beck, G. H., H. Bethe, and W. Riezler 1931. *Die Naturwissenschaften* 19, 39.

Becker O. 1923. "*Beiträge zur phänomenologischen Begründung der Geometrie und ihrer physikalischen Anwendungen*," *Jahrbuch für Philosophie und phänomenologische Forschung* 6, 385–560. Excerpts trans. in T. Kisiel and J. Kochelmans (eds.),

Phenomenology and the Natural Sciences. Evanston, IL: Northwestern University Press, 1970.

———. 1927. "Mathematische Existenz," *Jahrbuch für Philosophie und Phänomenologische Forschung* 8, 439–809.

Beiser, F. 1987. *The Fate of Reason*. Cambridge, MA: Harvard University Press.

Bell, D. 1990. *Husserl*. London: Routledge.

Ben-Menahem, Y. 2001. "Convention: Poincaré and Some of His Critics," *British Journal for Philosophy of Science* 52, 471–513.

Bergmann, P. 1949. "Non-linear Field Theories," *Physical Review* 75, 680–686.

———. 1961. "Observables in General Relativity," *Reviews in Modern Physics* 33, 510–514.

———. 1979. "Unitary Field Theory: Yesterday, Today, Tomorrow," in H. J. Treder (ed.), *Einstein-Centenarium 1979*. Berlin: Akademie Verlag, 62–73.

———. 1983. "Unitary Field Theories," in V. De Sabbata and E. Schmutzer (eds.), *Unified Field Theories of More than 4 Dimensions*. Singapore: World Scientific, 1–10.

Bernet, R., I. Kern, and E. Marbach. 1993. *An Introduction to Husserlian Phenomenology*. Evanston, IL: Northwestern University Press.

Bird, G. 1962. *Kant's Theory of Knowledge*. London: Routledge, Kegan Paul.

Birkhoff, G. D. 1923. *Relativity and Modern Physics*. Cambridge, MA: Harvard University Press.

Bondi, H. 1960. *Cosmology*, 2nd ed. Cambridge: Cambridge University Press.

Borel, A. 1986. "Hermann Weyl and Lie Groups," in K. Chandrasekharan (ed.), *Hermann Weyl 1885–1955, Centenary Lectures Delivered at the ETH Zürich*. Berlin: Springer, 53–82.

Born, M. 1969. *Albert Einstein-Max Born Briefwechsel 1916–1955*. München: Nyphenburger Verlag. Trans. by I. Born as *The Born-Einstein Letters*. New York: Walker, 1971.

Brading, K. A. 2002. "Which Symmetry? Noether, Weyl, and Conservation of Electric Charge," *Studies in History and Philosophy of Modern Physics* 33, 3–22.

Brading, K. A., and E. Castellani. 2003. *Symmetries in Physics: Philosophical Reflections*. Cambridge: Cambridge University Press.

Brading, K. A., and T. A. Ryckman. (forthcoming). "Causality as a Condition of Possible Experience: Hilbert and the Philosophical Background of Noether's Theorems."

Braithwaite, R. 1929. "Professor Eddington's Gifford Lectures," *Mind* n.s. 38, 409–435.

Bridgman, P. 1927. *The Logic of Modern Physics*. New York: Macmillan.

———. 1936. *The Nature of Physical Theory*. Princeton, NJ: Princeton University Press. Pagination according to repr. ed., New York: John Wiley and Sons, 1964.

Brown, H. R., and Brading, K. A. 2002. "General Covariance from the Perspective of Noether's Theorems," *Diálogos* 79, 59–86.

Brown, H. R., and Sypel, R. 1995. "On the Meaning of the Relativity Principle and Other Symmetries," *International Studies in the Philosophy of Science* 9, 235–253.

Campbell, N. 1931. "The Errors of Sir Arthur Eddington," *Philosophy* 6, 180–192.

Cao, T. Y. 1997. *Conceptual Development of Twentieth Century Field Theories*. New York: Cambridge University Press.

Cao, T. Y. (ed.) 1999. *Conceptual Foundations of Quantum Field Theory*. New York: Cambridge University Press.

Carnap, R. 1922. *Der Raum: Ein Beitrag zur Wissenschaftlehre. Kant-Studien Ergängzunghefte*, 56. Berlin: Verlag von Reuther and Reichard.

———. 1928. *Der Logische Aufbau der Welt*. Berlin: Weltkreis Verlag, 1928. Repr. ed., Hamburg: Felix Meiner, 1998. Trans. by R. A. George as *The Logical Structure of the World*. Berkeley, CA: University of California Press, 1969.

Carrier, M. 1990. "Constructing or Completing Physical Geometry? On the Relation between Theory and Evidence in Accounts of Space-Time Structure," *Philosophy of Science*, 57, 369–394.

Cartan, É. 1923. "*Sur les variétés à connexion affine et la théorie de la relativité généralisée*," *Annales de l'Ecole Normale Supérieure* 40, 352–412. Trans. by A. Magnon and A. Ashtekar as *On Manifolds with an Affine Connection and the Theory of General Relativity*. Napoli: Bibliopolis, 1986.

———. 1931. *Notice sur les travaux scientifiques*. Paris. Pagination according to repr. ed., Paris, Gauthier-Villars, 1974.

Cassirer, E. 1902. *Leibniz' System in seinen wissenschaftlichen Grundlagen*. Marburg: N. G. Elwert.

———. 1906. *Das Erkenntnisproblem in der Philosophie und Wissenschaft der neureren Zeit*, Bd. 1. Berlin: Verlag Bruno Cassirer.

———. 1907. *Das Erkenntnisproblem in der Philosophie und Wissenschaft der neureren Zeit*, Bd. 2. Berlin: Verlag Bruno Cassirer. Pagination according to *Dritte Auflage*, 1922.

———. 1910. *Substanzbegriff und Funktionsbegriff: Untersuchungen über die Grundfragen der Erkenntniskritik*. Berlin: Bruno Cassirer. Trans. by W. Swabey and M. Swabey in *Substance and Function and Einstein's Theory of Relativity*, 1923. Repr. ed., New York: Dover, 1953, xii, 3–346.

———. 1913. "*Erkenntnistheorie nebst den Grenzfragen der Logik*," *Jahrbücher der Philosophie* I, 1–59. Pagination according to repr. in Cassirer (1993), 3–76.

———. 1918. *Kants Leben und Lehre*, in E. Cassirer (hrsg.), *Immanuel Kants Werke*, Bd. XI, *Ergänzungsband*. Berlin: Bruno Cassirer. Trans. by J. Haden as *Kant's Life and Thought*. New Haven, CT: Yale University Press, 1981.

———. 1920. *Das Erkenntnisproblem in der Philosophie und Wissenschaft der neureren Zeit*, Bd. 3. Berlin: Bruno Cassirer.

———. 1921. *Zur Einsteinschen Relativitätstheorie: Erkenntnistheoretische Betrachtungen*. Berlin: Bruno Cassirer. Pagination as repr. in Cassirer (1957), 1–125. Trans. by W. Swabey and M. Swabey in *Substance and Function and Einstein's Theory of Relativity*, 1923. Repr. ed., New York: Dover, 1953, 347–460.

———. 1927. "*Erkenntnistheorie nebst den Grenzfragen der Logik und Denkpsychologie*', *Jahrbücher der Philosophie* 3, 31–92. Pagination according to repr. in Cassirer (1993), 77–153.

———. 1929. *Philosophie der Symbolische Formen*, Bd. 3. Berlin: Bruno Cassirer. Trans. by R. Manheim as *The Philosophy of Symbolic Forms*, vol. 3. New Haven, CT: Yale University Press, 1957.

———. 1931. "*Kant und das Problem der Metaphysik, Bemerkungen zu Martin Heideggers Kant-Interpretation*," Kant-Studien 36, 1–26.

———. 1937. *Determinismus und Indeterminismus in der moderne Physik: Historische und Systematische Studien zum Kausalproblem*. Göteborg: Wettergren and Kerbers. Pagination according to repr. in Cassirer (1957), 127–397.

———. 1939. "*Was Ist 'Subjektivismus'?*" *Theoria* V, 111–140. Pagination according to repr. in Cassirer (1993), 199–230.

———. 1957. *Zur modernen Physik*. Darmstadt: Wissenschaftliche Buchgesellschaft. Repr. ed., 1994.

———. 1993. *Erkenntnis, Begriff, Kultur*. R. A. Bast (hrsg.). Hamburg: Felix Meiner.

———. 1999. "*Ziele und Wege der Wirklichkeitserkenntnis*," Unpublished manuscript ca. 1936–1937, in K. C. Köhnke and J. M. Krois (eds.) *Ernst Cassirer: Nachgelassene Manuskripte und Texte*, Bd. 2. Hamburg: Felix Meiner, 118.

Chandrasekhar, S. 1935. "The Highly Collapsed Configurations of a Stellar Mass (Second Paper)," *Monthly Notices of the Royal Astronomical Society* 95, 207–225.

————. 1982. *Eddington: The Most Distinguished Astrophysicist of His Time*, and *Eddington: The Expositor and the Exponent of General Relativity* (Arthur Stanley Eddington Lectures, 1982). Cambridge: Cambridge University Press. Pagination as repr. in *Truth and Beauty: Aesthetics and Motivations of Science*. Chicago: University of Chicago Press, 1987.

Chern, S. S. 1996. "Finsler Geometry is Just Riemannian Geometry Without the Quadratic Restriction," *Notices of the American Mathematical Society* 43, 959–963.

Coolidge, J. 1940. *A History of Geometrical Methods*. Oxford: Clarendon Press.

Ciufolini, I., and J. Wheeler. 1995. *Gravitation and Inertia*. Princeton, NJ: Princeton University Press.

Clark, R. 1971. *Einstein: The Life and Times*. New York: World Publishing Company.

Clifford, W. K. 1870. "On the Space-Theory of Matter," *Proceedings of the Cambridge Philosophical Society* 2, 157–158.

————. 1875. "The Unseen Universe," *The Fortnightly Review* XVII (1 June 1875), 776–793.

————. 1878. "On the Nature of Things-in-Themselves," *Mind* 3, 57–67. Pagination as repr. in L. Stephen and F. Pollock (eds.), *Lectures and Essays*, vol. 2. London: Macmillan, 1879, 71–88.

Coffa, J. A. 1979. "Elective Affinities: Weyl and Reichenbach," in W. Salmon (ed.), *Hans Reichenbach: Logical Empiricist*. Dordrecht: D. Reidel, 267–304.

————. 1991. *The Semantic Tradition from Kant to Carnap: To the Vienna Station*. Ed. by L. Wessels. Cambridge: Cambridge University Press.

Cohen, H. 1902. *Logik der reinen Erkenntnis*. Berlin: Bruno Cassirer. Pagination according to *Dritte Auflage*, 1922.

Coleman, R., and H. Korté 2001. "Hermann Weyl: Mathematician, Physicist, Philosopher," in E. Scholz (ed.), *Hermann Weyl's Raum-Zeit-Materie and a General Introduction to His Scientific Work*. Basel: Birkhäuser, 157–386.

Davis, W. R. 1970. *Classical Fields, Particles, and the Theory of Relativity*. New York: Gordon and Breach.

Demopoulos, W., and M. Friedman. 1985. "Critical Notice: Bertrand Russell's *The Analysis of Matter*: Its Historical Context and Contemporary Interest," *Philosophy of Science* 52, 621–639.

Deppert, W., and K. Hübner (eds.) 1988. *Exact Sciences and their Philosophical Foundations: Proceedings of the International Hermann Weyl Congress Kiel 1985*. Frankfurt: Peter Lang.

De Sabbata, V., and E. Schmutzer (eds.). 1983. *Unified Field Theories of More than 4 Dimensions*. Singapore: World Scientific.

Dingle, H. 1937. "Modern Aristotelianism," *Nature* 140, 784–786.

————. 1954. *The Sources of Eddington's Philosophy*. Cambridge: Cambridge University Press.

Dingler, H. 1920. "Kritische Bemerkungen zu den Grundlagen der Relativitätstheorie," *Physikalische Zeitschrift* 21, 668–675.

Dirac, P. A. M. 1931. "Quantized Singularities in the Electromagnetic Field," *Proceedings of the Royal Society of London* A133, 60–72.

————. 1977. "Recollections of an Exciting Era," in C. Weiner (ed.), *Proceedings of the International School of Physics "Enrico Fermi,"* (Course LVII, History of Twentieth Century Physics). New York: Academic Press, 109–146.

————. 1982. "Early Years of Relativity," in G. Holton and Y. Elkana (eds.), *Albert Einstein: Historical Perspectives*. Princeton, NJ: Princeton University Press, 79–90.

DiSalle, R. 1993. "Helmholtz's Empiricist Philosophy of Mathematics: Between Laws of Perception and Laws of Nature," in D. Cahan (ed.), *Hermann von Helmholtz and the Foundations of Nineteenth-Century Science*. Berkeley, CA: University of California Press, 498–521.

Douglas, A. V. 1956a. "Forty Minutes with Einstein," *Journal of the Royal Astronomical Society of Canada* 50, 99–102.

———. 1956b. *The Life of Arthur Stanley Eddington*. London: Thomas Nelson and Sons.

Earman, J. 1989. *World Enough and Space-Time*. Cambridge, MA: MIT Press.

———. 2001. "Lambda: The Constant That Refuses to Die," *Archive for History of Exact Science* 55, 189–220.

———. 2003. "Once More General Covariance," unpublished manuscript.

Earman, J., and J. Eisenstaedt. 1999. "Einstein and Singularities," *Studies in the History and Philosophy of Modern Physics* 30B, 185–235.

Earman, J., and C. Glymour. 1978. "Lost in the Tensors: Einstein's Struggles with Covariance Principles, 1912–1916," *Studies in the History and Philosophy of Science* 9, 251–278.

Eddington, A. S. 1916. "Gravitation and the Principle of Relativity," *Nature* 98, 328–330.

———. 1918. *Report on the Relativity Theory of Gravitation*. London: Physical Society of London.

———. 1920a. *Report on the Relativity Theory of Gravitation*, 2nd ed. London: Fleetway Press.

———. 1920b. *Space, Time and Gravitation*. Cambridge: Cambridge University Press.

———. 1920c. "The Meaning of Matter and the Laws of Nature According to the Theory of Relativity," *Mind* n.s. 29, 145–158.

———. 1920d. "The Philosophical Aspect of the Theory of Relativity," *Mind* n.s. 29, 413–422.

———. 1921a. "A Generalization of Weyl's Theory of the Electromagnetic and Gravitational Fields," *Proceedings of the Royal Society of London* A99, 104–122.

———. 1921b. "The Relativity of Field and Matter," *Philosophical Magazine* 42, 800–806.

———. 1922. "The General Theory of Relativity," *Nature*, 109, 634–636.

———. 1923a. *The Mathematical Theory of Relativity*. Cambridge: Cambridge University Press.

———. 1923b. "Can Gravitation Be Explained?" *Scientia* 33, 315–324.

———. 1924. *The Mathematical Theory of Relativity*, 2nd ed. Cambridge: Cambridge University Press.

———. 1925a. "The Domain of Physical Science," in J. Needham (ed.), *Science, Religion and Reality*. New York: Macmillan, 187–218.

———. 1925b. *Relativitätstheorie in mathematischer Behandlung*. Berlin: J. Springer. German trans. of (1924).

———. 1926. "Universe: Electromagnetic-Gravitational Schemes," *Encyclopedia Britannica*, 13th ed. London, 907–908.

———. 1928. *The Nature of the Physical World*. New York: Macmillan.

———. 1930. *Science and the Unseen World*. New York: Macmillan.

———. 1933. *The Expanding Universe*. Cambridge: Cambridge University Press. Pagination according to repr. ed., 1987.

———. 1935a. *New Pathways in Science*. Cambridge: Cambridge University Press. Pagination according to repr. ed., Ann Arbor: University of Michigan Press, 1959.

———. 1935b. "On 'Relativistic Degeneracy,'" *Monthly Notices of the Royal Astronomical Society* 95, 194–206.

———. 1936. *The Relativity Theory of Protons and Electrons*. Cambridge: Cambridge University Press.

———. 1937. "The Reign of Relativity: 1915–1937." *Haldane Memorial Lecture*. London: Birkbeck College, University of London.

———. 1939a. *The Philosophy of Physical Science*. Cambridge: Cambridge University Press, and New York: Macmillan.

————. 1939b. "Lorentz-Invariance in Quantum Theory," *Proceedings of the Cambridge Philosophical Society* 35, 186–194.

————. 1942. "Lorentz-Invariance in Quantum Theory 2," *Proceedings of the Cambridge Philosophical Society* 38, 201–209.

————. 1946. *Fundamental Theory.* Cambridge: Cambridge University Press. Posthumously published, ed. by E. T. Whittaker.

Ehlers, J. 1973. "Survey of General Relativity," in W. Israel (ed.), *Relativity, Astrophysics and Cosmology.* Dordrecht: D. Reidel, 1–125.

————. 1988a. "Hermann Weyl's Contributions to the General Theory of Relativity," in Deppert and Hübner (1988), 83–105.

————. 1988b. *"Einführung in die Raum-Zeit-Struktur mittels Lichstrahlen und Teilchen,"* in J. Audretsch and K. Mainzer (eds.), *Philosophie und Physik der Raum-Zeit.* Mannheim: Bibliographische Institut, 145–162.

————. 1995. "Machian Ideas and General Relativity," in J. Barbour and H. Pfister (eds.), *Mach's Principle: From Newton's Bucket to Quantum Gravity.* Basel: Birkhäuser, 458–473.

Ehlers, J., F. Pirani, and A. Schild. 1972. "The Geometry of Free Fall and Light Propagation," in L. O'Raifeartaigh (ed.), *General Relativity: Papers in Honour of J. L. Synge.* Oxford: Clarendon Press, 63–84.

Einstein, A. 1911. *Über den Einfluß der Schwerkraft auf die Ausbreitung des Lichtes,"* *Annalen der Physik* 35, 898–908; repr. in Einstein (1993), 486–497. Trans. by W. Perrett and G. B. Jeffrey in Lorentz et al. (1923), 97–108.

————. 1914. *"Die formale Grundlage der allgemeinen Relativitätstheorie,"* *Königlich Preußische Akademie der Wissenschaften (Berlin) Sitzungsberichte. Physikalisch-Mathematische Klasse,* 1030–1085. Repr. in Einstein (1996), 72–128.

————. 1916a. *"Die Grundlage der allgemeinen Relativitätstheorie,"* *Annalen der Physik* 49, 769–822. Also issued as a *separatim, Die Grundlage der allgemeinen Relativitätstheorie.* Leipzig: J. Barth. Repr. in H. Lorentz et al. (1923), 72–124, and in Einstein, 1996, 284–337. Trans. by W. Perrett and G. B. Jeffrey as "The Foundation of the General Theory of Relativity," in Engl. trans. of Lorentz et al. (1923), 111–164.

————. 1916b. *"Hamiltonisches Prinzip und allgemeine Relativitätstheorie,"* *Preußische Akademie der Wissenschaflen (Berlin) Sitzungberichte. Physikalisch-Mathematische Klasse.* 1111–1116. Repr. in Lorentz et al. (1923), 125–129, and in Einstein (1996), 409–415. Trans. by W. Perrett and G. B. Jeffrey as "Hamilton's Principle and the General Theory of Relativity," in Engl. trans. of Lorentz et al. (1923), 165–173.

————. 1917a. *Über die spezielle und die allgemeine Relativitätstheorie.* Braunschweig: Vieweg. 23rd ed., 1988. Trans. by R. W. Lawson as Einstein (1961).

————. 1917b. "Kosmologische Betrachtungen zur allgemeinen Relativitäts theorie," *Preußische Akademie der Wissenschaflen (Berlin) Sitzungberichte. Physikalisch-Mathematische Klasse.* 142–152. Repr. in Lorentz et al. (1923), 130–139, and in Einstein (1996), 541–552. Trans. by W. Perrett and G. B. Jeffrey in Engl. trans. of Lorentz et al. (1923), 175–188.

————. 1918. *"Prinzipielles zur allgemeinen Relativitätstheorie,"* *Annalen der Physik* 55, 241–244. Repr. in Einstein (2002), 37–44.

————. 1919. *"Spielen Gravitationsfelder im Aufbau der materiellen Elementarteilchen eine wesentliche Rolle?"* *Preußische Akademie der Wissenschaflen (Berlin) Sitzungberichte. Physikalisch-Mathematische Klasse,* 349–356. Repr. in H. Lorentz et al. (1923), 140–146, and in Einstein (2002), 130–140. Trans. by W. Perrett and G. B. Jeffrey," Do Gravitational Fields Play an Essential Role in the Structure of the Elementary Particles of Matter?" in Engl. trans. of Lorentz et al. (1923), 189–198.

————. 1920. *"Bemerkungen,"* *Physikalische Zeitschrift* 21, 662.

————. 1921a. *"Geometrie und Erfahrung,"* *Preußische Akademie der Wissenschaften* (*Berlin*) *Sitzungberichte. Physikalisch-Mathematische Klasse.* 123–130. Issued separately in expanded form, Berlin: J. Springer. Repr. in Einstein (2002), 382–405. Expanded form trans. by W. Perrett and G. B. Jeffery as "Geometry and Experience," in *Sidelights on Relativity.* New York: E. P. Dutton, 1923, 235–256. Pagination according to repr., New York: Dover, 1983, 27–56.

————. 1921b. *"Über eine naheliegende Ergänzung des Fundamentes der allgemeinen Relativitätstheorie,"* *Preußische Akademie der Wissenschaften* (*Berlin*) *Sitzungberichte. Physikalisch-Mathematische Klasse,* 261–264. Repr. in Einstein (2002), 411–416.

————. 1922. *Vier Vorlesungen über Relativitätstheorie.* Braunschweig: Vieweg. Pagination according to 6th ed titled *Grundzüge der Relativitätstheorie,* 1990. Trans. by E. P. Adams as *The Meaning of Relativity.* Princeton, NJ: Princeton University Press, 1922. Pagination according to 5th ed., 1956.

————. 1923a. *"Zur allgemeinen Relataivitätstheorie,"* *Preußische Akademie der Wissenschaften* (*Berlin*) *Sitzungberichte. Physikalisch-Mathematische Klasse,* 32–38.

————. 1923b. *"Bemerkung zur meiner Arbeit 'Zur allgemeinen Relataivitätstheorie',"* *Preußische Akademie der Wissenschaften* (*Berlin*) *Sitzungberichte. Physikalisch-Mathematische Klasse,* 76–77.

————. 1923c. *"Zur affinen Feldtheorie,"* *Preußische Akademie der Wissenschaften* (*Berlin*) *Sitzungberichte. Physikalisch-Mathematische Klasse,* 137–140.

————. 1923d. "The Theory of the Affine Field," *Nature* 112, 448–449.

————. 1923e. *"Grundgedanken und Probleme der Relativitätstheorie,"* Lecture of 11 July 1923. Stockholm: Imprimerie Royale. Trans. as "Fundamental Ideas and Problems of the Theory of Relativity," in *Nobel Lectures—Physics, 1901–1921.* Amsterdam: Elsevier, 1967, 482–490.

————. 1924. *"Über den Äther,"* *Schweizerische naturforschende Gesellschaft, Verhandlungen* 105, 85–93. Trans. by S. W. Saunders as "On the Ether," in Saunders and Brown (1991), 13–20.

————. 1925a. *"Nichteuklidische Geometrie und Physik,"* *Die Neue Rundschau* 36, 16–20.

————. 1925b. *"Eddingtons Theorie und Hamiltonisches Prinzip,"* *Anhang* (appendix) to Eddington (1925b), 367–371.

————. 1925c. *"Einheitliche Feldtheorie von Gravitation und Elektrizität,"* *Akademie der Wissenschaften* (*Berlin*) *Sitzungberichte. Physikalisch-Mathematische Klasse,* 414–419.

————. 1925d. *"Die Relativitätstheorie,"* in E. Lecher (ed.), *Die Kultur der Gegenwart: Ihre Entwicklung und ihre Ziele,* 2nd rev. ed. Leipzig: Teubner, 784–797.

————. 1927. *"Zu Kaluzas Theorie des Zusammenhanges von Gravitation und Elektrizität,"* *Akademie der Wissenschaften* (*Berlin*) *Sitzungberichte. Physikalisch-Mathematische Klasse,* 26–30.

————. 1928. *"A propos de La Déduction relativiste de M. É. Meyerson,"* *Revue philosophique de las France et de l'étranger* 105, 161–166. Pagination according to trans. of original, incompletely published, German text by D. and M. Sipfle in the 1985 English trans. of Meyerson (1925), 252–256

————. 1929. *"Zur einheitliche Feldtheorie,"* *Akademie der Wissenschaften* (*Berlin*) *Sitzungberichte. Physikalisch-Mathematische Klasse,* 2–7.

————. 1932. *"Der gegenwärtige Stand der Relativitatstheorie,"* *Die Quelle Pedagogischer Führer* 82, 440–442.

————. 1933. *"Zur Methodik der theoretischen Physik,"* The Herbert Spencer Lecture, delivered at Oxford, 10 June 1933. In C. Seelig (hrsg.), *Mein Weltbild.* Amsterdam: Querido Verlag, 1934. Repr. ed., Frankfurt: Verlag Ullstein, 1988, 113–119. Trans. by S. Bargmann as "On the Method of Theoretical Physics," in Einstein (1983), 270–276.

————. 1936. "*Physik und Realität*," *Journal of the Franklin Institute* 221, 313–337. Trans. by S. Bargmann as "Physics and Reality," in Einstein (1983), 290–323.

————. 1949. "Autobiographical Notes," and "Replies to Criticism," in P. A. Schilpp (1949), 2–97, 663–688.

————. 1950. "On the Generalized Theory of Gravitation," *Scientific American* 184, no. 4, April 1950. Pagination as repr. in Einstein (1983), 341–356.

————. 1952. "Relativity and the Problem of Space," Pagination as repr. in Einstein (1961), appx. 5. New York: Crown Publishers, 135–157.

————. 1953. "Relativistic Theory of the Non-symmetrical Field," in Einstein (1956), 133–166.

————. 1956. *The Meaning of Relativity*. 5th ed. Trans. by E. P. Adams of Einstein (1922). Princeton, NJ: Princeton University Press.

————. 1961. *Relativity: The Special and General Theory*. Repr. 15th ed. New York: Crown, 1988. Trans. by R. W. Lawson of Einstein (1917).

————. 1983. *Ideas and Opinions*. New York: Crown.

————. 1993. *The Collected Papers of Albert Einstein*, vol. 3. M. Klein, A. Kox, J. Renn, and R. Schulmann (eds.). Princeton, NJ: Princeton University Press

————. 1996. *The Collected Papers of Albert Einstein*, vol. 6. A. Kox, M. Klein, and R. Schulmann (eds.). Princeton: Princeton University Press.

————. 1998. *The Collected Papers of Albert Einstein*, vol. 8, pts. A and B. R. Schulmann, A. Kox, M. Janssen, and J. Illy (eds.). Princeton, NJ: Princeton University Press.

————. 2002. *The Collected Papers of Albert Einstein*, vol. 7. M. Janssen, R. Schulmann, J. Illy, C. Lehner, and D. K. Buchwald (eds.). Princeton, NJ: Princeton University Press.

Einstein, A., and J. Grommer. 1923. "*Beweis der Nichtexistenz eines überall regulären zentrisch symmetrischen Felds der Feld-Theorie von Th. Kaluza*," *Scripta Universitatis atque Bibliothecae Hierosolymitanarum, Mathematica et Physica (Jerusalem)* 1, no. 7.

Eisenhart, L. P., and O. Veblen. 1922. "The Riemannian Geometry and its Generalization," *Proceedings of the National Academy of Sciences* 8, 19–23.

Emmet, D. 1945. *The Nature of Metaphysical Thinking*. London: Macmillan.

Feferman, S. 1988. "Weyl Vindicated: *Das Kontinuum* 70 years later," in *Atti del Congresso Temi e prospettive della logica e della filosofia della sciensa contemporanee, Casena 7–10 Gennaio, 1987*, t. I. Bologna: CLUEB, 59–93. Repr. in Feferman (1998), 249–283.

————. 1998. *In the Light of Logic*. New York: Oxford University Press.

Feigl, H. 1937–1938. "Moritz Schlick," *Erkenntuis* 7, 393–419. Trans. by P. Heath as "Moritz Schlick, A Memoir" in Schlick (1979a), xv–xxxviii.

————. 1956. "Some Major Issues and Developments in the Philosophy of Science of Logical Empricism," in H. Feigl and M. Scriven (eds.), *The Foundations of Science and the Concepts of Psychology and Psychoanalysis*. Minnesota Studies in the Philosophy of Science, vol. 1. Minneapolis: University of Minnesota Press, 3–37.

————. 1969. "The Origin and Spirit of Logical Positivism," in P. Achinstein and S. F. Barker (eds.), *The Legacy of Logical Positivism*. Baltimore: The Johns Hopkins University Press, 3–24.

————. 1975. "Russell and Schlick: A Remarkable Agreement on a Monistic Solution of the Mind-Body Problem," *Erkenntnis* 9, 11–34.

Feigl, H., and Brodbeck, M. (eds.). 1953. *Readings in the Philosophy of Science*. New York: Appleton-Century-Crofts, Inc.

Fine, A. 1996. *The Shaky Game: Einstein, Realism, and the Quantum Theory*. 2nd ed. Chicago: University of Chicago Press.

Fock, V. 1964. *The Theory of Space, Time and Gravitation*, 2nd rev. ed. New York: Pergamon Press.

Fölsing, A. 1997. *Albert Einstein*. Engl. trans. by E. Osers. New York: Viking.

Frank, P. 1917. *"Die Bedeutung der physikalsichen Erkenntnistheorie Machs für das Geistesleben der Gegenwart,"* *Die Naturwissenschaften* 5, 65–72. Trans. as "The Importance for our Times of Ernst Mach's Philosophy of Science," in Frank (1949a), 61–78.

———. 1934. *"La physique contemporaine manifest-t-elle une tendence a réintégrer un elément psychique?"* *Revue de Synthése* 8, 133–154; trans. as "Is There a Trend Today Towards Idealism in Physics?" in Frank (1949a), 122–137.

———. 1949a. *Modern Science and Its Philosophy*. Cambridge, MA: Harvard University Press.

———. 1949b. "Einstein, Mach, and Logical Positivism," in Schilpp (1949), 271–286.

Frankel, T. 1998. *The Geometry of Physics*. New York: Cambridge University Press.

French, A. P. 1968. *Special Relativity*. New York: W. W. Norton.

———. 1979. *Einstein: A Centenary Volume*. Cambridge, MA: Harvard University Press.

French, S. 2003. "Scribbling on the Blank Sheet: Eddington's Structuralist Conception of Objects," *Studies in History and Philosophy of Modern Physics* 34, 227–259.

Friedman, M. 1983. *Foundations of Space-Time Theories: Relativistic Physics and Philosophy of Science*. Princeton, NJ: Princeton University Press.

———. 1997. "Helmholtz's *Zeichentheorie* and Schlick's *Allgemeine Erkenntnislehre*: Early Logical Empiricism and Its Nineteenth-Century Background," *Philosophical Topics* 25, 19–50

———. 1999. *Reconsidering Logical Positivism*. Cambridge: Cambridge University Press.

———. 2000a. *A Parting of the Ways: Carnap, Cassirer, and Heidegger*. Chicago: Open Court.

———. 2000b. "Transcendental Philosophy and A Priori Knowledge: A Neo-Kantian Perspective," in P. Boghossian and C. Peacocke (eds.), *New Essays on the A Priori*. Oxford: Clarendon Press, 367–383.

———. 2001. *Dynamics of Reason*. Stanford, CA: CSLI Publications.

———. 2002. "Geometry as a Branch of Physics: Background and Context for Einstein's 'Geometry and Experience'," in D. B. Malament (ed.), *Reading Natural Philosophy: Essays in the History and Philosophy of Science and Mathematics*. Chicago: Open Court, 193–229.

Geroch, R. 1966. "Electromagnetism as an Aspect of Geometry? Already Unified Field Theory—The Null Field Case," *Annals of Physics* 36, 147–187.

———. 1978. *General Relativity from A to B*. Chicago: University of Chicago Press.

Geroch, R., and P. S. Jang. 1975. "Motion of a Body in General Relativity," *Journal of Mathematical Physics* 16, 65–67.

Ghins, M., and Budden, T. 2001. "The Principle of Equivalence," *Studies in History and Philosophy of Modern Physics* 32, 33–51.

Göckeler, M., and Schücker, T. 1987. *Differential Geometry, Gauge Theories, and Gravity*. Cambridge: Cambridge University Press.

Gross, D. 1999. "The triumph and the limitations of quantum field theory," in Cao, (1999), 56–67.

Grünbaum, A. 1973. *Philosophical Problems of Space and Time*. Dordrecht: D. Reidel.

Guyer, P. 1989. "The Rehabilitation of Transcendental Idealism?" in E. Schaper and W. Vossenkuhl (eds.), *Reading Kant*. Oxford: Basil Blackwell, 140–167.

Haag, R. 1992. *Local Quantum Physics: Fields Particles, Algebras*. Berlin: Springer Verlag.

Haas, A. 1920. "Die Physik als geometrische Notwendigkeit," *Die Naturwissenschaften* 8, 121–127.

Haldane, R. B. (Viscount). 1921. *The Reign of Relativity*. New Haven, CT: Yale University Press.

Hanna, R. 2001. *Kant and the Foundations of Analytic Philosophy.* Oxford: Clarendon Press.

Harward, A. 1922. "The Identical Relations in Einstein's Theory," *Philosophical Magazine* 44, 380–382.

Hass, A. 1920. "*Die Physik als geometrische Notwendigkeit,*" *Die Naturwissenschaften* 8, 121–127.

Havas, P. 1989. "The Early History of the 'Problem of Motion' in General Relativity," in Howard and Stachel (1989), 234–276.

Hawking, S., and G. Ellis. 1973. *The Large Scale Structure of Space-Time.* Cambridge: Cambridge University Press.

Hawkins, T. 2000. *Emergence of the Theory of Lie Groups: An Essay in the History of Mathematics 1869–1926.* New York: Springer Verlag.

Helmholtz, H. von. 1866. "*Über die tatsächliche Grundlagen der Geometrie,*" *Verhandlungen des naturhistorisch-medicinischen Vereins zu Heidelberg* 4, 197–202. Repr. in *Wissenschaftliche Abhandlungen,* Bd. 2. Leipzig: J. A. Barth, 1883, 610–617.

———. 1868. "*Über die Tatschen, die der Geometrie zum Grunde liegen,*" *Nachrichten von der Königliche Gesellschaft der Wissenschaften und der Georg-Augusts-Universität aus dem Jahre* 1868 no. 9, 193–221. Repr. in *Wissenschaftliche Abhandlungen,* Bd. 2. Leipzig: J. A. Barth, 1883, 618–639, and with "*Erläuterungen*" by P. Hertz in (1921), 38–69. Trans. of the latter by M. F. Lowe as "On the Facts Underlying Geometry" in (1977), 39–71.

———. 1870. "*Über die Ursprung und die Bedeutung der geometrischen Axiome,*" *Vorträge und Reden,* Bd. 2. *Dritte Auflage.* Braunschweig: Friedrich Vieweg und Sohn, 1884, 1–31. Repr. with "*Erläuterungen*" by M. Schlick in (1921), 1–37. Trans. of the latter by M. F. Lowe as "On the Origin and Significance of the Axioms of Geometry" in (1977), 1–38.

———. 1878. "*Die Tatsachen in der Wahrnehmung,*" *Vorträge und Reden,* Bd. 2. *Dritte Auflage.* Braunschweig: Friedrich Vieweg und Sohn, 1884, 215–247, 387–406. Repr. with "*Erläuterungen*" by M. Schlick in (1921), 109–175. Trans. of the latter by M. F. Lowe as "The Facts in Perception" in (1977), 115–185.

———. 1921. *Schriften zur Erkenntnistheorie.* P. Herz and M. Schlick (eds.). Berlin: J. Springer. Trans. by M. Lowe in (1977).

———. 1977. *Hermann von Helmholtz: Epistemological Writings.* R. S. Cohen and Y. Elkana (eds.). Trans. of (1921) by M. F. Lowe. Boston Studies in the Philosophy of Science, vol. 37. Dordrecht: D. Reidel.

Hempel, C. G. 2000. *Selected Philosophical Essays.* Ed. by R. Jeffrey. New York: Cambridge University Press.

Hentschel, K. 1986. "Die Korrespondenz Einstein-Schlick: Zum Verhältnis der Physik zur Philosophie," *Annals of Science* 43, 475–488.

———. 1990. *Interpretationen und Fehlinterpretationen der speziellen und der allgemeinen Relativitätstheorie durch Zeitgenossen Albert Einsteins.* Basel: Birkhäuser Verlag.

Hermann, A. (ed.). 1969. *Albert Einstein/Arnold Sommerfeld Briefwechsel.* Basel: Schwabe and Co.

Herneck, F. 1970. "*Moritz Schlick als Physiker,*" *Rostocker Philosophische Manuskripte,* Heft 8. Rostock, DDR: Universität Rostock.

———. 1976. *Einstein und sein Weltbild: Aufsätze und Vorträge.* Berlin: Buchverlag Der Morgen.

Hessenberg, G. 1917. "*Vektorielle Begründung der Differentialgeometrie,*" *Mathematische Annalen* 78, 187–217.

Hilbert, D. 1915. "*Die Grundlagen der Physik: Erste Mitteilung,*" *Nachrichten. Königliche Gesellschaft der Wissenschaften zu Göttingen. Mathematische-Physikalische Klasse,* 55–76.

———. 1917. "*Die Grundlagen der Physik: Zweite Mitteilung,*" Nachrichten. Königliche Gesellschaft der Wissenschaften zu Göttingen. Mathematische-Physikalische Klasse, 53–76.

Henrich, D. 1994. "On the Unity of Subjectivity," originally published in German in 1955. In *The Unity of Reason*, R. Velkley (ed.). Cambridge, MA: Harvard University Press, 17–54.

Hoddeson, L., M. Brown, M. Riordan, and M. Dresden (eds.). 1997. *The Rise of the Standard Model: Particle Physics in the 1960s and 1970s*. Cambridge: Cambridge University Press.

Hoefer, C. 2000. "Energy Conservation in GTR," *Studies in History and Philosophy of Modern Physics* 31, 187–226.

Hoffmann, B. 1972. *Albert Einstein: Creator and Rebel*. New York: Viking Press. With the assistance of H. Dukas.

Holzhey, H. 1992. "*Der Neukantianismus,*" in A. Hügli and P. Lübke (eds.), *Philosophie im 20. Jahrhundert*. Hamburg: Rowohlt-Verlag.

Howard, D. 1984. "Realism and Conventionalism in Einstein's Philosophy of Science: The Einstein-Schlick Correspondence," *Philosophia Naturalis* 21, 616–629.

———. 1990. "*Nicht Sein Kann Was Nicht Sein Darf,*" or, the Prehistory of EPR, 1909–1935: Einstein's Early Worries about the Quantum Mechanics of Composite Systems," in A. Miller (ed.), *Sixty-two Years of Uncertainty: Historical, Philosophical, Physical Inquiries into the Foundations of Quantum Physics*. New York: Plenum Press, 61–101.

———. 1994. "Einstein, Kant, and the Origins of Logical Empiricism," in W. Salmon and G. Wolters (eds.), *Language and the Structure of Scientific Theories*. Pittsburgh: University of Pittsburgh Press, 45–105.

Howard, D., and J. Stachel (eds.). 1989. *Einstein and the History of General Relativity*. Einstein Studies, vol. 1. Basel: Birkhäuser

Husserl, E. 1907. *Die Idee der Phänomenologie: Fünf Vorlesungen*, first published by W. Biemel (hrsg.). Haag: Martinus Nijhoff, 1950. 2 Auflage, 1958. Trans. by W. Alston and G. Naknnikian as *The Idea of Phenomenology*. The Hague: Martinus Nijhoff, 1964.

———. ca. 1908. "*Zur Auseinandersetzung meiner transzendentalen Phänomenologie mit Kants Transzendentalphilosophie,*" in E. Husserl (1956), 381–395.

———. 1910–1911. "*Philosophie als strenge Wissenschaft,*" Logos, I: 289–341. Trans. by Q. Lauer as "Philosophy as Rigorous Science," in E. Husserl, *Phenomenology and the Crisis of Philosophy*. New York: Harper and Row, 71–147.

———. 1913. *Ideen zu einer reinen Phänomenologischen Philosophie* (Jahrbuch für Philosophie und phänomenologische Forschung) Bd. I, Halle. Augmented text (with later insertions) repr. as Husserl (1950).

———. 1921. *Logische Untersuchungen*, Bd. 2, Teil 2. Zweite Auflage. Halle: Max Niemeyer. Trans. by J. N. Findlay as *Logical Investigations*, vol. 2, pt. 2. London: Routledge and Kegan Paul, 1970.

———. 1922. *Logische Untersuchungen*, Bd. I: *Prolegomena zur reinen Logik, Dritte unveränderte Auflage* (i.e., from the 2nd ed. of 1913). Halle: Max Niemeyer. Trans. by J. N. Findlay as *Logical Investigations*, vol. 1, *Prolegonema to Pure Logic*. London: Routledge and Kegan Paul, 1970.

———. 1931. "Author's Preface to the English Edition," in *Ideas: General Introduction to Pure Phenomenology*. Trans. of Husserl (1913) by W. Boyce Gibson. London: Allen and Unwin, 1931.

———. 1936–1937. *Die Krisis der europäischen Wissenschaften und die tranzendentale Phänomenologie. Eine Einleitung in die phänomenologische Philosophie*. W. Biemel (hrsg). 2. Auflage. Haag: Martins Nijhoff, 1962. Parts published in *Philosophia*,

1, 1936, 77–176. Trans. by D. Carr as *The Crisis of European Sciences and Tran-scendental Phenomenology*. Evanston, IL: Northwestern University Press.

———. 1950. *Ideen zu einer reinen Phänomenologie und Phänomenologischen Philosophie. Erstes Buch. Allgemeine Einführung in die Reine Phänomenologie. Neue, auf Grund der handschriften Zusätze des Verfassers erweiterte Auglage.* W. Biemel (hrsg.) Haag: Martinus Nijhoff. Trans. by F. Kersten as *Ideas Pertaining to a Pure Phenomenology and to a Phenomenological Philosophy. First Book.* The Hague: Martinus Nijhoff, 1983.

———. 1956. *Erste Philosophie (1923/1924). Erster Teil.* R. Boehm (hrsg.). The Hague: Martinus Nijhoff.

Ihmig, K.-N. 2001. *Grundzüge einer Philosophie der Wissenschaften bei Ernst Cassirer.* Darmstadt: Wissenschaftliche Buchgesellschaft.

Jeffreys, H. 1941. "Epistemology and Modern Physics," *Philosophical Magazine* 32, 177–205.

Joshi, P. 1993. *Global Aspects in Gravitation and Cosmology.* Oxford: Clarendon Press.

Kaku, M. 1993. *Quantum Field Theory.* New York: Oxford University Press.

Kaluza, T. 1921. "Zum Unitätsproblem der Physik," *Preußische Akademie der Wissenschaften (Berlin). Sitzungsberichte. Physikalisch-Mathematische Klasse,* 966–972. Trans. as "On the Problem of Unity in Physics," in De Sabbata and Schmutzer (1983), 427–433.

Kamlah, A. 1979. "*Erlauterungen zur 'Axiomatik der relativistischen Raum-Zeit-Lehre" bis §19,*" in A. Kamlah and M. Reichenbach (eds.), *Hans Reichenback: Gesammelte Werke,* vol. 3. Braunschweig: F. Vieweg and Sohn, 431–464.

Kant, I. 1781/1787. *Critique of Pure Reason.* First ("A") and second ("B") editions. Trans. and ed. by P. Guyer and A. Wood. Cambridge: Cambridge University Press, 1999.

Kasner, E. 1921. "Einstein's Theory of Gravitation: Determination of the Field by Light Signals," *American Journal of Mathematics* 43, 20–28.

Kerszberg, P. 1989. *The Invented Universe.* Oxford: Clarendon Press.

Kilmister, C. W. 1994. *Eddington's Search for a Fundamental Theory: A Key to the Universe.* Cambridge: Cambridge University Press.

Koenigsberger, L. 1906. *Hermann von Helmholtz.* Oxford: Oxford University Press. Trans. by F. Welby from original German ed., 1903. Repr. ed., New York: Dover, 1965.

Kretschmann, E. 1917. "*Über den physikalischen Sinn der Relativitätspostulate: A. Einstein's neue und seine ursprüngliche Relativitätstheorie,*" *Annalen der Physik* 53, 575–614.

Krois, J. M. 1987. *Cassirer: Symbolic Forms and History.* New Haven, CT: Yale University Press.

Ladyman, J. 1998. "What Is Structural Realism?" *Studies in History and Philosophy of Science* 29, 409–424.

Landau, L. D., and E. M. Lifshitz. 1975. *The Classical Theory of Fields.* 4th rev. Engl. ed. Oxford: Pergamon Press.

Lange, I. 1996. *Arnold Zweig, Beatrice Zweig, Helene Weyl: Komm her, Wir lieben Dich. Briefe einer ungewöhnlichen Freundschaft.* Berlin: Aufhan-Verlag.

Laugwitz, D. 1958. "*Über eine Vermutung von Hermann Weyl zum Raumproblem,*" *Archiv der Mathematik* 9, 128–133.

———. 1965. *Differential and Riemannian Geometry.* New York: Academic Press.

Levi-Civita, T. 1917. "*Nozione di parallelismo in una varietà qualungue e consequente specificazione geometrica della curvature Riemanniana,*" *Rendiconti del Circolo mathematico di Palermo* 43, 173–205.

Lie, S. 1890. *Theorie der Transformationsgruppen,* unter Mitwirkung von Dr. F. Engel, pt. 3. Repr. of Leipzig edition of 1888–1893. New York: Chelsea Publishing, 1970.

Lorentz, H., A. Einstein, H. Minkowski, and H. Weyl. 1923. *Das Relativitätsprinzip. Fünften Auflage.* Stuttgart: B. G. Teubner. Translated by W. Perrett and G. B. Jeffrey as *The Principle of Relativity*. London: Meuthen, 1923. Repr., New York, Dover, 1952.

Lovejoy, A. O. 1930. *The Revolt Against Dualism*. New York: W. W. Norton and Open Court Publishing Co.

McCrae, W. 1991. "Arthur Stanley Eddington," *Scientific American* June, 92–97.

Maidens, A. 1998. "Symmetry Groups, Absolute Objects and Action Principles in General Relativity," *Studies in the History and Philosophy of Modern Physics* 29, 245–272.

Majer, U. 1995. "Geometry, Intuition and Experience: From Kant to Husserl," *Erkenntnis* 42, 261–285.

Mancosu, P. 1998. *From Brouwer to Hilbert: The Debate on the Foundations of Mathematics in the 1920's*. New York and Oxford: Oxford University Press.

Mancosu, P., and T. A. Ryckman 2002. "Mathematics and Phenomenology: The Correspondence between O. Becker and H. Weyl," *Philosophia Mathematica* 10, 130–202.

———. 2004. "Geometry, Physics, and Phenomenology: Four Letters of O. Becker to H. Weyl," in V. Peckhaus (ed.), *Die Philosophie und die Mathematik: Oskar Becker in der mathematischen Grundlagendiskussion*. Munich: Wilhelm Fink Verlag, 71–136.

Matthews, H. 1969. "Strawson on Transcendental Idealism," *Philosophical Quarterly* 19, 204–220. Pagination as repr. in R.C.S. Walker (ed.), *Kant on Pure Reason* Oxford: Oxford University Press, 1982, 132–149.

Mehra, J., and H. Rechenberg. 1982. *The Historical Development of Quantum Theory*, vol. 2. New York: Springer-Verlag.

Merleau-Ponty, J. 1965. *Philosophie et théorie physique chez Eddington*. Paris: Belles-Lettres.

Meyerson, É. 1908. *Identité et Réalité*. Paris: Alcan. Trans. by K. Loewenberg as *Identity and Reality*. London: Allen and Unwin, 1930. Repr., New York, Dover, 1962.

———. 1921. *De l'Explication dans les Sciences*. Paris: Payot. Trans. by M. and D. Sipfle as *Explanation in the Sciences*. Dordrecht: Kluwer, 1991.

———. 1925. *La Déduction Relativiste*. Paris: Payot. Trans. by D. and M. Sipfle as *The Relativistic Deduction*. Dordrecht: D. Reidel, 1985.

Mielke, E., and F. Hehl. 1988. "*Die Entwicklung der Eichtheorien: Marginalien zu deren Wissenschaftsgeschichte*," in Deppert and Hübner (1988), 191–231.

Mills, R. 1989. "Gauge Fields," *American Journal of Physics* 57, 191–231.

Minkowski, H. 1908. "*Raum und Zeit*," with "*Anmerkungen*" by A. Sommerfeld, in Lorentz et al. (1923), 54–71. Trans. by W. Perrett and G. B. Jeffrey as "Space and Time," in Engl. trans. of Lorentz et al. (1923), 73–96.

Misner, C., and J. A. Wheeler. 1962. "Classical Physics as Geometry," in Wheeler (1962), 225–307.

Misner, C., K. Thorne, and J. A. Wheeler. 1973. *Gravitation*. San Francisco: W. A. Freeman.

Moriyasu, K. 1983. *An Elementary Primer for Gauge Theory*. Singapore: World Scientific.

Moynahan, G. 2003. "Hermann Cohen's *Das Prinzip der Infinitesimalmethode*, Ernst Cassirer, and the Politics of Science in Wilhelmine Germany," *Perspectives in Science*, 11, 35–75.

Müller, A. 1923. "*Der Sinn der physikalischen Axiomatik*," *Physikalische Zeitschrift* 24, 444–450.

Neurath, O. 1929. *Wissenschaftliche Weltauffassung—Der Wiener Kreis*. Co-authored with R. Carnap and H. Hahn. Wien: Veröffentlichungen des Vereins Ernst Mach. Pagination according to repr. in R. Hegselmann (hrsg.), *Otto Neurath Wissenschaftliche Weltauffassung, Sozialismus und Logischer Empirismus*. Frankfurt am Main: Suhrkamp, 1979, 81–101.

Newman, M.H.A. 1928. "Mr. Russell's Causal Theory of Perception," *Mind* n.s. 37, 137–148.

Norton, J. D. 1984. "How Einstein Found His Field Equations, 1912–1915," *Historical Studies in the Physical Sciences* 14, 253–316, pagination as repr. in Howard and Stachel (1989), 101–159.

———. 1985. "What Was Einstein's Principle of Equivalence?" *Studies in History and Philosophy of Science* 16; pagination as repr. in Howard and Stachel (1989), 5–47.

———. 1992a. "Einstein, Nordström and the Early Demise of Scalar, Lorentz-Covariant Theories of Gravitation," *Archive for History of Exact Science* 45, 17–94.

———. 1992b. "The Physical Content of General Covariance," in J. Eisenstaedt and A. Kox (eds.), *Studies in the History of General Relativity*. Basel, Boston, Berlin: Birkhäuser, 281–315.

———. 1993. "General Covariance and the Foundations of General Relativity: Eight Decades of Dispute," *Reports on Progress in Physics* 56, 791–858.

Ohanian, H. C. 1977. "What Is the Principle of Equivalence?" *American Journal of Physics* 45, 903–909.

Ohanian, H. C., and R. Ruffini. 1994. *Gravitation and Spacetime*, 2nd ed. New York: Norton.

O'Raifeartaigh, L. 1997. *The Dawning of Gauge Theory*. Princeton Series in Physics. Princeton, NJ: Princeton University Press.

O'Raifeartaigh, L., and N. Straumann. 2000. "Gauge Theory: Historical Origins and Some Modern Developments," *Reviews of Modern Physics* 72, 1–23.

Orth, E. and H. Holzhey 1995. *Neukantianismus: Perspektiven und Probleme*. Würzburg: Königshausen und Neumann GmbH.

Pais, A. 1982. *"Subtle Is the Lord...": The Science and the Life of Albert Einstein*. New York: Oxford University Press.

Pauli, W., Jr. 1919a. *"Merkurperihelbewegung und Stralenablenkung in Weyls Gravitationstheorie," Verhandlungen der Deutschen Physikalischen Gesellschaft* 21, 742–750.

———. 1919b. *"Zur Theorie der Gravitation und der Elektrizitätat von Hermann Weyl," Physikalischen Zeitschrift* 20, 457–467.

———. 1921. *"Relativitätstheorie," Encyklopädie der mathematischen Wissenschaften*, vol. 19. Leipzig: B. G. Teubner. Pagination according to trans. as *The Theory of Relativity*. New York: Pergamon Press, 1958.

———. 1926. Review of Eddington (1924), *Die Naturwissenschaften* 13, 273–274.

———. 1979. *Wolfgang Pauli Wissenschaftlicher Briefwechsel*, Bd. 1, 1919–29. A. Hermann (ed.). New York: Springer-Verlag.

Peckhaus, V. 1990. *Hilbertprogramm und Kritische Philosophie*. Göttingen: Vandenhoeck und Ruprecht.

Pedoe, D. 1988. *Geometry: A Comprehensive Course*. New York: Dover.

Philpse, H. 1995. "Transcendental Idealism," in Smith and Smith (1995), 239–322.

Planck, M. 1908. *"Die Einheit des physikalischen Weltbildes,"* as repr. in Planck (1934), 1–32.

———. 1910. *Acht Vorlesungen über Theoretische Physik*. Leipzig: S. Hirzel.

———. 1934. *Wege zur physikalischen Erkenntnis*. Leipzig: S. Hirzel.

Poincaré, H. 1902. *La Science et L'Hypothèse*. Paris: Flammarion. Repr., 1968. Trans. by G. B. Halsted as *Science and Hypothesis* in H. Poincaré, *The Foundations of Science*. New York: The Science Press, 1929.

———. 1906. *La Valeur de la Science*. Paris: Flammarion. Trans. by G. Halsted as *The Value of Science*. New York: Dover, 1958.

Prugovečki, E. 1995. *Principles of Quantum General Relativity*. Singapore: World Scientific Publishers.

Putnam, H. 1963. "An Examination of Grünbaum's Philosophy of Geometry," pagination as repr. in his *Philosophical Papers*, 2nd ed., vol. 1. New York: Cambridge University Press, 1979, 193–129.

Quine, W. V. 1995. *From Stimulus to Science*. Cambridge, MA: Harvard University Press.

Rainich, G. 1925. "Electrodynamics in the General Relativity Theory," *Transactions of the American Mathematical Society* 27, 106–136.

Redhead, M. 2002. "The Interpretation of Gauge Symmetry," in M. Kuhlmann, H. Lyre, and A. Wayne (eds.), *Ontological Aspects of Quantum Field Theory*. London: World Scientific, 281–301.

Reichenbach, H. 1916–1917. "*Der Begriff der Wahrscheinlichkeit für die mathematische Darstellung der Wirklichkeit*," *Zeitschrift für Philosophie und Philosophische Kritik* 161, 209–239 (pt. 1); 162, 98–112 (pt. 2); 222–239 (pt. 3); 163, 86–98 (pt. 4).

———. 1920. *Relativitätstheorie und Erkenntnis A Priori*. Berlin: J. Springer. Trans. by M. Reichenbach as *The Theory of Relativity and A Priori Knowledge*. Berkeley: University of California Press, 1965.

———. 1921. "*Erwiderung auf H. Dinglers Kritik an der Relativitätstheorie*," *Physicalische Zeitschrift* 22, 379–384.

———. 1922a. Review of Helmholtz 1921, *Zeitschrift für angewandte Psychologie* 20, 421–422.

———. 1922b. "*La Signification philosophique de la théorie de la relativité*," *Revue Philosophique* 94, 5–61.

———. 1922c. "*Der gegenwärtige Stand der Relativititätsdiskussion*," *Logos* 10, 316–378. Trans., with omissions, by M. Reichenbach as "The Present State of the Discussion on Relativity: A Critical Investigation," in (1978), 3–47.

———. 1924. *Axiomatik der relativistischen Raum-Zeit-Lehre*. Braunschweig: F. Vieweg, Trans. by M. Reichenbach as *Axiomatization of the Theory of Relativity*. Berkeley: University of California Press, 1969.

———. 1925. "*Über die physikalischen Konsequenzen der relativistichen Axiomatik*," *Zeitschrift für Physik* 34, 32–48.

———. 1927. "*Lichtgeschwindigkeit und Gleichzeitigkeit*," *Annalen der Philosophie* 6, 128–144.

———. 1928. *Philosophie der Raum-Zeit-Lehre*. Berlin: Walter de Gruyter. Trans., with omission of an appendix, by M. Reichenbach and J. Freund as *Philosophy of Space and Time*. New York: Dover, 1958.

———. 1978. *Hans Reichenbach: Selected Writings, 1909–1953*, vol. 2. M. Reichenbach and R. Cohen (eds.). Dordrecht: D. Reidel.

Rickert, H. 1921. *Der Gegenstand der Erkenntnis. Einführung in die Transzendental-Philosophie. Vierte und Fünfte Auflage*. Tübingen: Verlag J.C.B. Mohr (Paul Siebeck).

Ritchie, A. D. 1948. *Reflections on the Philosophy of Sir Arthur Eddington*. Cambridge: Cambridge University Press.

Robb, A. 1914. *A Theory of Time and Space*. Cambridge: Cambridge University Press.

Rovelli, C. 1997. "Halfway Through the Woods: Contemporary Research on Space and Time," in J. Earman and J. Norton (eds.), *The Cosmos of Science: Essays of Exploration*. Pittsburgh: University of Pittsburgh Press, 180–223.

———. 2001. "Quantum Spacetime: What Do We Know?" in C. Callender and N. Huggett (eds.), *Physics Meets Philosophy at the Planck Scale*. New York: Cambridge University Press, 101–122.

Rowe, D. 2001. "Einstein Meets Hilbert: At the Crossroads of Physics and Mathematics," *Physics in Perspective* 3, 379–424.

———. 2002. "Einstein's Gravitational Field Equations and the Bianchi Identities," *The Mathematical Intelligencer* 24, 57–66.

Russell, B. A. W. 1912. *The Principles of Philosophy*. Oxford: Oxford University Press.

————. 1919. *Introduction to Mathematical Philosophy*. London: Kegan Paul.

————. 1926. "Relativity: Philosophical Consequences." *Encyclopedia Britannica*, 13th ed. London, 331–332.

————. 1927. *The Analysis of Matter*. New York: Macmillan.

————. 1931. *The Scientific Outlook*. London: Allen and Unwin.

————. 1948. *Human Knowledge: Its Scope and Limits*. New York: Simon and Schuster.

Ryckman, T. A. 1991. "*Conditio Sine Qua Non: Zuordnung* in the Early Epistemologies of Cassirer and Schlick," *Synthese* 88, 57–95.

————. 1992. "'P(oint)-C(oincidence) Thinking': The Ironical Attachment of Logical Empiricism to General Relativity (and Some Lingering Consequences)," *Studies in History and Philosophy of Science* 23, 471–497.

————. 1994. "Weyl, Reichenbach and the Epistemology of Geometry," *Studies in the History and Philosophy of Modern Physics* 25, 831–870.

————. 2003a. "The Philosophical Roots of the Gauge Principle: Weyl and Transcendental Phenomenological Idealism," in Brading and Castellani (2003), 61–88.

————. 2003b. "Surplus Structure from the Standpoint of Transcendental Idealism: The 'World Geometries' of Weyl and Eddington," *Perspectives on Science* 11, 76–106.

————. forthcoming a. "Logical Empiricism and the Philosophy of Physics," in A. Richardson and T. E. Uebel (eds.), *The Cambridge Companion to Logical Empiricism*. New York: Cambridge University Press.

————. forthcoming b. "Husserl and Carnap," in R. Creath and M. Friedman (eds.), *The Cambridge Companion to Carnap*. New York: Cambridge University Press.

Rynasiewicz, R. 1999. "Kretschmann's Analysis of Covariance and Relativity Principles," in H. Goenner, J. Renn, J. Ritter, and T. Sauer (eds.), *The Expanding Worlds of General Relativity*. Basel: Birkhäuser, 431–462.

Saunders, S., and H. R. Brown. 1991. *The Philosophy of Vacuum*. Oxford: Clarendon Press.

Scheibe, E. 1988. "Hermann Weyl and the Nature of Space Time," in Deppert and Hübner (1988), 61–82.

Schilpp, P. A., ed. 1949. *Albert Einstein: Philosopher-Scientist*. Evanston, IL: Northwestern University Press.

Schlick, M. 1904. *Über die Reflection des Lichtes in einer inhomogenen Schicht*. Ph.D. Dissertation. Berlin University.

————. 1915. "*Die philosophische Bedeutung des Relativitätsprinzips*," *Zeitschrift für Philosophie und philosophische Kritik* 159, 129–175. Trans. by P. Heath as "The Philosophical Significance of the Principle of Relativity," in Schlick (1979a), 153–189.

————. 1917. "*Raum und Zeit in der gegenwärtigen Physik*," *Die Naturwissenschaften* 5, 161–167, 177–186. Expanded and published separately. Berlin: J. Springer. For trans., see (1922a).

————. 1918. *Allgemeine Erkenntnislehre*. Berlin: J. Springer.

————. 1921a. "*Kritizistische oder Empiristische Deutung der neuen Physik*," *Kant-Studien*, 26, 96–111. Trans. by P. Heath as "Critical or Empiricist Interpretation of Modern Physics," in Schlick (1979a), 322–334.

————. 1921b. "*Helmholtz als Erkenntnistheoretiker*." Trans. by P. Heath as "Helmholtz the Epistemologist," in Schlick (1979a), 335–342.

————. 1922a. *Raum und Zeit in der gegenwärtigen Physik. Vierte Auflage*. Berlin: J. Springer. Trans. by H. Brose and P. Heath as *Space and Time in Contemporary Physics*, in Schlick (1979a), 207–269.

————. 1922b. "*Die Relativitätstheorie in der Philosophie*," *Verhandlungen der Gesellschaft Deutscher Naturforscher und Ärzte*, Leipzig, 58–69. Trans. by P. Heath as "The Theory of Relativity in Philosophy," in Schlick (1979a), 343–353.

———. 1925. *Allgemeine Erkenntnislehre. Zweite Auflage*. Berlin: J. Springer. Trans. by A. Blumberg as *The General Theory of Knowledge*. La Salle, IL: Open Court, 1985.

———. 1929. "Rezension von Reichenbach (1928)," *Die Naturwissenschaften* 17, 549.

———. 1934. "Über das Fundament der Erkenntnis," *Erkenntnis* 4, 79–99. Trans. by P. Heath as "The Foundation of Knowledge," in Schlick (1979b), 370–387.

———. 1935. *Form and Content. An Introduction to Philosophical Thinking*. Three lectures. London, 1932. Corrected repr. in Schlick (1979b), 285–369.

———. 1979a. *Moritiz Schlick: Philosophical Papers*, vol. 1. H. L. Mulder and B. van de Velde-Schlick (eds.). Dordrecht: D. Reidel.

———. 1979b. *Moritiz Schlick: Philosophical Papers*, vol. 2. H. L. Mulder and B. van de Velde-Schlick (eds.). Dordrecht: D. Reidel.

Scholz, E. 1994. "Hermann Weyl's Contributions to Geometry, 1917–1923," in S. Chikara, S. Mitsua, and J. Dauben (eds.), *The Intersection of History and Mathematics*. Basel: Birkhäuser, 203–230.

———. 1995. "Hermann Weyl's 'Purely Infinitesimal Geometry,'" in *Proceedings of the International Congress of Mathematics, Zürich 1994*. Basel: Birkhäuser, 1592–1603.

———. 1999. "Weyl and the Theory of Connections," in J. Gray (ed.), *The Symbolic Universe: Geometry and Physics 1890–1930*. Oxford: Oxford University Press, 260–284.

———. 2001. "Weyls Infinitesimalgeometrie, 1917–1925," in E. Scholz (ed.), *Hermann Weyl's Raum-Zeit-Materie and a General Introduction to His Scientific Work*. Basel: Birkhäuser, 48–104.

Schouten, J. A. 1919. *Die direkte Analysis zur neueren Relativitätstheorie*. Amsterdam: Johannes Müller. (Paper appeared in 1918 but has publication date of 1919.)

———. 1922. "Über die verschiedenen Arten der Übertragung in einer n-dimensionalen Mannigfaltigkeit, die einer Differentialgeometrie zugrunde gelegt werden können," *Mathematische Zeitschrift* 13, 56–81.

Schouten, J. A., and D. J. Struik. 1921. "On Some Properties of General Manifolds Relating to Einstein's Theory of Gravitation," *American Journal of Methematics* 43, 213–216.

———. 1924. "Note on A. E. Harwood's 'The Identical Relations in Einstein's Theory,'" *Philosophical Magazine* 47, 584–585.

Schrödinger, E. 1937. "World Structure," *Nature* 140, 742–744.

———. 1938. "Sur la théorie du monde d'Eddington," *Nuovo Cimento* 15, 246–254.

———. 1939. "The Proper Vibrations of the Expanding Universe," *Physica* 6, 899–912.

———. 1950. *Space-Time Structure*. Cambridge: Cambridge University Press.

Schuhmann, K. 1977. *Husserl-Chronik*. Den Haag: Martinus Nijhoff.

———. 1994. *Edmund Husserl Briefwechsel*, Bd. 5. Ed. in collaboration with E. Schuhmann. Dordrecht: Kluwer.

———. 1996. *Edmund Husserl Briefwechsel*, Bd. 7. Ed. in colloboration with E. Schuhmann. Dordrecht: Kluwer.

Seelig, C. 1960. *Albert Einstein*. Zürich: Europa Verlag.

Shankland, R. 1964. "The Michelson-Morley Experiment," *American Journal of Physics* 32, 16–35.

Sharpe, R. W. 1997. *Differential Geometry: Cartan's Generalization of Klein's Erlangen Problem*. New York: Springer Verlag.

Sigurdsson, S. 1991. *Herman Weyl, Mathematics and Physics, 1900–27*. Ph.D. Dissertation. Cambridge, MA: Harvard University.

Sklar, L. 1985. "*Facts, Conventions, and Assumptions in the Theory of Spacetime*," as repr. in his *Philosophy and Spacetime Physics*. Berkeley: University of California Press, 73–147.

Slater, N. 1957. *The Development and Meaning of Eddington's "Fundamental Theory."* Cambridge: Cambridge University Press.

Sluga, H. 1993. *Heidegger's Crisis.* Cambridge, MA: Harvard University Press.

Smith, B., and D. W. Smith (eds.). 1995. *The Cambridge Companion to Husserl.* Cambridge: Cambridge University Press.

Smolin, L. 1992. "Space and Time in the Quantum Universe," in A. Ashtekar and J. Stachel (eds.), *Conceptual Problems of Quantum Gravity.* Basel: Birkhäuser, 228–288.

———. 2003. "How Far Are We from the Quantum Theory of Gravity?" http:// xxx.lanl.arXiv:hep-th/0303185v2, 11 April 2003.

Sommerfeld, A. 1949. "To Albert Einstein's Seventieth Birthday," in Schilpp (1949), 99–105.

Speziali, P. (ed.). 1972. *Albert Einstein/Michele Besso Correspondence, 1903–1955.* Paris: Hermann.

Spiegelberg, H. 1982. *The Phenomenological Movement.* 3rd rev. and enl. ed., with collaboration of K. Schuhmann. The Hague: Mouton.

Stachel, J. 1986a. "Eddington and Einstein," in E. Ullman-Margalit (ed.), *The Prism of Science.* Dordrecht: D. Reidel, 225–250.

———. 1986b. "What a Physicist Can Learn from the Discovery of General Relativity," in R. Ruffini (ed.), *Proceedings of the Fourth Marcel Grossmann Meeting on General Relativity.* Amsterdam: Elsevier Science Publishers, 1857–1862.

———. 1989a. "Einstein's Search for General Covariance, 1912–1915," in Howard and Stachel (1989), 63–100. Based on a paper circulating privately since 1980.

———. 1989b. "The Rigidly Rotating Disk as the 'Missing Link' in the History of Relativity," in Howard and Stachel (1989), 48–62.

———. 1992a. "Einstein and Quantum Mechanics," in A. Ashtekar and J. Stachel (eds.), *Conceptual Problems of Quantum Gravity.* Boston: Birkhäuser, 13–42.

———. 1992b. "The Cauchy Problem in General Relativity—the Early Years," in J. Eisenstaedt and A. Kox (eds), *Studies in the History of General Relativity.* Boston: Birkhäuser, 407–418.

———. 1993a. "The Meaning of General Covariance: The Hole Story," in J. Earman, A. Janis, G. Massey, and N. Rescher (eds.), *Philosophical Problems of the Internal and External Worlds: Essays on the Philosophy of Adolf Grünbaum.* Pittsburgh: University of Pittsburgh Press, 129–160.

———. 1993b. "The Other Einstein: Einstein Contra the Field Theory," *Science in Context* 6, 275–290.

———. 1996. "Introduction," in Einstein (1996), xv–xxv.

Stebbing, L. S. 1937. *Philosophy and the Physicists.* London: Methuen and. Co.

Straumann, N. 1987. "*Zum Ursprung der Eichtheorien bei Hermann Weyl,*" *Physikalischen Blätter* 43, 414–421.

———. 2001. "*Ursprünge der Eichtheorien,*" in E. Scholz (ed.), *Hermann Weyl's Raum-Zeit-Materie and a General Introduction to His Scientific Work.* Basel: Birkhäuser, 138–155.

Struik, D. J. 1989. "Schouten, Levi-Città, and the Emergence of Tensor Calculus," in D. Rowe and J. McCleary (eds.), *The History of Modern Mathematics*, vol. 2. Boston: Academic Press, 99–105.

Tonietti, T. 1988. "Four Letters of E. Husserl to H. Weyl and Their Context," in Deppert et al. (1988), 343–384.

Torretti, R. 1978. *Philosophy of Geometry from Riemann to Poincaré.* Dordrecht: D. Reidel. Pagination according to corrected repr. (1984).

———. 1983. *Relativity and Geometry.* Oxford: Pergamon.

Trautman, A. 1966. "The General Theory of Relativity," *Soviet Physics Uspekhi* 89, 319–336.

————. 1980a. "Fibre Bundles, Gauge Fields, and Gravitation," in A. Held (ed.), *General Relativity and Gravitation: One Hundred Years after the Birth of Albert Einstein*, vol. 1. New York: Plenum Press, 287–308.

————. 1980b. "Generalities on Geometric Theories of Gravitation," in P. G. Bergmann and V. De Sabbata (eds.), *Cosmology and Gravitation: Spin, Torsion, Rotation, and Supergravity*. New York: Plenum Press, 1–4.

Uebel, T. E. 1992. *Overcoming Logical Positivism from Within*. Amsterdam: Rodopi.

Utiyama, R. 1956. "Invariant Theoretical Interpretation of Interaction," *Physical Review* 101, 1597–1607.

van Dalen, D. 1984. "Four Letters from E. Husserl to H. Weyl," *Husserl Studies* 1, 1–12.

Veblen, O. 1923. "Geometry and Physics," *Science* 57, no. 1466 (February 2), 129–139.

Vizgin, V. 1994. *Unified Field Theories in the First Third of the 20th Century*. As trans. from Russian by J. B. Barbour. Basel: Birkäuser Verlag.

Von Laue, M. 1921. *Die Relativitätstheorie. Zweiter Band: Die Allgemeine Relativitätstheorie und Einsteins Lehre von der Schwerkraft*. Braunschweig: Vieweg.

Wald, R. 1984. *General Relativity*. Chicago: University of Chicago Press.

Weinberg, S. 1972. *Gravitation and Cosmology*. New York: Wiley.

————. 1993. *Dreams of a Final Theory*. New York: Pantheon.

————. 1995. *The Quantum Theory of Fields*, vol. 1. New York: Cambridge University Press.

Weitzenböck, R. 1920. "*Über die Wirkungsfunktion in der Weyl'schen Physik,*" *Akademie der Wissenschaften in Wien. Sitzungsberichte. Abteilung IIa. Mathematisch-naturwissenschaftliche Klasse* 129, 683–708.

Welton, D. 2000. *The Other Husserl: The Horizons of Transcendental Philosophy*. Bloomington, IN: Indiana University Press.

Weyl, H. 1913. *Die Idee der Riemannschen Fläche*. Leipzig: B. G. Teubner.

————. 1917. "*Zur Gravitationstheorie,*" *Annalen der Physik* 54, 117–145; repr. in (1968), vol. 1, 670–698.

————. 1918a. *Das Kontinuum: Kritische Untersuchungen über die Grundlagen der Analysis*. Leipzig: Veit. Trans. by S. Pollard and T. Bowl as *The Continuum: A Critical Examination of the Foundations of Analysis*. New York: Dover, 1987.

————. 1918b. *Raum-Zeit-Materie*. Berlin: J. Springer.

————. 1918c. "*Gravitation und Elektrizität,*" *Preußische Akademie der Wissenschaften (Berlin) Sitzungsberichte. Physikalisch-Mathematische Klasse*, 465–480; repr. in (1968), vol. 2, 29–42. Trans. in O'Raifeartaigh (1997), 24–37.

————. 1918d. "*Reine Infinitesimalgeometrie,*" *Mathematische Zeitschrift* 2, 384–411; repr. in (1968), vol. 2, 1–28.

————. 1919a. "*Eine Neue Erweiterung der Relativitätstheorie,*" *Annalen der Physik* 59, 101–133; repr. in (1968), vol. 2, 55–87.

————. 1919b. *Raum-Zeit-Materie. 3 Auflage*. Berlin: J. Springer.

————. 1920a. "*Das Verältnis der kausalen zur statischen Betrachtungsweise in der Physik,*" *Schweiserische Medizinische Wochenschrift*, 737–741; repr. in (1968), vol. 2, 113–122.

————. 1920b. "*Elektrizität und Gravitation,*" *Physikalische Zeitschrift* 21, 649–651; repr. in (1968), vol. 2, 141–142.

————. 1921a. *Raum-Zeit-Materie. 4 Auflage*. Berlin: J. Springer. Trans. by H. L. Brose as *Space-Time-Matter*. London: Meuthen, 1921. Repr. ed., New York: Dover, 1953.

————. 1921b. "*Über die physikalischen Grundlagen der erweiterten Relativitätstheorie,*" *Physikalische Zeitschrift* 22, 473–480; repr. in (1968), vol. 2, 99–112.

————. 1921c. "*Zur Infinitesimalgeometrie: Einordnung der projektiven und der konformen Auffassung,*" *Nachrichten. Königlichen Gesellschaft der Wissenschaften zu Göttingen, Mathematische-Physikalische Klasse*, 99–112; repr. in (1968), vol. 2, 195–207.

————. 1921d. "*Feld und Materie*," *Annalen der Physik* 65, 541–563; repr. in (1968), vol. 2, 237–259.

————. 1921e. "Electricity and Gravitation," Trans. by R. W. Lawson. *Nature* 106, 800–802; repr. in (1968), vol. 2, 260–262.

————. 1921f. "*Über die neue Grundlagenkrise der Mathematik*," *Mathematische Zeitschrift* 10, 39–79; repr. in (1968), vol. 2, 143–179. Trans. by B. Müller as "On the New Foundational Crisis of Mathematics," in Mancosu (1998), 86–122.

————. 1921g. "*Das Raumproblem*," *Jahresberichte der Deutschen Mathematikvereinigung* 31, 205–221; repr. in (1968), vol. 2, 328–344.

————. 1922a. "*Die Einzigartigkeit der Pythagoreischen Massbestimmung*," *Mathematische Zeitschrift* 12, 114–146; repr. in (1968), vol. 2, 263–295.

————. 1922b. "*Die Relativitätstheorie auf der Naturforscherversammlung in Bad Nauheim*," *Jahresbericht der Deutschen Mathematische-Vereinigung*, 51–63; repr. in (1968), vol. 2, 315–327.

————. 1922c. "*Rezension von R. Carnap 'Der Raum: Ein Beitrag zur Wissenschaftslehre,'*" *Jahrbuch über die Fortschritte der Mathematik*, 631–632.

————. 1923a. *Raum-Zeit-Materie. 5 Auflage*. Berlin: J. Springer.

————. 1923b. *Mathematische Analyse des Raumproblems: Vorlesungen gehalten in Barcelona und Madrid*. Berlin: J. Springer.

————. 1923c. "*Rezension von Schlick 'Allgemeine Erkenntnislehre,'*" *Jahrbuch über die Fortschritte der Mathematik*, 59–62.

————. 1924a. "*Was ist Materie*," *Die Naturwissenschaften* 12, 561–568, 585–593, 604–611; and separately issued, *Was Ist Materie? Zwei Aufsätze zur Naturphilosophie*. Berlin: J. Springer; repr. in (1968), vol. 2, 486–510.

————. 1924b. "*Rezension von Hans Reichenbach, 'Axiomatik der relativistischen Raum-Zeit-Lehre,'*" *Deutsche Literaturzeitung* 30, cols. 2122–2127.

————. 1925a. "*Die Heutige Erkenntnislage in der Mathematik*," *Symposium* 1, 1–32; repr. in (1968), vol. 2, 511–542; trans. by B. Müller as "The Current Epistemological Situation in Mathematics," in Mancosu (1998), 123–142.

————. 1925b. *Riemanns geometrische Ideen, ihre Auswirkung und ihre Verknüpfung mit der Gruppentheorie*. K. Chandrasekharan (ed.). Berlin: Springer Verlag, 1988 (1st publ.).

————. 1926. "*Philosophie der Mathematik und Naturwissenschaft*," in A. Baeumler and M. Schröter (eds.), *Handbuch der Philosophie*, Abt. 2. Munich: R. Oldenbourg. Separate monographic publication dated 1927.

————. 1928a. "*Diskussionbemerkungen zu dem zweiten Hilberschen Vortrag über die Grundlagen der Mathematik*," *Abhandlungen aus dem mathematischen Seminar der Hamburgischen Universität* 6, 86–88. Trans. by S. Bauer-Mengelberg and D. Føllesdal, in J. Van Heijenoort (ed.), *From Frege to Gödel*. Cambridge, MA: Harvard University Press, 1967, 482–484.

————. 1928b. *Gruppentheorie und Quantenmechanik*. Leipzig: S. Hirzel. Trans. of 2nd ed. by H. P. Robertson as *The Theory of Groups and Quantum Mechanics*. London: Methuen, 1931.

————. 1929a. "*Elektron and Gravitation*," *Zeitschrift für Physik* 56, 330–352. Repr. in (1968), vol. 3, 245–267. Trans. in O'Raifeartaigh (1997) 121–144.

————. 1929b. "On the Foundations of Infinitesimal Geometry," *Bulletin of the American Mathematical Society* 35, 716–725. Repr. in (1968), vol. 3, 121–144.

————. 1930. "*Axiomatik*," *Vorlesungen der Winter Semester 1930*, Göttingen Universität.

————. 1931. "*Geometry und Physik*," *Die Naturwissenschaften* 19, 49–58; repr. in (1968), vol. 3, 336–345.

————. 1934. "*Universum und Atom*," *Die Naturwissenschaften* 22, 145–194; repr. in (1968), vol. 3, 420–424.

———. 1938. "Cartan on Groups and Differential Geometry," *Bulletin of the American Mathematical Society* 44, 598–601.

———. 1940. "The Ghost of Modality" in M. Farber (ed.), *Philosophical Essays in Memory of Edmund Husserl*. Cambridge, MA: Harvard University Press; repr. in (1968), vol. 3, 684–709.

———. 1948. "*In memoriam* Helene Weyl," in Lange (1996), 379–390.

———. 1949a. *Philosophy of Mathematics and Natural Science*. Rev. augmented Engl. ed., of 1926/1927 based on the trans. of O. Helmer. Princeton, NJ: Princeton University Press.

———. 1949b. "Relativity Theory as a Stimulus in Mathematical Research," *Proceedings of the American Philsophical Society* 93, 535–541; repr. in (1968), vol. 4, 206–216.

———. 1951. "*50 Jahre Relativitätstheorie*," *Die Naturwissenschaften* 38, 73–83; repr. in (1968), vol. 4, 421–431.

———. 1952. *Symmetry*. Princeton, NJ: Princeton University Press.

———. 1954. "Address on the Unity of Knowledge Delivered at the Bicentennial Conference of Columbia University," as repr. in (1968), vol. 4, 623–630.

———. 1955. "*Erkenntnis und Besinnung*" ("*Vortrag, gehalten an der Universität Lausanne im Mai 1954*"), *Studia Philosophia* (*Jarbuch der Schwiezerischen Philosophischen Gesellschaft*) 15, 153–171; repr. in (1968), vol. 4, 631–649.

———. 1968. *Gesammelte Abhandlungen*. Vols. 1–4. Ed. by K. Chandrasekharan. Berlin: Springer Verlag.

———. 1985. "Axiomatic Versus Constructive Procedures in Mathematics," ed. by T. Tonietti, *The Mathematical Intelligencer* 7, 10–17.

Wheeler, J. A. 1962. *Geometrodynamics*. New York: Academic Press.

———. 1990. *A Journey into Gravitation and Space-Time*. San Francisco: W. H. Freeman.

Whitehead, A. N. 1925. *Science and the Modern World*. New York: Macmillan.

Whittaker, E. T. 1942. "Some Disputed Questions in the Philosophy of the Physical Sciences," *Philosophical Magazine* 33, 353–366.

———. 1945. "Eddington's Theory of the Constants of Nature," *The Mathematical Gazette* 29, 137–144.

Witt-Hansen, J. 1958. *Exposition and Critique of the Conceptions of Eddington Concerning the Philosophy of Physical Science*. Copenhagen: G. E. C. Gad.

Wood, A. 1957. *Bertrand Russell: The Passionate Sceptic*. London: Allen and Unwin.

Worrall, J. 1989. "Structural Realism: The Best of Both Worlds," *Dialectica* 43, 99–124.

Yang, C. N. 1977. "Magnetic Monopoles, Fiber Bundles, and Gauge Fields," *Annals of the New York Academy of Sciences* 294, 86–97.

———. 1980. "Einstein's Impact on Theoretical Physics," *Physics Today* 30, 42–49.

———. 1985. "Hermann Weyl's Contributions to Physics," in K. Chandrasekharan (ed.), *Hermann Weyl 1885–1985: Centenary Lectures at the ETH Zürich*. Berlin: Springer-Verlag, 7–21.

Yolton, J. 1960. *The Philosophy of Science of A. S. Eddington*. The Hague: M. Nijhoff.

Zahar, E. 1989. *Einstein's Revolution: A Study in Heuristics*. La Salle, IL: Open Court.

Zeeman, E. G. 1967. "The Topology of Minkowski Space-Time," *Topology* 6, 161–170.

INDEX

Hawking, Stephen, 21, 269 n.141
Hegel, Georg Wilhelm Friedrich, 239
Heidegger, Martin, 26, 128
Hellinger, Ernst, 111
Helmholtz, Hermann von, 8, 52, 56–57,
    67–75, 154
Hempel, Carl Gustav, 11
Hertz, Heinrich, 40, 67
Hertz, Paul, 67, 70
Hessenberg, Gerhard, 150
Hilbert, David, 7, 9, 29, 45, 111, 119,
    138, 248, 269 n.136
  axiomatic method of, 84, 97, 109
  choice of Riemann scalar, 165
  finitism of, 114
  on general covariance, 258 n.148
Hoefer, Carl, 282 n.106
Hoffmann, Banesh, 18, 265 n.64
holism, epistemological, 64
Howard, Don A., 59, 255 n.53, 259 n.43
Hume, David, 25
Husserl, Edmund, 49, 79, 108, 110–128,
    136–144, 176, 232–233
  concept of *a priori*, 143
  conception of phenomenology as
    doctrine of essence, 142
  essential insight *(Wesenserschauung)*,
    139
  formal ontology, 143
  Ideen I, 118, 122–128
  Logical Investigations, 112, 118–119,
    137
  method of essential analysis
    (Wesensanalyse), 139–142
  method of phenomenological
    reduction, 122, 126
  "Philosophy as Rigorous Science", 111,
    120–122
  "principle of all principles", 125
  regional ontology, 118, 143–144
  rejoinder to Schlick, 113
  transcendental turn of, 119

ideal, regulative, 34, 44
idealism
  critical, 41
  postulate of transcendental, 6
  subjective, 127, 217
  transcendental, 6, 11, 231–234, 240,
    287 n.53
  transcendental-phenomenological, 109

argument for in *Ideen I*, 122–128
fundamental thesis of, 117, 148
method of essential analysis in,
    117, 149
vs. pre-epistemological "natural
    attitude", 120–121, 145
ideas, of reason, 45–46
  as synthetic principle of unity, 27,
    43, 45
instrumentalism, 6, 200
intentionality, 119, 123. *See also*
    idealism, transcendental-
    phenomenological
intuition
  categorical, 173, 271 n.44
  form of outer (Helmholtz), 72
  forms of, 147
  inner, 74
  phenomenological, 148
  pure, 44, 48
  sensible vs. categorical, 113
invariance. *See also* symmetry
  diffeomorphism, 21, 23, 254 n.39
  gauge, 78, 80, 84, 109, 147, 173
  Lorentz, 279 n.31
  principle, meta-empirical status of, 15
  principle of general, 22

Jeffreys, Harold, 178
Joseph, Friederike Bertha Helene. *See*
    Weyl, Hella

Kaluza, Theodore, 78, 221–222
Kant, Immanuel, 10, 72, 74, 115,
    155, 240
  Critique of Pure Reason, 5, 177
    Transcendental Aesthetic, 14, 37,
      48, 133, 241
    Transcendental Analytic, 37, 186
    Transcendental Logic, 26
  doctrine of pure intuition, 44
  ideas, regulative, 46
  on schematism, transcendental, 25
  on sensibility and understanding,
    faculties of, 14, 24–25
Kant Gesellschaft, 47
Kilmister, Clive, 178, 283 n.139
Klein, Felix, 40
König, Rudolf, 111
Koyré, Alexander, 11
Kretschmann, Erich, 17, 82